Sensors for Machine Condition Monitoring, Diagnostics, Prognostics and Maintenance

Sensors for Machine Condition Monitoring, Diagnostics, Prognostics and Maintenance

Guest Editors

Janis Terpenny
Thomas Kurfess
Vittal Prabhu
Dazhong Wu

MDPI

Basel • Beijing • Wuhan • Barcelona • Belgrade • Novi Sad • Cluj • Manchester

Guest Editors
Janis Terpenny
Department of Systems Engineering and Operations Research and the Department of Mechanical Engineering
George Mason University
Fairfax
USA

Thomas Kurfess
George W. Woodruff School of Mechanical Engineering
Georgia Institute of Technology
Atlanta
USA

Vittal Prabhu
Harold and Inge Marcus Department of Industrial and Manufacturing Engineering
Pennsylvania State University
University Park
USA

Dazhong Wu
Department of Mechanical and Aerospace Engineering
University of Central Florida
Orlando
USA

Editorial Office
MDPI AG
Grosspeteranlage 5
4052 Basel, Switzerland

This is a reprint of the Special Issue, published open access by the journal *Sensors* (ISSN 1424-8220), freely accessible at: https://www.mdpi.com/journal/sensors/special_issues/MCM_DPM.

For citation purposes, cite each article independently as indicated on the article page online and as indicated below:

Lastname, A.A.; Lastname, B.B. Article Title. *Journal Name* **Year**, *Volume Number*, Page Range.

ISBN 978-3-7258-5461-5 (Hbk)
ISBN 978-3-7258-5462-2 (PDF)
https://doi.org/10.3390/books978-3-7258-5462-2

Contents

About the Editors . . . . **vii**

Prince Waqas Khan and Yung-Cheol Byun
Multi-Fault Detection and Classification of Wind Turbines Using Stacking Classifier
Reprinted from: *Sensors* **2022**, *22*, 6955, https://doi.org/10.3390/s22186955 . . . . **1**

Amol Kulkarni, Janis Terpenny and Vittaldas Prabhu
Leveraging Active Learning for Failure Mode Acquisition
Reprinted from: *Sensors* **2023**, *23*, 2818, https://doi.org/10.3390/s23052818 . . . . **18**

Florina Maria Șerdean, Mihai Dan Șerdean and Silviu-Dan Mândru
Maintenance Cost Minimization for an Agricultural Harvesting Gripper
Reprinted from: *Sensors* **2023**, *23*, 4103, https://doi.org/10.3390/s23084103 . . . . **38**

Lukas Krupp, Christian Wiede, Joachim Friedhoff and Anton Grabmaier
Explainable Remaining Tool Life Prediction for Individualized Production Using Automated Machine Learning
Reprinted from: *Sensors* **2023**, *23*, 8523, https://doi.org/10.3390/s23208523 . . . . **51**

Leroy Anozie, Bodo Fink, Christoph M. Friedrich, Christoph Engels
Monitoring Flow-Forming Processes Using Design of Experiments and a Machine Learning Approach Based on Randomized-Supervised Time Series Forest and Recursive Feature Elimination
Reprinted from: *Sensors* **2024**, *24*, 1527, https://doi.org/10.3390/s24051527 . . . . **79**

Simone Mari, Giovanni Bucci, Fabrizio Ciancetta, Edoardo Fiorucci and Andrea Fioravanti
Impact of Measurement Uncertainty on Fault Diagnosis Systems: A Case Study on Electrical Faults in Induction Motors
Reprinted from: *Sensors* **2024**, *24*, 5263, https://doi.org/10.3390/s24165263 . . . . **105**

Wenbo Lu, Junpeng Xue, Wei Pu, Hongyang Chen, Kelei Wang and Ran Jia
A Calculation Method of Bearing Balls Rotational Vectors Based on Binocular Vision Three-Dimensional Coordinates Measurement
Reprinted from: *Sensors* **2024**, *24*, 6499, https://doi.org/10.3390/s24196499 . . . . **121**

Yuepeng Zhang, Jun Wu, Bo Gao, Linzhong Xia, Chen Lu, Hui Wang and Guangzhong Cao
Fault Types and Diagnostic Methods of Manipulator Robots: A Review
Reprinted from: *Sensors* **2025**, *25*, 1716, https://doi.org/10.3390/s25061716 . . . . **141**

Sylwia Werbinska-Wojciechowska and Rafal Rogowski
Proactive Maintenance of Pump Systems Operating in the Mining Industry—A Systematic Review
Reprinted from: *Sensors* **2025**, *25*, 2365, https://doi.org/10.3390/s25082365 . . . . **169**

Ryszard Błażej, Leszek Jurdziak and Aleksandra Rzeszowska
Sensor-Based Diagnostics for Conveyor Belt Condition Monitoring and Predictive Refurbishment
Reprinted from: *Sensors* **2025**, *25*, 3459, https://doi.org/10.3390/s25113459 . . . . **236**

About the Editors

Janis Terpenny

Janis Terpenny is a program director at the National Science Foundation (NSF) in the division of Civil, Mechanical, and Manufacturing Innovation (CMMI). She is also a Professor of Mechanical Engineering and Systems Engineering & Operations Research at George Mason University. Her research focuses on smart integrated systems, processes for design and manufacturing, and engineering education. Her work has been supported by more than USD 16M in research funding and she has published in 200+ peer-reviewed journal and conference papers. Previously, Janis served as the Dean of Engineering at the University of Tennessee, Department Head at Penn State, Department Chair at Iowa State, Director of the NSF Center for e-Design, Program Director for NSF's Division of Undergraduate Education, and as a Professor at Virginia Tech and the University of Massachusetts. She has served as the first technology thrust lead for the Digital Manufacturing and Design Innovation Institute (DMDII), one of the nation's first Manufacturing USA institutes (now MxD). She has industry work experience with General Electric (GE). She is a fellow and member of ASEE, ASME, and IISE, and a member of AAAS, Alpha Pi Mu, INFORMS, SME, and Tau Beta Pi. She is a Governor on ASME's Board of Governors. Janis has received several awards for research and teaching innovation. She has served as an NSF Advance Professor at Virginia Tech and has been an innovator and leader of many initiatives for the professional growth and success of women and other underrepresented groups in engineering.

Thomas Kurfess

Thomas Kurfess is the Chief Manufacturing Officer of the Georgia Institute of Technology and the Executive Director of the Georgia Tech Manufacturing Institute. At Georgia Tech he is the HUSCO/Ramirez Distinguished Chair in Fluid Power and Motion Control and the Regents' Professor of Mechanical Engineering. He also serves as the Chief Technology Officer at the National Center for Manufacturing Sciences. He received his S.B., S.M., and Ph.D. degrees in mechanical engineering from M.I.T. in 1986, 1987, and 1989, respectively. He also received an S.M. degree from M.I.T. in electrical engineering and computer science in 1988. During 2019-2021 he was on leave serving as the Chief Manufacturing Officer at the Oak Ridge National Laboratory (ORNL), where he was responsible for strategic planning for ORNL's advanced manufacturing. He was also the Founding Director for the Manufacturing Science Division at ORNL. During 2012-2013 he served as the Assistant Director for Advanced Manufacturing at the Office of Science and Technology Policy in the Executive Office of the President of the United States of America, where he was responsible for coordinating federal advanced manufacturing R&D. He served as the President of the American Society of Mechanical Engineers in 2023-2024 and was President of the Society of Manufacturing Engineers in 2018. His research focuses on the design and development of advanced manufacturing systems, targeting secure digital manufacturing, additive and subtractive processes, and large-scale production enterprises. He is an elected member of the National Academy of Engineering and is a Fellow of ASME, AAAS, and SME.

Vittal Prabhu

Vittal Prabhu is a Professor & Charles and Enid Schneider Faculty Chair in Service Enterprise Engineering in the Marcus Department of Industrial and Manufacturing Engineering at Penn State University. Over the last 30 years, he has made significant technical contributions in the area of distributed control systems, with a focus on manufacturing and service systems. He has created an innovative methodology for controlling discrete-event systems by formulating these problems as continuous variable control problems. Dr. Prabhu's recent technical contributions include unified supply chain servo control algorithms for simultaneous real-time control of processing capacity, production, maintenance, inventory, distribution, and energy in manufacturing supply chains. He has collaborated with industry and faculty from several departments and colleges in PSU's Consortium for Digital Enterprises (CODE). Since 2013, he has been serving as the Director of Service Systems Engineering program at Penn State.

Dazhong Wu

Dazhong Wu is an Associate Professor in the Department of Mechanical and Aerospace Engineering at the University of Central Florida (UCF). He received his Ph.D. in Mechanical Engineering from Georgia Tech. His research focuses on machine learning-enabled smart manufacturing, additive manufacturing processes, prognostics and health monitoring. He has published over 70 journal articles with over 10,000 citations. He received the *Journal of Manufacturing Systems* Best Paper Award in 2019 and 2021, as well as the *Journal of Manufacturing Systems* Outstanding Associate Editor Award in 2025. He was selected as one of the Most Influential Professors in Smart Manufacturing by the *Smart Manufacturing Magazine*. He serves as an Associate Editor for the *Journal of Manufacturing Systems*, the *Journal of Intelligent Manufacturing*, and *IEEE Transactions on Industrial Informatics*.

sensors

MDPI

Article

Multi-Fault Detection and Classification of Wind Turbines Using Stacking Classifier

Prince Waqas Khan and Yung-Cheol Byun *

Department of Computer Engineering, Jeju National University, Jeju-si 63243, Korea
* Correspondence: ycb@jejunu.ac.kr

Abstract: Wind turbines are widely used worldwide to generate clean, renewable energy. The biggest issue with a wind turbine is reducing failures and downtime, which lowers costs associated with operations and maintenance. Wind turbines' consistency and timely maintenance can enhance their performance and dependability. Still, the traditional routine configuration makes detecting faults of wind turbines difficult. Supervisory control and data acquisition (SCADA) produces reliable and affordable quality data for the health condition of wind turbine operations. For wind power to be sufficiently reliable, it is crucial to retrieve useful information from SCADA successfully. This article proposes a new AdaBoost, K-nearest neighbors, and logistic regression-based stacking ensemble (AKL-SE) classifier to classify the faults of the wind turbine condition monitoring system. A stacking ensemble classifier integrates different classification models to enhance the model's accuracy. We have used three classifiers, AdaBoost, K-nearest neighbors, and logistic regression, as base models to make output. The output of these three classifiers is used as input in the logistic regression classifier's meta-model. To improve the data validity, SCADA data are first preprocessed by cleaning and removing any abnormal data. Next, the Pearson correlation coefficient was used to choose the input variables. The Stacking Ensemble classifier was trained using these parameters. The analysis demonstrates that the suggested method successfully identifies faults in wind turbines when applied to local 3 MW wind turbines. The proposed approach shows the potential for effective wind energy use, which could encourage the use of clean energy.

Keywords: wind turbines; fault detection; stacking ensemble classifier; AdaBoost; K-nearest neighbors; logistic regression

1. Introduction

The global energy system is recognized as a significant contributor to greenhouse gas emissions. Sustainable development of low carbon emissions is required for decarbonization. The need for modern energy has taken an essential step toward renewable energy. Due to its positive benefits, wind power is attracting a significant amount of interest from investors in renewable energy sources. Wind energy is a clean, pollution-free renewable energy source with excellent growth potential [1]. Wind's kinetic energy does not produce carbon dioxide and provides clean, reliable electricity. Over the last few years, the installed capacity of wind power has grown exponentially. However, most wind farms are located in remote areas with challenging environments such as mountains, deserts, or oceans. Many components of wind turbines, such as fans, bearings, gears, and generators, are at risk of failure, leading to higher maintenance and operating costs for wind turbines. Continuous health monitoring and maintenance can improve the reliability of wind turbines and prevent catastrophic accidents. Therefore, monitoring the condition of wind turbines and identifying faults is necessary and valuable for maintenance and planning, which can reduce economic losses and promote the development of the wind power industry.

Due to high-quality complex structures, high maintenance costs, and an increase in wind turbine installations worldwide, wind turbine operation and maintenance (O&M)

technologies are now required [2]. In addition, offshore wind power sources, especially floating wind turbines, are rapidly evolving, significantly increasing the hassle and cost of O&M. Operating and maintenance costs make up a large portion of the total annual cost of wind turbines. For a new turbine, the cost of O&M can easily be as high as 20–25 percent of the total cost of an aircraft producing kWh during the turbine's life. If the turbine is relatively new, the component may be only 10–15 percent, but it may increase by at least 20–35 percent during the turbine's life. As a result, more attention is paid to O&M as manufacturers try to significantly reduce costs by designing new turbines that require less service testing and shorter turbine downtime. The consistency and timely maintenance of wind turbines can enhance their reliability and performance [3]. Much research has focused on monitoring control and early detection of faults through Supervisory Control and Data Acquisition (SCADA). Detecting the condition or malfunction of wind turbines is an inexpensive solution to reduce maintenance costs and revenue loss due to component malfunction. The design of debugging deployment turbines has a more significant impact on life expectancy. Older turbine gearboxes did not take advantage of advances in engineering and were still at risk of decay and wear and tear. Several artificial intelligence solutions have been proposed that allow technicians to predict and identify turbine faults, assist in diagnosis, and determine when safety precautions should be taken. Machine learning solutions have become popular in many fields, and their use in wind turbines has yielded excellent results.

This paper proposes a new AdaBoost, K-nearest neighbors, and logistic regression-based stacking ensemble (AKL-SE) classifier to classify the faults of the wind turbine condition monitoring system. The work is based on the pre-identified status of the fault obtained from the condition monitoring system (CMS). Using the proposed system, the need and expenses of CMS can be cut off. A stacking ensemble classifier combines different classification models to enhance the model's accuracy. We have used three classifiers, AdaBoost, K-nearest neighbors, and logistic regression, as base models to make output. The output of these three classifiers is used as input in the meta-model [4]. To improve the data validity, SCADA data are first preprocessed by cleaning and removing any abnormal data. Then, using the Pearson correlation coefficient, the input variables are chosen. The stacking ensemble classifier was then trained using these parameters. The analysis demonstrates the effectiveness of the proposed method in identifying wind turbine anomalies after being applied to several 3 MW wind turbines. In the eleven months of data, there were more than forty different types of faults, but only a small fraction occurred frequently enough for a classifier to be trained to identify them. We classify five different types of faults: feeding faults, excitation errors, generator heating faults, malfunctioning air cooling, and mains failure. The proposed AKL-SE outperforms the state-of-the-art methods for intelligent multi-fault classification in the wind turbine.

1.1. Contributions

The significant contributions of this article can be summarized as follow

- An AdaBoost, K-nearest neighbors, and logistic regression-based stacking ensemble (AKL-SE) classifier is introduced.
- We integrated SCADA data with status data for fault detection and employed data and predictive analytics techniques for data obtained from wind turbines.
- We performed a comparison with state-of-the-art ML models and different combinations of ensemble models.

1.2. Article Organization

The remainder of the article is arranged as follows. Section 2 presents the review and analyses of related publications. Section 3 presents the proposed methodology and explanation of the models used. Section 4 explains the exploratory data analysis. Section 5 provides the results and also covers the comparison with the latest machine learning models. Finally, we conclude this article in the last conclusion section.

2. Literature Review

Wind power is an essential renewable energy source due to technological advances and cheapness. Wind turbines (WT) are essential for wind power generation systems [5,6]. Defects in various parts of wind turbines should be eliminated for better performance. Vibration testing is an integral part of monitoring the condition of wind turbine components [7]. In the article by Bodla et al. [8], a test was performed using average and incorrect WT vibration data to monitor low position. The purpose is to monitor the condition of the wind turbine for immediate fault predictions so that the turbine can be configured for immediate better performance and longer service life. Quality monitoring and evaluation are essential for proper functioning wind power generation systems. In the study by Malik et al. [9], health conditions were assessed for three different fault types using K-nearest neighbors (KNN) algorithms. To decompose the raw signal, they used a separate estimation of the wave function called the Meyer wavelet function. The wind generator used 21 full-featured templates to classify imbalance errors. They compared the proposed method to a multilayer perceptron (MLP). The proposed approach results and various comparisons can serve as an essential tool for WTG error detection.

The approaches used in the study by Lima et al. [10] enable information mining from easily accessible SCADA data in order to identify potential abnormalities when wind turbines run. With this system monitoring the turbine parts, it is feasible to see problems before they arise and take appropriate action, minimizing downtime and maintenance expenses. A customized deep learning model was suggested by Chatterjee et al. [11] for anomaly prediction and transparent decision-making. It was extended to transfer learning from one domain to another, eliminating the need for training in the new domain.

The converter can quickly fail as a vital component of the transmission system. In order to isolate vulnerable sums during the integration process, the best model parameters for each SVM can be found by repeating BSA optimization in the article by Zheng et al. [12]. To improve fault detection accuracy for wind turbines, a method of correction of fault correlation with wave change and compression detection theory has been proposed. It also uses the AdaBoost-SVM wind turbine converter for diagnostics. Based on the compression observation theory, the estimated coefficient of the wave is quantitatively reduced to obtain the measured signal. The target of the orthogonal adjustment is the algorithm tracking vector and then the error multiplier vector. AdaBoost-SVM's advanced classification for error detection has introduced error detection vectors.

The study by Wu et al. [13] presented a wind turbine diagnostic method based on the XGBoost and ReliefF algorithms that used SCADA data to improve the accuracy of wind turbine diagnostics. This article provides solutions to wind turbine malfunction. The advantage of high accuracy in the algorithm was verified by comparison with the state-of-the-art models.

Wind energy is converted into electrical energy by a rotating blade attached to a generator [14]. Due to environmental conditions and extensive construction, the blades suffer from many defects and lack of productivity [15]. The time of inactivity can be reduced if their condition is checked periodically using monitoring techniques. Feature extraction, feature selection, and error classification have been considered machine learning issues. The study by Joshua et al. [16] provided an explanation based on the algorithm of vibration signals for analyzing wind turbine aircraft conditions. A model was created using the data formatting technique from the vibration data obtained. A logistic model tree (LMT) algorithm was used to learn and classify different aircraft locations. Samples were tested ten times under cross-verification, and 90.33 percent accurate classification samples were found. The error rate was relatively low and can be considered blade debugging. Therefore, the tree logistics model is available mainly for monitoring wind turbine blades to reduce idle time and provide continuous wind power. Renewable energy sources such as wind power are plentiful. The reliability of wind turbines is essential to maximize wind power. Vibration signals on the rotating side of wind turbines were average, not Gaussian, and instability and malfunction patterns were generally minimal. With these issues in mind,

The article by Wenyi et al. [17] designed and proposed a method for detecting wind turbine faults based on the grouping of the diagonal spectrum and binary tree support-vector machine (SVM). First, the input properties of the diagonal spectrum vibration machine were considered vectors. Second, a self-regulating neural network mapping feature was introduced to create error feature group sampling and a cluster binary tree. Numerous error classifications were made for practice and sample testing. The diagnostic analysis results confirm that this paper method is effective and efficient. Impact classification effects were RBF neural network methods, and higher accuracy could be achieved in better and fewer sample cases than for traditional SVM methods.

The study by Jiang et al. [18] focused on a survey of the proposed architecture and the multifaceted properties of low-turn vibration signals for troubleshooting WT gearboxes under different operating conditions. Unlike traditional methods, when the emission and classification of features are individually designed, the design's purpose was to automatically determine the characteristics of the desired error of the vibration signal. The proposed multiscale convolutional neural network system was evaluated by testing the test machine on the WT gearbox. Experimental results and extensive comparative analysis of traditional CNN and traditional multidimensional extractors demonstrated the effectiveness of the proposed method. The accuracy could be enhanced by increasing the features [19]. They verified the scale of their proposed MSCNN with an accurate mass WT gearbox. In addition, they further explored analytical methods based on unbalanced multidimensional presentations to significantly reduce the impact of the distribution of diagonal data between standard and inaccurate data, leading to learning algorithms for error diagnosis to significantly improve performance.

Modern debugging and rating systems have become necessary to achieve wind turbines' desired reliability and efficiency [20]. In the work by Vidal et al. [21], sampling frequency was increased, and database multi-error detection monitoring and current sensor classification (in all commercial wind turbines) monitoring controls and data acquisition (SCADA) systems were monitored without the use of special equipment. A high-quality wind turbine was used. They understood that there were several defects in wind turbine actuators and sensors. First, the SCADA measurement groups were preset by a feature change based on analyzing multiple key components and sample base openings. Then, classification based on 10-fold cross-verification SVM was applied. The result was a uniquely trained classification that can solve all the errors analyzed by calculating only a set of features from the data for evaluation. As a result, their proposed approach is better than the other methods.

Time series data on the monitoring and debugging of wind turbines and other power systems are widely used [22], where long-term reliability is essential for developing classification features. Lei et al. [23] identified errors in time series signals based on long shorts. They introduced a new method using the long short-term memory (LSTM) model to learn features directly from multiple variable time series data and gain long-term possibilities through recurring behaviors and LSTM gate mechanisms. Errors can be efficiently classified by one or more sensors using raw time-series signals, and the performance of modern technologies is more efficient. Further, the sustainability of the proposed structure can be verified by testing it on smaller datasets with limited data. This could be enhanced by using a CNN as part of preparations to acquire local features to enhance the functionality of the proposed structure.

Classification of multi-fault detection is a challenging task due to weak faults [24], especially in wind turbine gearboxes with different gears and bearings [25]. The study by Teng et al. [26] analyzed the vibration signals coming from a real multi-fault wind turbine gearbox with catastrophic failure. The complex waveform can provide a multiscale enveloping spectrogram for simultaneous decomposition and distortion of signals. Using this method, testing multiscale enveloping spectrogram disks on different scales can quickly determine the faulty characteristics of the mounted holder under compression force. Unplanned or unresponsive maintenance of wind turbines due to component failure can lose a significant amount of time and revenue [27]. For this purpose, it is necessary to maintain it before it is needed. By constantly monitoring the health of the turbine, it is possible to eliminate the need for periodic inspections, identify underlying defects, and adjust maintenance schedules as needed. Efforts have been made to develop a condition monitoring system (CMS) based on detecting expensive vibration [28,29] or oil analysis sensors [30,31] in turbines. Instead, critical analyses of existing data from the turbine's SCADA system can provide essential insights into turbine performance at a low cost. The study by Leahy et al. [32] examines a new method for classifying and predicting turbine faults based on SCADA data. The data were taken from a SCADA system in southeastern Ireland. Error classification works on three levels: distinguishing between error/error-free operations and classifying a specific error. Error and warning data were analyzed using simple and forced curves to identify the duration of filtering and error activation. The results were good and showed that perfect memory and accuracy could differentiate between error and error-free operations, but the F1 score dropped due to poor accuracy.

3. Methodology

Classification of multi-fault detection is a challenging task due to weak faults [24]. Different machine learning algorithms have their positive and negative sides [33,34]. We have proposed a stacking ensemble classifier that utilizes different classification models to enhance model accuracy. We have proposed a new AdaBoost, K-nearest neighbors, and logistic regression-based stacking ensemble (AKL-SE) classifier. Figure 1 shows the flow diagram of the proposed methodology. We combine SCADA data with the status data obtained from the wind energy converter (WEC). WEC status data corresponds to a status message directly related to the turbine itself [35]. We preprocessed the batch data and applied feature engineering techniques. Then we used a set of parameters to train the stacking ensemble classifier. Finally, we used different evaluation metrics for the validation to classify the fault.

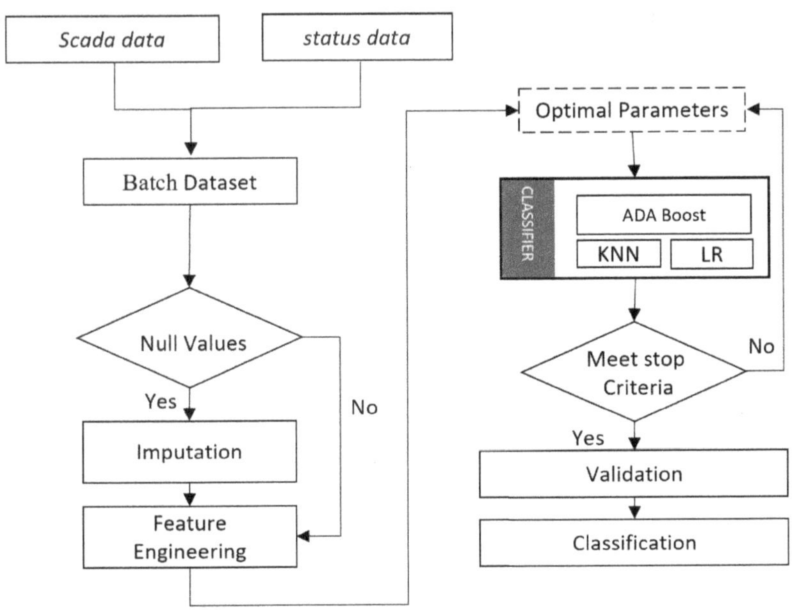

Figure 1. Flow diagram.

3.1. Stacking Ensemble Classification

Stacking ensemble classifier methods is a method for different classification models to enhance the model's accuracy [36]. Stacking is one of the ensemble methods. We have used three classifiers for the meta-models at level 0 to make output. The output of these three classifiers is used as input in level 1 of the meta-model classifier. Figure 2 shows the base and meta-models of the proposed AdaBoost, K-nearest neighbors, and logistic regression-based stacking ensemble (AKL-SE) classifier model.

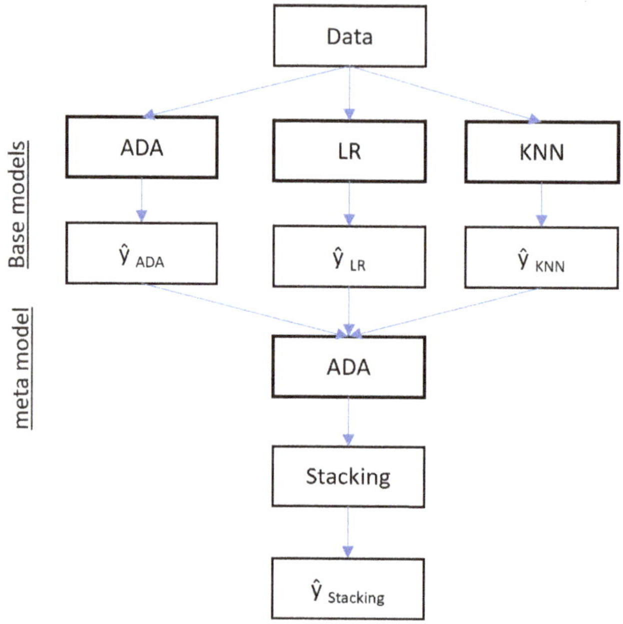

Figure 2. Stacking ensemble classifier model.

We used the "pycaret" Python library to obtain the training accuracy of different models. Then we selected the top three models and used them for stacking. Table 1 summarizes the hyperparameter settings for the base models: AdaBoost, K-nearest neighbors, and logistic regression. The AdaBoost classifier is used as a meta-model.

Table 1. Hyperparameter settings for the base and meta-models.

Sr #	Parameter	AdaBoost	Logistic Regression	KNN
1	base estimators	none	-	-
2	learning rate	1.0	-	-
3	n estimators/jobs	50	none	−1
4	random state	none	none	-
5	leaf size	-	-	30
6	max iter	-	100	-
7	n neighbors	-	-	5
8	c/p	-	1.0	2
9	weights	none	none	uniform
10	penalty	-	12	-

3.2. *AdaBoost*

AdaBoost is short for Adaptive Boosting. It is a statistical classification meta-algorithm [37]. AdaBoost works by systematically applying a weak classification to the training dataset that has been reconsidered by re-weighting [38]. It can be used with many other types of learning algorithms to improve performance. Algorithm 1 explains the pseudocode for AdaBoost training, where we assign every observation x_i an initial weight value and n represents the total number of observations. The first weak decision model is trained, which is usually a decision tree model for every observation. If the prediction is correct, w_i is increased; otherwise, it is decreased. Then, we train another weak model where greater-weight observations are given more priority.

Algorithm 1 Pseudo code for AdaBoost training

```
Require: Initialize weight to x_i
Require: Initialize W_i = 1/n
  Train first weak decision tree model
  for Each observation x_i do
    if pred ≠ correct then
      W_i ++
    else
      W_i --
    end if
  end for
  Train second weak model with greater weights
  return pred
```

3.3. *Logistic Regression*

Logistic regression is a classification algorithm that is used to assign observations to a separate class group. In contrast to linear regression, which produces continuous numeric values, logistic regression translates its output into the use of the logistic sigmoid function. A probability value can be mapped to two or more separate classes. The goal is to choose parameters that maximize likelihood (Equation (1)) and the partial derivative of the log-likelihood (Equation (2)) corresponding to each parameter [39].

$$p(y_i = 1|{x_i}^T w) = g\left({x_i}^T w\right) = 1/\left(1 + exp\left(-{x_i}^T w\right)\right) \tag{1}$$

$$L(w) = \sum_{i=1}^{n} y_i \log\left(g\left({x_i}^T w\right)\right) + (1 - y_i) \log\left(1 - g\left({x_i}^T w\right)\right). \tag{2}$$

where $p(y_i)$ represents the likelihood function and $L(w)$ represents the log-likelihood function [40]. The training record is denoted by a matrix $X = (x_1, x_2, ... x_n)^T$, and $X = (x_i^1, x_i^2, ... x_i^m)^T$ shows each row vector. Corresponding labels to each record are represented by Y, where $Y = (y_1, y_2, ... y_n)^T$. The parameters $W = (w_1, w_2, ... w_n)^T$ can be optimized by training the model through $L(w)$.

3.4. *K-Nearest Neighbor*

K-nearest neighbors (KNN) is a supervised analytics algorithm used for classification and regression problems. The rule is to find the number of pre-determined training samples close to the new point and predict the training samples' labels [41]. Algorithm 2 represents the pseudocode for K-nearest neighbors training. It starts with the initializing of $k, func, target, data_t$, where k represents the user-defined constant, the number of closest training data, $data_t$ represents all training data points, $target$ represents a new point, and $func$ represents functions used to get the target label. The first step is calculating the Euclidean distance between the new point and all training data. The next step is to pick the top-K closest training data. The most common label of these labels, as a result, is picked and returned.

Figure 3 represents the structure of the proposed methodology. A 3 MW direct-drive turbine in the South of Ireland that provides power to a sizable manufacturing facility close to the coast provided us with benchmark SCADA data. Data come from the turbine SCADA system in the form of 10-minute operational and instantaneous alarm system data.

Algorithm 2 Pseudo code for K-nearest neighbors training

Require: Initialize $k, func, target, data_t$
Require: Initialize *neighbors*=[]
  Train first weak decision tree model
  for Each observation $data_t$ **do**
    distance = euclidean distance $(data[: -1], target)$
    calculate euclidean distance
    append neighbors
  end for
  pick the top-K closest training data
  take the most common label of these labels
  return labels

In order to properly train a classifier, the data must be correctly labeled [32]. In this paper, we classified samples as specific fault or fault-free. We labeled the data according to the faults, where status number 62 represents feeding fault or load shedding, fault 80 represents excitation error due to an overvoltage DC-link, fault 228 is for a timeout warning message or malfunction in air cooling, Fault 60 is for mains failure and start delay, and fault nine is for generator heating due to the hygrostat inverter. After performing feature engineering and exploratory data analysis, the data were divided into two parts. One was for training the staking ensemble classifier, and the other was for testing. We used different evaluation metrics, including accuracy, area under the curve (AUC), recall, orecision, and F1 score, to evaluate the accuracy of the proposed model.

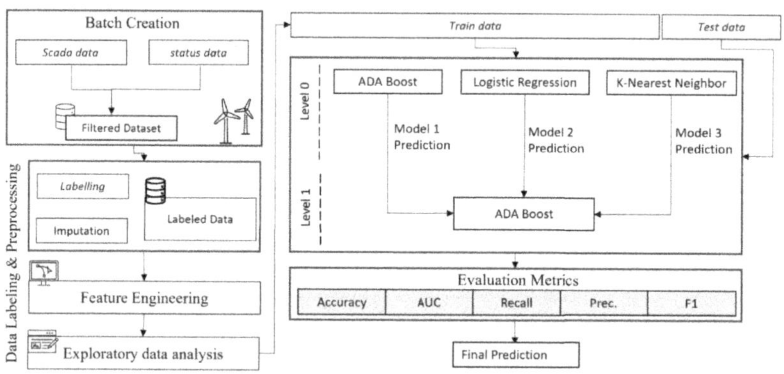

Figure 3. Structure of the proposed methodology.

4. Data Analysis

This section covers the exploratory data analysis of SCADA data. The data used in this study were gathered from a 3 MW direct-drive turbine that powers a sizable manufacturing facility close to the coast in the south of Ireland [42]. Data come from the turbine SCADA system in the form of 10-min operational and instantaneous alarm system data. The data cover 11 months from May 2014 to April 2015.

Figure 4 shows the relationship between power and wind speed, where x-axis represents the wind speed in m/s, and the y-axis represents the active power in kwh. An increment in power is observed with the increase in wind speed. However, there are some outliers where the power is zero, even with a high wind speed.

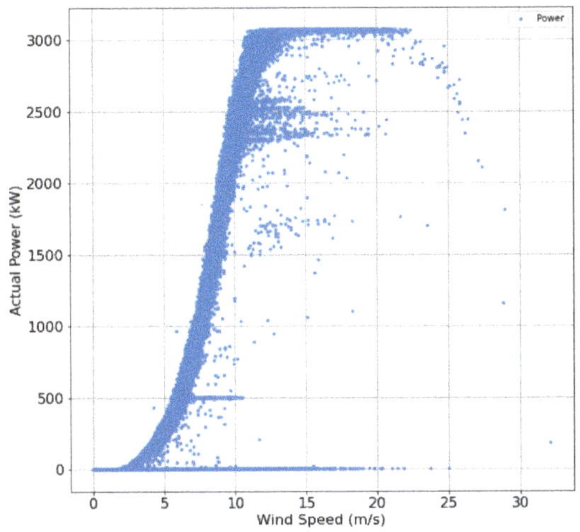

Figure 4. Relation ship between power and wind speed.

Two distinct datasets, WEC Status data and RTU Status data, save the number of normal and abnormal operation states. The WEC (wind energy converter) status information relates to a message that is specific to the turbine. At the point of connection to the grid, RTU data corresponds to power control data, such as active and reactive power. There are "main status" and "sub-status" codes assigned to each status. Any main WEC status code above 0 denotes erroneous behavior but does not necessarily indicate a fault; for example, status code 2 denotes "lack of wind". Only active and reactive power set-points are addressed by statuses in RTU; for example, status 100:82 corresponds to limiting the active power output to 82 percent of its actual current output.

Figure 5 represents the count of all statuses from the wind energy converter in SCADA data. Any main status code of zero indicates that the turbine is generating or spinning up to generate, while anything above zero represents either faulty or reduced operation. The turbine is available to operate but not producing power due to weather, grid, or other events, e.g., status code 2—"lack of wind".

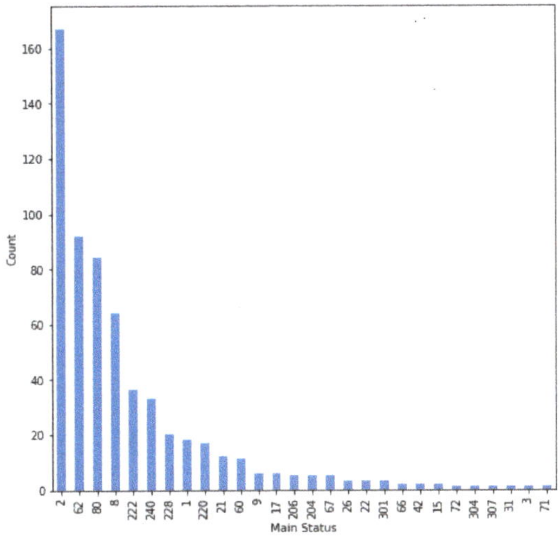

Figure 5. Count of all status readings.

Any "main status" above zero denotes unusual or possibly unusual behavior, but it is not always a fault. In the eleven months of data, there were more than forty different types of faults, but only a small fraction was common enough for a classifier to be trained to identify them. Figure 6 shows the count of faulty status readings. Status number 62 represents a feeding fault or load shedding. Fault 80 represents excitation error due to an overvoltage DC-link. Fault 228 is for a timeout warning message or a malfunction in air cooling. Fault 60 is for mains failure and start delay, and fault nine is for generator heating due to the hygrostat inverter.

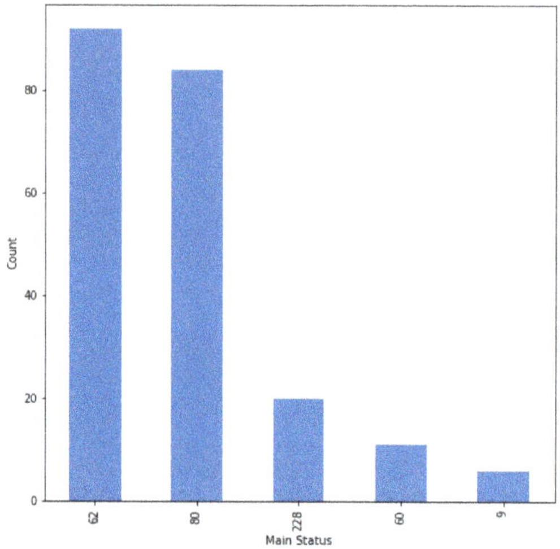

Figure 6. Count of faulty status.

Figure 7 shows the relationship between power and wind speed concerning non-faulty and all individual fault operations. Blue dots represent no faults, orange dots represent fault 62, green dots represent fault 80, red dots represent fault 228, purple dots represent fault 60, and brown dots represent fault 9.

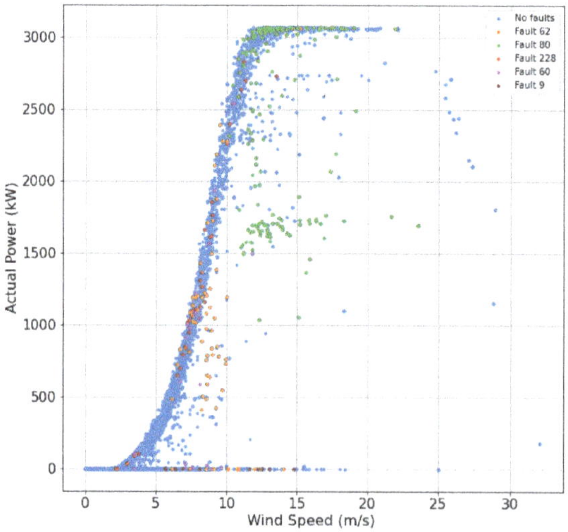

Figure 7. Power and wind speed with respect to non-faulty status and all individual faults.

Figure 8 shows the count of power. The power in the range of 0–300 has the highest count, whereas 1900–2100 has the lowest count.

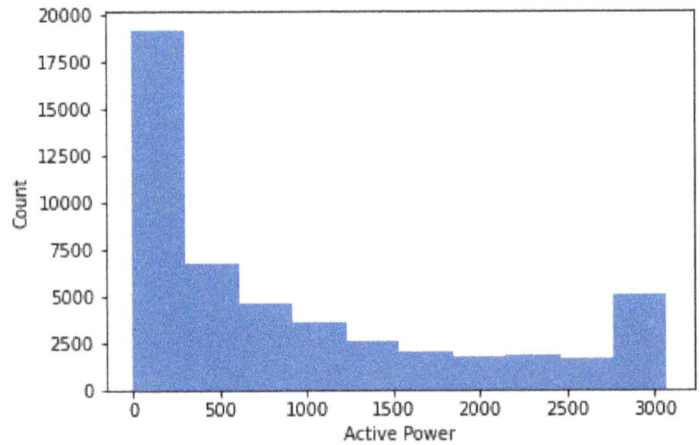

Figure 8. Count of power according to active power range.

Figure 9 shows the count of records where wind speed was recorded but the power was zero. The count is at its maximum in the wind speed range of 0–2.5 m/s and at its minimum in the range of 22.5–25 m/s.

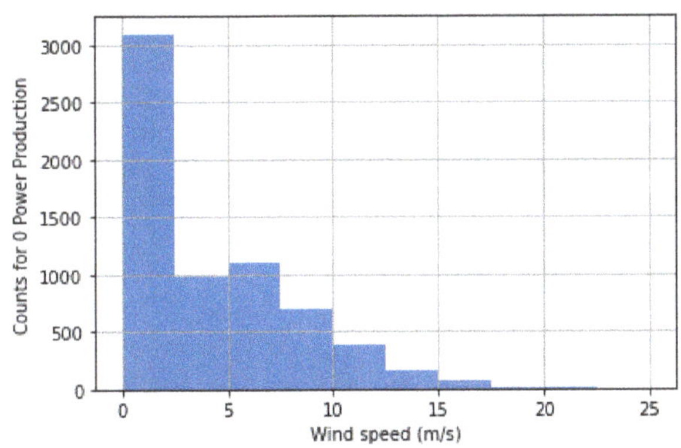

Figure 9. Count of records where wind speed was recorded, but power was zero.

Figure 10 shows the power recorded according to each month. May has the lowest power recorded.

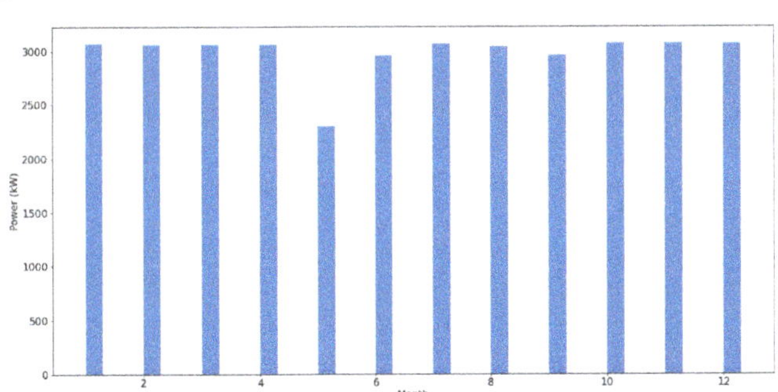

Figure 10. Power recorded according to each month.

5. Results

This section consists of different evaluation metrics. Different tabular and graphical ways are used to evaluate the performance of the proposed model. We have used a correlation diagram (Appendix A), confusion matrix, and receiver operating characteristics curve for a graphical representation of performance. Accuracy, precision, recall, and F1 score are used to compare with other state-of-the-art classifiers.

5.1. Confusion Matrix

The proposed model is interpretable. The engineers and technicians in the wind industry can trust the model's accuracy. We have used a confusion matrix to show the classification visually. A classification problem's prediction outcomes are compiled in a confusion matrix. Count values are used to summarize the number of accurate and inaccurate predictions for each class. Figure 11 shows the confusion matrix obtained after fitting the proposed model. The x-axis represents the actual faults, and the y-axis represents the predicted faults. The diagonal of the matrix shows the true predictions.

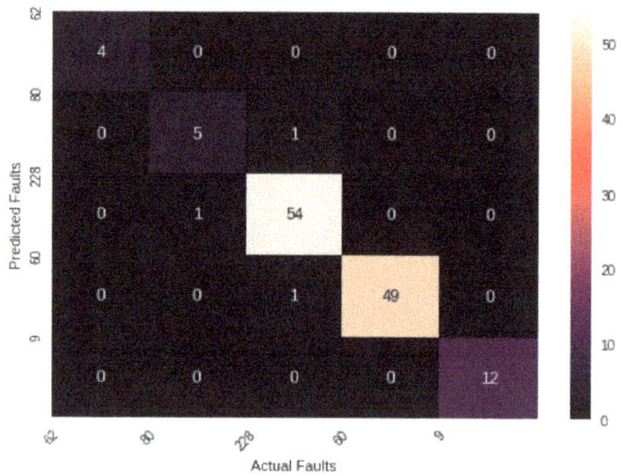

Figure 11. Confusion matrix of faults.

5.2. Receiver Operating Characteristics

For multiclass problems, ROC curves can be plotted using one class versus the rest [43]. To see how well a model performs between sensitivity and specificity, we use a ROC (receiver operating characteristics) plot. Sensitivity is the ability to correctly recognize entries that belong to the positive class. Accurately identifying entries that belong in the negative class is referred to as specificity. Figure 12 shows the ROC curve for multiclass faults, which compares each fault with the rest of the faults.

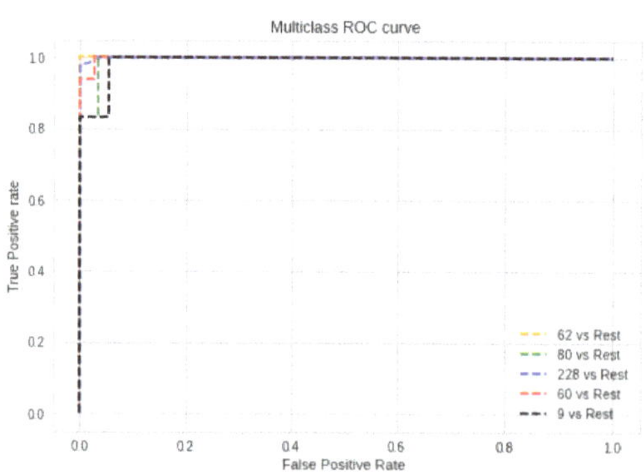

Figure 12. ROC curve for multiclass faults.

5.3. Precision

Precision is the number of classified correct outputs, or the exactness of the model. It is calculated by using Equation (3), where T_p represents true positive values and F_p represents false positive values.

$$Precision = \frac{T_p}{T_p + F_p} \tag{3}$$

5.4. Recall

Recall is the measure of our model correctly identifying true positives. It is calculated by using Equation (4), where T_p represents true positive values and F_n represents false negative values.

$$Recall = \frac{T_p}{T_p + F_n} \tag{4}$$

5.5. Accuracy

Accuracy is the percentage of correctly predicted outputs. It measures how many positive and negative observations were correctly classified. It is calculated by using Equation (5), where T_p represents true positive values, T_n represents true negative values, F_p represents false positive values, and F_n represents false negative values.

$$Accuracy = \frac{T_p + T_n}{T_p + F_p + F_n + T_n} \tag{5}$$

5.6. F1 Score

F1 score is the average of precision and recall. It combines precision and recall into one metric by calculating the harmonic mean between those two. It is calculated by using Equation (6), where T_p represents true positive values, F_p represents false positive values, and F_n represents false negative values.

$$F_1 = \frac{2T_p}{2T_p + F_p + F_n} \tag{6}$$

We compared our proposed model with other state-of-the-art machine learning classifiers. We compare the proposed stacking ensemble classifier with AdaBoost, KNN, logistic regression, quadratic discriminant analysis, naive Bayes, and decision tree classifier. Some of the machine learning and deep learning models perform better individually; however, a stacking ensemble learns the most effective way to integrate the predictions from various effective deep learning or machine learning models. We have compared the performance of individual models with the proposed model. Table 2 shows a comparison of accuracy, precision, recall, and F1 score. AdaBoost, KNN, and logistic regression performed best as compared to other models, so we used these three classifiers and ensembled them using the stacking ensemble method. Machine learning models can perform reasonably well on their own, but in this study, we discovered that combining several of the best-performing ML models can produce even better results. The best feature selection, for which we used a confusion matrix, is a further factor in increasing accuracy. The proposed model performs better than the individual models when measured against a variety of evaluation metrics.

Table 2. Comparison of evaluation matrices to other models.

Sr #	Model	Accuracy	Precision	Recall	F1 Score
1	AdaBoost	0.95	0.91	0.95	0.93
2	K-Nearest Neighbors	0.97	0.97	0.96	0.96
3	Logistic Regression	0.96	0.97	0.96	0.96
4	Quadratic Discriminant Analysis	0.88	0.88	0.88	0.87
5	Naive Bayes	0.68	0.77	0.68	0.65
6	Decision Tree Classifier	0.65	0.66	0.65	0.62
7	Recurrent Neural Network	0.72	0.74	0.71	0.72
8	Stacking Classifier	0.98	0.98	0.98	0.97

6. Conclusions

With a focus on renewable energy sources, the reliability and efficiency of wind turbines have become an essential issue in enhancing the total installed capacity of wind turbines. Improved performance requires upgrades in various parts of wind turbines. This paper demonstrates a wind turbine fault detection method based on the AdaBoost, K-nearest neighbors, and logistic regression-based stacking ensemble (AKL-SE) classifier. The anomaly identification model proposed in this paper can identify the fault state of a wind turbine. A new stacking classifier-based ensemble classification method is proposed for fault diagnosis. We have used data from the supervisory control and data acquisition (SCADA) system to improve wind turbine fault detection accuracy. This article provides a solution to identify multi faults in a wind turbine. We used three classification models, AdaBoost, K-nearest neighbors, and logistic regression, as the primary format for generating output. The output of these three classifications was used as the input to the meta-model of the logistic regression classifier. We compared our proposed model with other advanced machine learning classifications. We compared the proposed stack ensemble classifier with AdaBoost, a CNN, logistic regression, quadratic discrimination analysis, naive Bayes, a recurrent neural network, and decision tree classifier. Accuracy, precision, recall, and F1 scores were also compared. Since AdaBoost, KNN, and logistic regression worked better than other models, we combined these three classifications using the stacking ensemble method.

The data used for this study were obtained from a single wind turbine; however, the proposed approach is transferable and generalizable to multiple wind farms. The proposed method is more robust as compared to the traditional approach. However, some improvements need to be made to enhance trustworthiness. In the future, optimization algorithms can be used to enhance accuracy.

Author Contributions: Conceptualization, P.W.K.; formal analysis, P.W.K.; funding acquisition, Y.-C.B.; investigation, Y.-C.B.; methodology, P.W.K.; project administration, Y.-C.B.; supervision, Y.-C.B. All authors have read and agreed to the published version of the manuscript.

Funding: This research was supported by the 2022 scientific promotion program funded by Jeju National University.

Institutional Review Board Statement: Not applicable.

Informed Consent Statement: Not applicable.

Data Availability Statement: Not applicable.

Acknowledgments: The authors would like to show their gratitude to Dr. Kevin Leahy from the School of Engineering, University College Cork, Ireland, for his cooperation in this research.

Conflicts of Interest: The authors declare no conflict of interest.

Abbreviations

The following abbreviations are used in this manuscript:

SCADA	Supervisory control and data acquisition
O&M	Operation and maintenance
ML	Machine learning
WT	Wind turbines
RES	Renewable energy sources
LR	Logistic regression
MLP	Multi-layer perceptron
SVM	Support-vector machine
LSTM	Long short-term memory
CMS	Condition monitoring system
WEC	Wind energy converter
KNN	K-nearest neighbors
AUC	Area under the curve
ROC	Receiver operating characteristics

Appendix A

A correlation plot of all the variables is the first step in any factor analysis to evaluate whether some variables are useful or too correlated with others. Figure A1 shows the Correlation diagram of features.

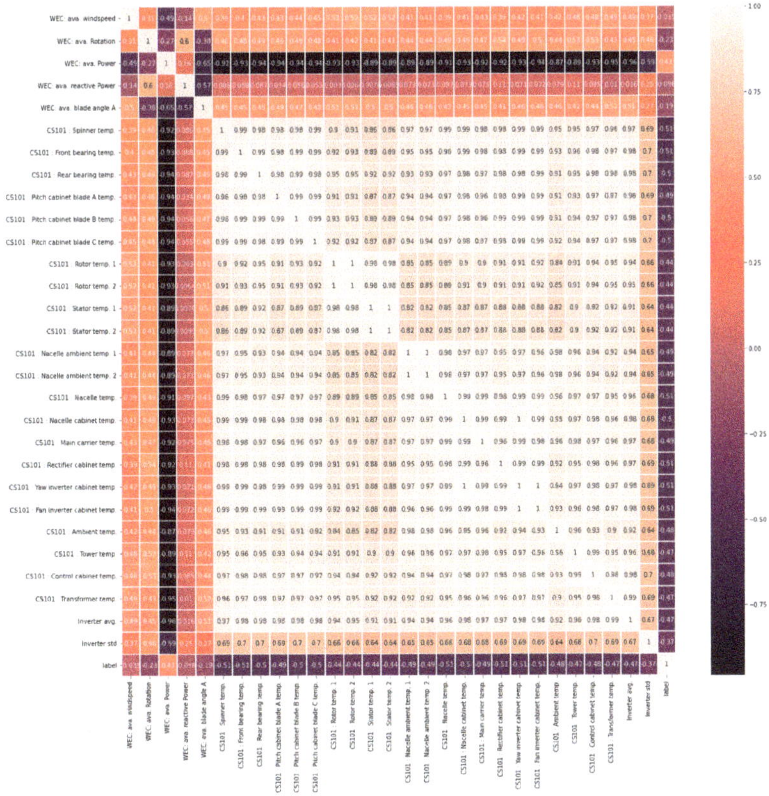

Figure A1. Correlation diagram of features.

References

1. Xiang, L.; Yang, X.; Hu, A.; Su, H.; Wang, P. Condition monitoring and anomaly detection of wind turbine based on cascaded and bidirectional deep learning networks. *Appl. Energy* **2022**, *305*, 117925. [CrossRef]
2. Zhu, Y.; Zhu, C.; Tan, J.; Tan, Y.; Rao, L. Anomaly detection and condition monitoring of wind turbine gearbox based on LSTM-FS and transfer learning. *Renew. Energy* **2022**, *189*, 90–103. [CrossRef]
3. Zhang, Y.; Zheng, H.; Liu, J.; Zhao, J.; Sun, P. An anomaly identification model for wind turbine state parameters. *J. Clean. Prod.* **2018**, *195*, 1214–1227. [CrossRef]
4. Khan, P.W.; Byun, Y.C. Adaptive error curve learning ensemble model for improving energy consumption forecasting. *Comput. Mater. Contin* **2021**, *69*, 1893–1913. [CrossRef]
5. Li, H.; Chen, Z. Overview of different wind generator systems and their comparisons. *IET Renew. Power Gener.* **2008**, *2*, 123–138. [CrossRef]
6. Wang, Y.; Ma, X.; Qian, P. Wind turbine fault detection and identification through PCA-based optimal variable selection. *IEEE Trans. Sustain. Energy* **2018**, *9*, 1627–1635. [CrossRef]
7. Chen, X.; Yang, Y.; Cui, Z.; Shen, J. Vibration fault diagnosis of wind turbines based on variational mode decomposition and energy entropy. *Energy* **2019**, *174*, 1100–1109. [CrossRef]
8. Bodla, M.K.; Malik, S.M.; Rasheed, M.T.; Numan, M.; Ali, M.Z.; Brima, J.B. Logistic regression and feature extraction based fault diagnosis of main bearing of wind turbines. In Proceedings of the 2016 IEEE 11th Conference on Industrial Electronics and Applications (ICIEA), Hefei, China, 5–7 June 2016; pp. 1628–1633.
9. Malik, H. Wavelet and Hilbert Huang transform based wind turbine imbalance fault classification model using k-nearest neighbour algorithm. *Int. J. Renew. Energy Technol.* **2018**, *9*, 66–83. [CrossRef]
10. Lima, L.; Blatt, A.; Fujise, J. Wind turbine failure prediction using SCADA data. In *Proceedings of the Journal of Physics: Conference Series*; IOP Publishing: Bristol, UK, 2020; Volume 1618, p. 022017.
11. Chatterjee, J.; Dethlefs, N. Deep learning with knowledge transfer for explainable anomaly prediction in wind turbines. *Wind Energy* **2020**, *23*, 1693–1710. [CrossRef]
12. Zheng, X.X.; Peng, P. Fault diagnosis of wind power converters based on compressed sensing theory and weight constrained Adaboost-SVM. *J. Power Electron.* **2019**, *19*, 443–453.
13. Wu, Z.; Wang, X.; Jiang, B. Fault diagnosis for wind turbines based on ReliefF and eXtreme gradient boosting. *Appl. Sci.* **2020**, *10*, 3258. [CrossRef]
14. Downey, A.; Ubertini, F.; Laflamme, S. Algorithm for damage detection in wind turbine blades using a hybrid dense sensor network with feature level data fusion. *J. Wind. Eng. Ind. Aerodyn.* **2017**, *168*, 288–296. [CrossRef]
15. Kushwah, K.; Sahoo, S.; Joshuva, A. Health Monitoring of Wind Turbine Blades Through Vibration Signal Using Machine Learning Techniques. In Proceedings of the International Conference on Computing and Communication Systems, Vellore, India, 9–11 December 2021; Springer: Berlin/Heidelberg, Germany, 2021; pp. 239–247.
16. Joshuva, A.; Deenadayalan, G.; Sivakumar, S.; Sathishkumar, R.; Vishnuvardhan, R. Logistic model tree classifier for condition monitoring of wind turbine blades. *Int. J. Recent Technol. Eng.* **2019**, *8*, 202–209.
17. Wenyi, L.; Zhenfeng, W.; Jiguang, H.; Guangfeng, W. Wind turbine fault diagnosis method based on diagonal spectrum and clustering binary tree SVM. *Renew. Energy* **2013**, *50*, 1–6. [CrossRef]
18. Jiang, G.; He, H.; Yan, J.; Xie, P. Multiscale convolutional neural networks for fault diagnosis of wind turbine gearbox. *IEEE Trans. Ind. Electron.* **2018**, *66*, 3196–3207. [CrossRef]
19. Rahimilarki, R.; Gao, Z.; Jin, N.; Zhang, A. Convolutional neural network fault classification based on time-series analysis for benchmark wind turbine machine. *Renew. Energy* **2022**, *185*, 916–931. [CrossRef]
20. Li, Y.; Huang, X.; Tee, K.F.; Li, Q.; Wu, X.P. Comparative study of onshore and offshore wind characteristics and wind energy potentials: A case study for southeast coastal region of China. *Sustain. Energy Technol. Assess.* **2020**, *39*, 100711. [CrossRef]
21. Vidal, Y.; Pozo, F.; Tutivén, C. Wind turbine multi-fault detection and classification based on SCADA data. *Energies* **2018**, *11*, 3018. [CrossRef]
22. Miele, E.S.; Bonacina, F.; Corsini, A. Deep anomaly detection in horizontal axis wind turbines using Graph Convolutional Autoencoders for Multivariate Time series. *Energy AI* **2022**, *8*, 100145. [CrossRef]
23. Lei, J.; Liu, C.; Jiang, D. Fault diagnosis of wind turbine based on Long Short-term memory networks. *Renew. Energy* **2019**, *133*, 422–432. [CrossRef]
24. He, X.; Zhou, X.; Yu, W.; Hou, Y.; Mechefske, C.K. Adaptive variational mode decomposition and its application to multi-fault detection using mechanical vibration signals. *ISA Trans.* **2021**, *111*, 360–375. [CrossRef] [PubMed]
25. Joshuva, A.; Kumar, R.S.; Sivakumar, S.; Deenadayalan, G.; Vishnuvardhan, R. An insight on VMD for diagnosing wind turbine blade faults using C4. 5 as feature selection and discriminating through multilayer perceptron. *Alex. Eng. J.* **2020**, *59*, 3863–3879. [CrossRef]
26. Teng, W.; Ding, X.; Zhang, X.; Liu, Y.; Ma, Z. Multi-fault detection and failure analysis of wind turbine gearbox using complex wavelet transform. *Renew. Energy* **2016**, *93*, 591–598. [CrossRef]
27. Mishnaevsky, L., Jr.; Thomsen, K. Costs of repair of wind turbine blades: Influence of technology aspects. *Wind Energy* **2020**, *23*, 2247–2255. [CrossRef]

28. Ou, Y.; Tatsis, K.E.; Dertimanis, V.K.; Spiridonakos, M.D.; Chatzi, E.N. Vibration-based monitoring of a small-scale wind turbine blade under varying climate conditions. Part I: An experimental benchmark. *Struct. Control. Health Monit.* **2021**, *28*, e2660. [CrossRef] [PubMed]
29. Zhang, Y.; Hutchinson, P.; Lieven, N.A.; Nunez-Yanez, J. Adaptive event-triggered anomaly detection in compressed vibration data. *Mech. Syst. Signal Process.* **2019**, *122*, 480–501. [CrossRef]
30. Coronado, D.; Wenske, J. Monitoring the oil of wind-turbine gearboxes: main degradation indicators and detection methods. *Machines* **2018**, *6*, 25. [CrossRef]
31. Bie, Y.; Liu, X.; Xu, T.; Zhu, Z.; Li, Z. A review of the application of oil analysis in condition monitoring and life prediction of wind turbine gearboxes. *Insight-Non-Destr. Test. Cond. Monit.* **2021**, *63*, 289–301. [CrossRef]
32. Leahy, K.; Hu, R.L.; Konstantakopoulos, I.C.; Spanos, C.J.; Agogino, A.M. Diagnosing wind turbine faults using machine learning techniques applied to operational data. In Proceedings of the 2016 IEEE International Conference on Prognostics and Health Management (Icphm), Ottawa, ON, Canada, 20–22 June 2016; pp. 1–8.
33. Khan, P.W.; Kim, Y.; Byun, Y.C.; Lee, S.J. Influencing Factors Evaluation of Machine Learning-Based Energy Consumption Prediction. *Energies* **2021**, *14*, 7167. [CrossRef]
34. Khan, P.W.; Byun, Y.C. Analysis of factors affecting machine learning-based energy prediction. In Proceedings of the KIIT Conference, 2021; pp. 33–35.
35. Pang, Y.; He, Q.; Jiang, G.; Xie, P. Spatio-temporal fusion neural network for multi-class fault diagnosis of wind turbines based on SCADA data. *Renew. Energy* **2020**, *161*, 510–524. [CrossRef]
36. Khan, P.W.; Byun, Y.C.; Lee, S.J. Optimal Photovoltaic Panel Direction and Tilt Angle Prediction Using Stacking Ensemble Learning. *Front. Energy Res.* **2022**, *10*, 865413. [CrossRef]
37. Lu, W.; Liu, J.; Fu, X.; Yang, J.; Jones, S. Integrating machine learning into path analysis for quantifying behavioral pathways in bicycle-motor vehicle crashes. *Accid. Anal. Prev.* **2022**, *168*, 106622. [CrossRef] [PubMed]
38. Guo, H.; Li, Y.; Li, Y.; Liu, X.; Li, J. BPSO-Adaboost-KNN ensemble learning algorithm for multi-class imbalanced data classification. *Eng. Appl. Artif. Intell.* **2016**, *49*, 176–193.
39. Piech, C. Logistic Regression. 2017. Available online: https://pdfs.semanticscholar.org/8a27/7cf63806ee25977bb8a59fa511e591 8d2cfe.pdf (accessed on 7 May 2022).
40. Jiang, Y.; Hamer, J.; Wang, C.; Jiang, X.; Kim, M.; Song, Y.; Xia, Y.; Mohammed, N.; Sadat, M.N.; Wang, S. SecureLR: Secure logistic regression model via a hybrid cryptographic protocol. *IEEE/ACM Trans. Comput. Biol. Bioinform.* **2018**, *16*, 113–123. [CrossRef]
41. Niño-Adan, I.; Landa-Torres, I.; Portillo, E.; Manjarres, D. Influence of statistical feature normalisation methods on K-Nearest Neighbours and K-Means in the context of industry 4.0. *Eng. Appl. Artif. Intell.* **2022**, *111*, 104807. [CrossRef]
42. Leahy, K.; Hu, R.L.; Konstantakopoulos, I.C.; Spanos, C.J.; Agogino, A.M.; O'Sullivan, D.T. Diagnosing and predicting wind turbine faults from SCADA data using support vector machines. *Int. J. Progn. Health Manag.* **2018**, *9*. [CrossRef]
43. Yansari, R.T.; Mirzarezaee, M.; Sadeghi, M.; Araabi, B.N. A new survival analysis model in adjuvant Tamoxifen-treated breast cancer patients using manifold-based semi-supervised learning. *J. Comput. Sci.* **2022**, *61*, 101645. [CrossRef]

sensors

MDPI

Article

Leveraging Active Learning for Failure Mode Acquisition

Amol Kulkarni [1], Janis Terpenny [2,*] and Vittaldas Prabhu [1]

[1] Department of Industrial and Manufacturing Engineering, The Pennsylvania State University, University Park, State College, PA 16802, USA
[2] Department of Systems Engineering and Operations Research, George Mason University, Fairfax, VA 22030, USA
* Correspondence: terpenny@gmu.edu

Abstract: Identifying failure modes is an important task to improve the design and reliability of a product and can also serve as a key input in sensor selection for predictive maintenance. Failure mode acquisition typically relies on experts or simulations which require significant computing resources. With the recent advances in Natural Language Processing (NLP), efforts have been made to automate this process. However, it is not only time consuming, but extremely challenging to obtain maintenance records that list failure modes. Unsupervised learning methods such as topic modeling, clustering, and community detection are promising approaches for automatic processing of maintenance records to identify failure modes. However, the nascent state of NLP tools combined with incompleteness and inaccuracies of typical maintenance records pose significant technical challenges. As a step towards addressing these challenges, this paper proposes a framework in which online active learning is used to identify failure modes from maintenance records. Active learning provides a semi-supervised machine learning approach, allowing for a human in the training stage of the model. The hypothesis of this paper is that the use of a human to annotate part of the data and train a machine learning model to annotate the rest is more efficient than training unsupervised learning models. Results demonstrate that the model is trained with annotating less than ten percent of the total available data. The framework is able to achieve ninety percent (90%) accuracy in the identification of failure modes in test cases with an F-1 score of 0.89. This paper also demonstrates the effectiveness of the proposed framework with both qualitative and quantitative measures.

Keywords: fault-mode acquisition; maintenance records; active learning; human-in-the-loop learning

1. Introduction

To improve product quality and reliability, it is necessary to implement appropriate actions to avoid faults. This is typically completed through a process known as Failure Mode and Effects Analysis (FMEA). A critical step in FMEA is to identify different failure modes, associated product subsystems, and components where faults occur. The traditional approach to improving reliability has been to perform finite element analysis to simulate and predict different failures [1–4]. Typically, failure modes are acquired through historical documents, bill of materials, and failure analysis reports. This can be a very labor intensive and time-consuming task that is both inefficient and prone to errors because of incomplete data, differences in recording maintenance information, or the lack of a standard vocabulary for maintenance records among different departments within the industry [5]. Recent advances in Natural Language Processing (NLP) are enabling automated processing of large volumes of textual data. For example, asset rich industries such as mining require substantive upfront investment specifically to acquire heavy mobile equipment. In 2022, the capital expenditure of the twenty leading miners was expected to reach USD 70.4 billion [6]. Similar situations exist in other industrial sectors such as aviation, power, and energy. Clearly, it is necessary to maintain assets over a long-life cycle to ensure return on investment along with asset maintenance data stored in company maintenance systems [7].

There are existing studies that leverage the information from maintenance logs to identify failure modes to construct a component–failure matrix using various machine learning techniques [5,8–11]. However, most of the studies focus on techniques related to a specific domain. In the cases where unsupervised learning techniques are employed, the number of clusters is typically fixed, making it necessary to provide the number of clusters or topics present within the documents. Active learning is a semi-supervised machine learning approach, which involves a human in the training stage of the model. This paper proposes to bridge this gap with the use of an online active learning method for failure mode identification. The key contributions of the paper include:

1. An uncertainty-based online active learning model for identifying and classifying faults in maintenance records;
2. Identifying some of the shortcomings in current state-of-the-art techniques related to classifying or clustering maintenance records;
3. A proposed standard vocabulary for fault modes across specific industries.

The following section provides a review of the literature including recent developments to extract failure modes using NLP along with a brief overview of the issues with maintenance texts. The next section presents the research framework and different querying techniques in online active learning. The Case Study section demonstrates the advantage of using the proposed active learning model over NLP techniques on mining equipment maintenance data. The paper concludes with a discussion of findings, benefits of the framework, and proposed future work.

2. Literature Review

Natural Language Processing (NLP)-based methods have been used in the past to obtain failure modes. Extraction of standard failure modes using failure mode classification from a failure–experience matrix [12] and bill of materials with recorded failure information [13] have been successfully demonstrated. A classification method using tree kernel-based support vector machine to identify bearing failures [14] based on a set category of failure modes has also been accomplished. However, these classification methods pre-suppose the type of failure modes, which is not applicable to complex systems, such as industrial robots, wind turbines, and more. Such complex systems are likely to have more failure modes than the set of failure modes under consideration.

Unsupervised learning methods such as cluster analysis have been studied in the past during the conceptual design phase of systems. For example, "K-means" clustering combined with a row addition method based on artificially established failure modes and their frequency was studied in rotating mechanical systems [15] to develop the Component–Failure (CF) matrix. The K-means clustering algorithm has also been used on preprocessed software failure text using the representative samples as cluster labels [16]. Processes to extract failure from maintenance texts using hierarchical clustering and similar distribution-based clustering methods have been explored [17]. Researchers have also explored clustering historical FMEA data to extract failure modes and convert the structured data in FMEA sheets into an evolving tree algorithm [18]. Apart from supervised and unsupervised learning, ontology-based methods have also been proposed where the failure modes are extracted from historical data on maintenance and services [11] as well as repairing verbatim data [19].

While promising, the methods described above do not account for the non-uniform description of failure mode data or the lack of standard vocabulary for FMEA data. Although efforts have been made to effectively maintain and utilize a standard failure mode vocabulary [13], a systematic approach to standardize failure modes has not been established. To tackle these issues, a frequency itemset mining [5] approach was proposed to construct the CF matrix. This approach, however, is computationally expensive as frequency set mining is resource-intensive and time-consuming. Frequency itemset mining has also been known to develop associations or rules that are a product of chance.

Most clustering algorithms, such as K-means and hierarchical agglomerative clustering algorithms, are similarity-based. These algorithms represent documents as a vector space model (VSM) [20], where a vector of length 'V' represents each document. The weight of a word typically calculated using TF-IDF (Term Frequency Inverse Document Frequency) represents the corresponding elements of the vector. Due to the sparsity of short texts, most of the words will have TF = 1, which means TF is ineffective in calculating the weights. This representation results in high time and space complexity, creating dimensionality issues in short texts.

More recently, the focus has been on utilizing a topic modeling approach to extract hidden failure information from maintenance records [8,21]. Topic modeling has become a widely used text clustering approach to discover abstract topics and infer hidden semantic structures from a collection of documents with techniques that can be categorized as clustering, classification, or probabilistic.

The most widely used approach in topic modeling is the Latent Dirichlet Allocation (LDA) model which has been used to improve user experience [22], summarize and provide an overview of discovered themes to recommend books [23], and analyze research trends [24]. LDA is a generative probabilistic model with three levels of hierarchy [25] for collections of discrete data such as texts. The finite mixture of topics serves as an underlying model for each collection of items, which in turn is modeled as an infinite mixture over an underlying set of probabilities. While LDA works well on long text data, it has been criticized as not being suitable for inferring topics from short texts [26], having issues with scaling [27], and ignoring co-occurrence relations [28]. These issues make it unsuitable to infer topics (failure modes) from maintenance records. It also requires that the number of topics that are specified as a parameter be supplied to the model.

To overcome the co-occurrence limitation of LDA, a Hierarchical Latent Tree Model (HLTM) has been proposed [29]. HLTM clusters documents by treating words at the bottom level as observed binary variables and words at other levels as latent binary variables representing co-occurrence patterns. While efficient when compared to LDA, this would be unsuitable for short-text clustering. A latent topic text representation model, based on the assumption that words in the same topic follow a Gaussian distribution, was proposed to overcome the limitations of vector representation models and semantic models [30]. To discover topics, it aims to represent text as a probability distribution, causing it to perform better than LDA. However, this model is likely to perform less coherently when short texts are considered due to sparsity and lack of structure. Another proposed method that clusters texts thematically is based on Formal Concept Analysis (FCA), organizing text in the form of a concept lattice for topic modeling of Twitter data [31]. FCA also facilitates the detection of new topics based on the information coming from previous topics. However, this approach does not generalize well. There are numerous approaches that serve as an alternative to LDA, yet limitations persist.

With the advent of deep-learning and sentence embedding techniques such as BERT (bi-directional encoder representation from Transformers), GloVe (Global Vectors for Word Representation) doc2Vec, and others, short text-classification has made significant progress in recent years. For example, a recent study [32], sought to identify misinformation pertaining to COVID-19 by performing sentiment analysis utilizing Twitter data. The study focuses on attitudes, sentiments, and how misinformation affects "E-learning". The noise in textual data is removed using a denoising auto-encoder and an attentional feature fusion mechanism is used to combine multi-level features of ELM-AE (Extreme Learning Machine—Auto Encoder) with LSTM (Long Short-Term Memory). The authors achieved an F-1 score of 0.945, which is a higher F-1 score than other state-of-the-art approaches. A similar study [33] focused on short text classification using a feature fusion framework based on BERT. The short text defined in this case pertained to Twitter data, which have

180–200 characters. However, the short text considered focused on maintenance records which have less than 40 characters, making it much more difficult to classify.

2.1. Active Learning

Active learning, also referred to as "query learning" or "optimal experimental design", is a subset of machine learning where a learning algorithm can interact with the user to label unknown data points with desired outputs. It is a type of semi-supervised machine learning algorithm. The algorithm selects a subset of samples to be labeled next from a pool of unlabeled data. The fundamental belief is that if the algorithm can choose the observations it wants to learn from, it can perform better than the traditional learning algorithms with substantially fewer observations [34]. The observations under consideration can be categorized into three subsets. If N is the total number of observations under consideration, during each iteration i, N can be broken up into three subsets:

- $(N–K)_i$: observations where the label is known;
- K_i: observations where the label is unknown;
- C_i: A subset of Ki that is chosen to be labeled.

The decision whether to query a specific label depends on whether the gain from querying the label is greater than the cost of obtaining that information. This practice can take a few different forms depending on the budget limit and other factors, namely: 1. membership query synthesis, 2. pool-based sampling, and 3. Stream-based selective sampling.

2.1.1. Membership Query Analysis

Membership Query Analysis was the first active learning scenario to be investigated [35]. In this scenario, the algorithm may request a label for any unlabeled observation in the input space, including queries generated instead of the ones sampled from an underlying distribution. For finite problem domains, query synthesis is often tractable and efficient [36]. This idea has also been extended to regression learning tasks [37] for predicting the absolute coordinates of a robot hand.

The scenario described here is not applicable to all cases since it involves the generation of synthetic data. In this method, the active learner creates its own examples for labeling. This method is suitable for situations where it is easy to generate an example of data. For instance, some have tried to classify handwritten characters using neural networks using human oracles [38]. The queries generated by the learner often contained unrecognizable symbols and were mostly artificial hybrid characters that had no natural semantic. To overcome this limitation, an innovative and promising approach using a "robot scientist" [39] has been investigated to discover metabolic pathways in yeast saccharomyces cerevisiae (Brewer's yeast). This approach is promising in domains where humans are not the oracles.

2.1.2. Pool-Based Sampling

Most real-world datasets have many unlabeled data. In pool-based active learning [40], the algorithm is trained on a fully labeled part of the data. This helps the algorithm to determine the instances that would be most beneficial to be added to the training set in the next iteration. An informativeness measure evaluates all the data before a query or a set of queries is carried out. It is one of the most popular active learning scenarios and has been studied for various applications such as image classification and retrieval [40–43]. The downside of this algorithm is the amount of memory required.

2.1.3. Stream-Based Selective Sampling

This scenario assumes that obtaining unlabeled data is inexpensive. It first selects samples from the actual distribution before deciding whether to request the oracle for labels. Stream-based selective sampling, alternatively known as sequential active learning, draws out each unlabeled instance one at a time from the data before making the decision to query. If the underlying distribution is non-uniform or unknown, the queries will be sensible, coming from the real distribution [34].

There are different strategies to help the algorithm arrive at a decision on whether to carry out a query. One approach is to use an informativeness measure and make a biased random decision by favoring more informative observations to be queried [44]. Another approach is to select the observations that are still ambiguous to the algorithm. A naïve strategy approach is to set a threshold for the informativeness measure and to select observations that are above the threshold.

Active learning has been increasingly used in the realm of fault diagnosis [45–47]. A novel probabilistic framework, based on active learning, was used as an online strategy for SHM [47]. It was found that active learning was more efficient and accurate compared to traditional machine learning algorithms. The authors used vibration data from different sensors on a Z24 bridge and simulated vibration measurements from an aircraft engine to classify faults. Similarly, an active learning method based on uncertainty and complexity was proposed for gearbox fault diagnosis [46]. The authors used empirical mode decomposition–singular value decomposition (EMD-SVD) to obtain feature vectors from sensor signals, and then used active learning combined with Random Forest to train and classify faults in a gearbox. The results from both studies showed that active learning models outperformed traditional machine learning models. While the use of sensor data to identify and classify fault modes is a novel approach, it does not deal with the classification of textual data. The authors make an assumption that sensor data is readily available to identify faults. This is not the case for most industries, which are still hesitant to invest in expensive sensors for health monitoring. Hence, this paper suggests the use of maintenance records, which are maintained by most industries and require little to no investment in additional hardware for fault mode acquisition.

To recap, although there are many approaches that have been proposed for topic modeling, LDA continues to be the most frequently used algorithm. The methods that serve as alternatives to LDA are not well-suited for short-text clustering. The next section provides details of issues with short-text clustering, specifically maintenance text, which is the focus of this paper. It will also provide a brief overview of the issues with the typical text pre-processing steps/pipeline that exist in the NLP realm. To support the discussion, a summary of the literature review is provided in Table 1 below.

Table 1. Summary of Literature.

	Title	Year	Objective	Method
[15]	Failure Mode Analysis of Mechanical Systems at Conceptual Design Stage	2006	To determine the criticality of failure modes	K-means Clustering
[22]	A correlated topic model of science	2007	To improve user experience by using identified topics as guides	Correlation Topic Modeling
[17]	A case study of failure mode analysis with text mining methods	2007	To automate failure mode extraction	Multiple methods
[18]	Clustering and visualization of failure modes using an evolving tree	2015	To cluster and visualize failure modes in FMEA document as tree structures	Evolving Tree
[14]	A Preliminary Study of Clinical Abbreviation Disambiguation in Real Time	2015	To recognize clinical abbreviations	Word sense disambiguation
[19]	A data and ontology-driven text mining-based construction of reliability model to analyze and predict component failures	2016	To enhance the reliability estimation of automobiles using textual data	Ontology
[31]	A step forward for Topic Detection in Twitter: An FCA-based approach	2016	To improve topic detection process	Formal Concept Analysis

Table 1. *Cont.*

	Title	Year	Objective	Method
[11]	Knowledge management of automobile system failures through development of failure knowledge ontology from maintenance experience	2017	To enhance knowledge management of automobile system failures to aid in design and maintenance of automobiles	Ontology
[23]	Unsupervised Topic Modelling in a Book Recommender System for New Users	2017	To summarize the overview of discovered themes for book recommendation	LDA
[24]	Analyzing research trends in personal information privacy using topic modeling	2017	To analyze the research trends between 1972 and 2015 on personal information privacy to offer direction for future research	LDA
[29]	Latent tree models for hierarchical topic detection	2017	To model patterns of word co-occurrence and co-occurrence of those patterns to overcome LDA limitation	HLTM
[30]	Latent Topic Text Representation Learning on Statistical Manifolds	2018	To provide effective text representation and text measurement with latent topics	Gaussian Mixture Model
[47]	Probabilistic active learning: An online framework for structural health monitoring	2019	Fault mode Classification using sensor signals	Online Probabilistic Active Learning
[5]	A data-driven approach for constructing the component-failure mode matrix for FMEA	2020	Failure mode extraction of automobile seat module to build a Component–Failure Matrix for FMEA	Association Rule Mining
[21]	A Framework Based on K-Means Clustering and Topic Modeling for Analyzing Unstructured Manufacturing Capability Data	2020	To discover patterns in manufacturing capability corpus with clustering suppliers' capability	K-means Clustering and Topic Modeling
[45]	An Applicable Predictive Maintenance Framework for the Absence of Run-to-Failure Data	2021	Predictive maintenance framework to identify and classify faults using sensor signals	Autoencoder and simple linear regression
[32]	Topic Modeling and Sentiment Analysis of Online Education in the COVID-19 Era Using Social Networks Based Datasets	2022	To identify misinformation related to COVID-19 as it pertains to E-Learning	Attentional Feature Fusion with ELM-AE and LSTM

3. Overview of Maintenance Text Issues

A typical NLP pre-processing step involves:

1. Tokenization: Tokenization is the fundamental step in NLP which is a process of segmenting texts into tokens from sentences into words and characters. It is possible to use the token occurrences of a document as a vector representation. This process converts unstructured text into structured data for analysis.
2. Removing stop words: The stop word removal process involves removing common language articles, pronouns, and prepositions such as 'and', 'the', or 'to' in English. This process not only frees up space but also improves processing time for any ML model. Stop words can be removed by performing a lookup operation from a predefined list. It is important to note that there is no universal list of stop words. This is true especially when it comes to technical text. The trend in the past few years has shifted from using a long list of stop words to not removing any stop words at

all. Because this process decontextualizes the sentences, it is typically not suitable to perform sentiment analysis.

3. Stemming: This is the process of reducing a word to its root form. Stemming typically removes prefixes and suffixes, generating the stem of the word. It is a relatively fast and simple process. The issue with stemming is that it can generate stems that are not actual words, thus affecting the accuracy of NLP.
4. Lemmatization: The base of a word in its base form or dictionary form is called a lemma. The process of reducing words to lemma is termed lemmatization. The primary difference with stemming is that lemmatization takes context and grammar into consideration. The process itself requires access to a dictionary or a knowledge base mapping words to their lemma forms. This process is slower and computationally more expensive compared to stemming.

While helpful, these steps also cause several issues with tailoring Natural Language Processing tools to expert-driven text data [48]. The pre-processing step may cause issues such as removing asset identifiers (ABB IRB 140) or result in reversing the meaning of certain orders when stop words are removed. This is problematic. When the goal is to construct a CF matrix, it becomes imperative to identify not only the failure modes but to identify the sub-system and components where the issue occurs. Therefore, Technical Language Processing (TLP) encourages semi-supervised models with human-in-the-loop to tailor NLP tools for technical text data. One approach to address this issue is to use active learning, which needs an oracle to label the data iteratively. A text classification model can then be trained to recognize the failure modes and label them on incoming data. This requires only a small amount of input from the user. Models have demonstrated classification with sufficient accuracy [37,41,42,49–51]. In the case of this paper, the focus is mainly on the extraction of failure modes for the purpose of sensor selection. This makes the components in which the issue occurs of less importance.

Technical texts are typically short texts that average about four to five words per row. This makes it difficult to use certain NLP tools such as TF-IDF as it requires large volumes of text and repeating words to gain meaningful information. The vectorization of high-dimensional data results in sparsity, requiring high computing and storage capacity. This makes it challenging to determine the number of clusters, and the clusters obtained will be difficult to label, which is the goal of this paper.

There are several clustering methods that have been proposed to deal with the issue of sparsity in vector representation for short-text clustering [52–55]. To enrich the representation of short text, most of the proposed approaches mentioned above use external sources such as Wikipedia or Wordnet, which sometimes lead to inconsistency. To deal with this issue, a corpus-based topic diffusion technique has been proposed which focuses on finding words that do not appear in the text that are related to its content [56].

Hybrid LDA models that learn vector representation were proposed to overcome the short-text clustering limitation of LDA models [57]. In a study comparing short text classification with non-negative matrix factorization (NMF), it was shown that NMF outperforms LDA [58]. Another extension of the LDA model is the Gibbs Sampling for Dirichlet Multinomial Mixture (GSDMM) model, which extends LDA with a Dirichlet Multinomial Mixture [59]. GSDMM has been shown to perform better on short texts compared to other hybrid LDA models. GSDMM not only infers the number of topics present in a document, but it also achieves a good balance between completeness and homogeneity. GSDMM assumes that the same symmetric distribution of Dirichlet priors is given to all words, which is not realistic. To overcome the limitation of GSDMM, a method that combines Chi-square statistics with GSDMM has been proposed and shown to perform well on railway maintenance data [60]. However, the stability of the model was shown to last for a shorter duration.

In summary, it is challenging to adapt the NLP text pre-processing pipeline for maintenance data. Inconsistency in performance is one of the issues with clustering models that utilize external resources to deal with the sparsity of short texts. Hybrid LDA models outperform traditional LDA models in short-text clustering. However, hybrid LDA models do not extend to maintenance texts.

4. Research Framework

Figure 1 provides an overview of the research framework. As shown, it begins with a sensor fault mapping process, which takes either the maintenance logs or FMEA data as input, whichever is available. Note that it is better to utilize maintenance logs as input if the system is deployed and operational since there is actual failure data. In case the system is still in the design phase, it is better to use FMEA data, which provides anticipated failure data.

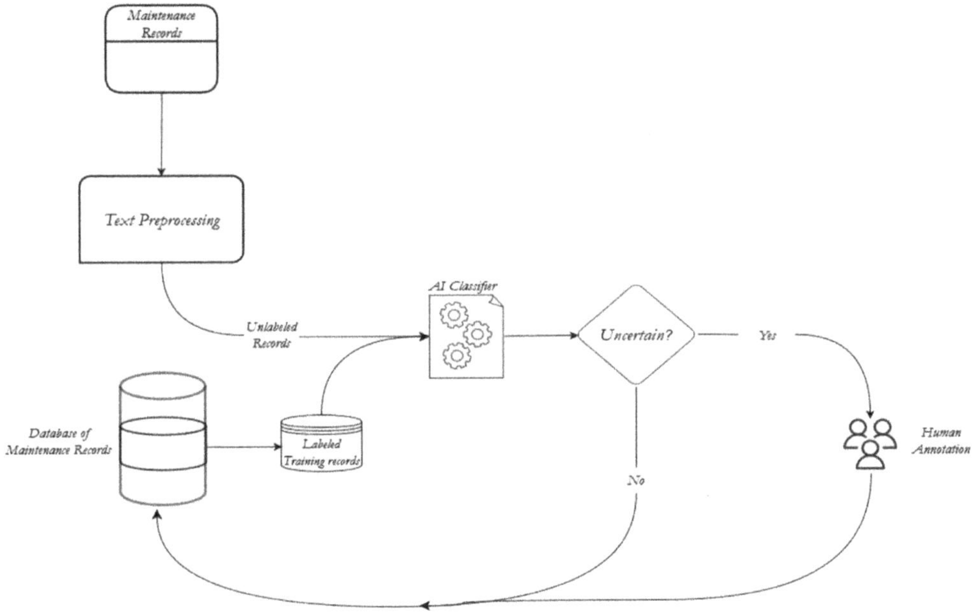

Figure 1. Research Framework.

Exploratory analysis of the raw data is conducted with the construction of word clouds, n-gram analysis, and a co-occurrence network. During the text processing stage, all special characters are removed, and any misspelled words are corrected. The pre-processed text is then clustered using non-parametric Bayesian algorithms such as GS-DMM and a hierarchical LDA model to extract the failure modes. To establish a baseline model, a similarity-based Hierarchical agglomerative clustering algorithm is used. The algorithm uses soft cosine measurement to measure the similarity between sentences. Topic modeling using LDA is also performed to illustrate its performance on short texts, specifically maintenance data. Finally, clustering using sentence embedding to identify failure modes is carried out. To support this, a supervised learning algorithm for text classification is used by an active learning model. The performance of unsupervised models is compared using Normalized Mutual Information (NMI) and Adjusted Rand Index (ARI). The supervised and active learning models are compared based on classification accuracy.

5. Case Study

This section demonstrates the extraction of failure modes for an operational system using maintenance records on several excavators collected from the Mobile Mining Equipment Reliability Database [61]. The maintenance logs used in this study were collected from eight excavators where the asset code has been de-identified. The data were obtained via the prognostics data library [62]. The details of how the data were collected are described by

Hodkiewicz and Ho [63]. The dataset contains records of maintenance services performed on eight similarly sized excavators/assets at different mine sites across Australia over a period of ten years. The asset types have been de-identified to an uppercase letter (A, B, C, D, and E). The dataset has five variables. A brief description of each variable is provided in Table 2. Type of maintenance is a categorical variable with six levels as follows:

- PM01: unplanned or breakdown maintenance;
- PM02: planned or preventive maintenance;
- PM06: accidental damage;
- PM13: Repair of a spare for an asset;
- PM04 and PM05: unknown.

Table 2. Different fields in the dataset.

Fields/Variables	Description	Data Type
BscStartDate	Date of commencement of maintenance work	Date
Asset	De-identified asset type	Categorical
OriginalShorttext	Description of maintenance work needed	String/Unstructured text
PM Type	Maintenance work type	Categorical
Cost	Cost in Australian dollars	Float

For the analysis, PM01, PM13, PM04, and PM05 are considered with a focus on identifying failure modes for sensor selection in predictive maintenance. The occurrences of each of the above-mentioned categories are shown in Figure 2. A marginal distribution plot of the cost of each asset over time is provided in Appendix A.

Figure 2. Maintenance type for different assets.

Word clouds are powerful techniques to represent textual data. The size and color of each word indicate its frequency and importance. The word cloud generated using maintenance data (Figure 3) provides a glimpse into the dominating failures of the asset. This makes it easier to expect clusters containing certain failure modes. This process also helps in identifying the extent to which the documents need to be pre-processed.

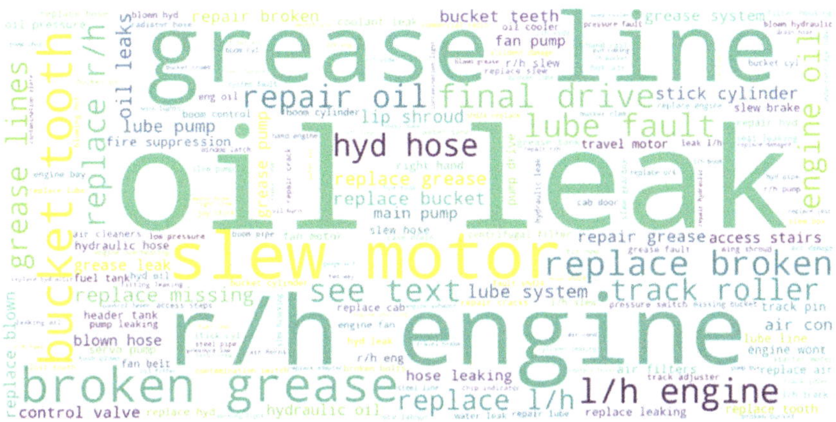

Figure 3. Word cloud of maintenance records.

Most of the failure modes present in Figure 3 are associated with lubricants or hydraulic fluids, indicating that lubricant leaking is a major problem in these excavators. The component that experiences frequent maintenance issues appears to be the engine on the right-hand side or 'r/h engine'. Some of the terms are repetitive such as 'grease line' v/s 'grease lines' indicating the plural form of the line is considered a separate word. It might be possible to remove this discrepancy by stemming the word to its base form. However, doing so will make it difficult to interpret the results of the clusters to identify the cluster label or failure mode. There are some terms such as repair grease which do not create this problem. To ensure that these two terms occur together, it is necessary to conduct n-gram analysis on the data.

N-gram refers to a continuous sequence of n words. If the number of words is two, it is called as bigram, for three words, it is called a trigram, and so on. If the most frequently occurring words are considered in isolation, as shown in Figure 4, it would be difficult to recognize the context in which a word was used. It becomes somewhat easier to understand the texts if bigrams are considered, shown in Figure 5. However, it is still difficult to discern the failure modes for some of the entries, such as 'rh engine', which simply lists the name of the component where the fault is present. Considering the first step is to identify the failure modes within the document, at least three words need to be considered, shown in Figure 6. There are some repetitive observations that refer to the same failure mode but are worded differently, such as 'repair broken grease' and 'broken grease line'. Some of the words refer to the same failure mode but also list the component where the fault can be observed. A co-occurrence network provides insight into which words are most likely to appear together. The higher the likelihood, the more prominent the edge between the two words. A co-occurrence network with a forced atlas 2 layout (a force-directed layout used for network spatialization) was created using Gephi, as shown in Figure 7. The size of the labels represents the degree of the associated nodes in the network. The co-occurrence network confirms most of the findings in bigrams that most of the components or lubricants need to be replaced. The major issue that these machines have is oil leaks as they are more likely to occur together. Another issue that can be inferred from this is that most of the issues appear in the 'rh engine' than in the 'lh engine'.

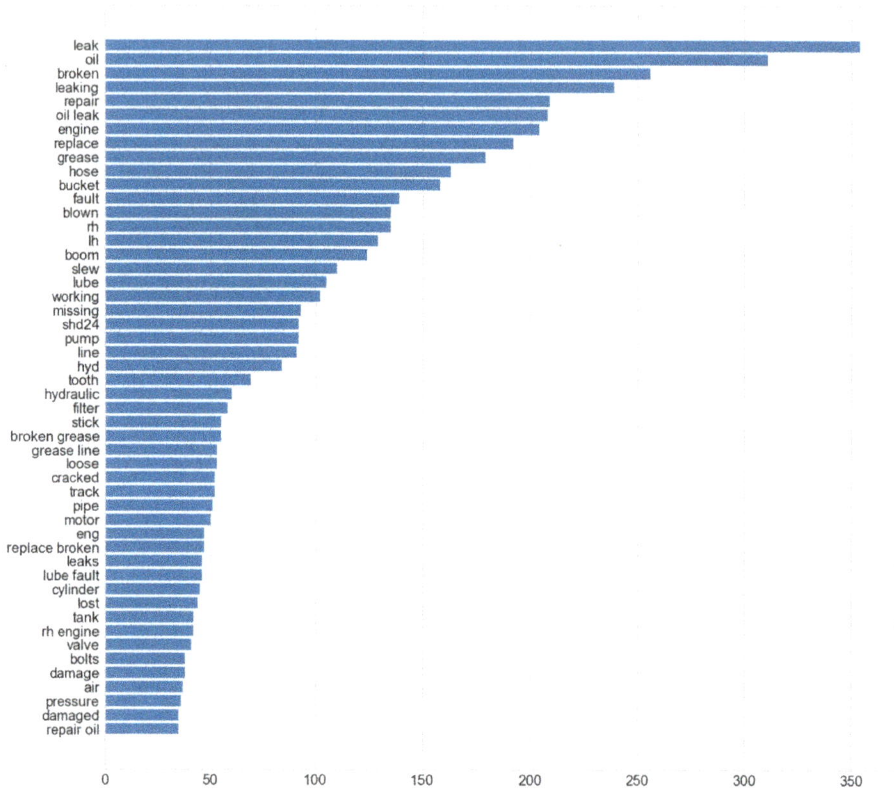

Figure 4. Frequently occurring words in excavator maintenance records.

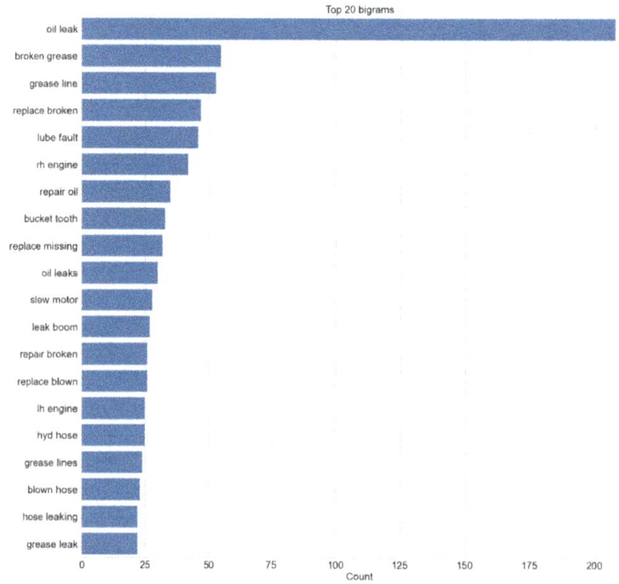

Figure 5. Top 20 bigrams in maintenance records.

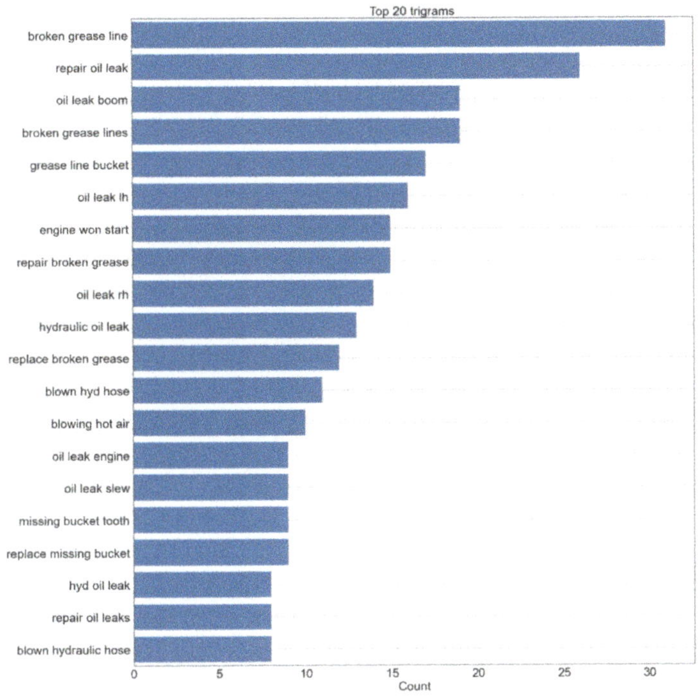

Figure 6. Top 20 trigrams in maintenance records.

5.1. Tokenization

Tokenization is a process that makes it easier to remove punctuations and numbers which typically do not provide much information in unstructured textual data. As an example, consider '#1 Swing Motor contamination switch'. After tokenization, it becomes '['#', '1', 'Swing', 'Motor', 'contamination', 'switch']'. It is important to note that tokenization can trigger certain complications when it comes to technical data. In the case of abbreviated words, the period following the word should be part of the same token and not a separate token. This can be problematic as technical texts often have hyphens, parentheses, and other punctuation marks. Care should be taken so that a phrase with a single meaning with white space, such as 'New York', is not split into two different words.

The emerging trend seems to be to forgo tokenization entirely and develop the NLP algorithms at the character level, which makes it easier to avoid the pitfalls of tokenization. The nuances of the tokenization schemes are not enough of a crucial issue to be dealt with at the development stage of an NLP model. However, adjusting the tokenization scheme is often the easiest way to improve the model for failure modes that are worded differently, such as repair broken grease and broken grease line.

5.2. Text-Preprocessing

Certain words such as 'cyl', 'hyd', and 'eng' were changed to not count as separate words from 'cylinder', 'hydraulic', and 'engine', respectively. The misspelled words were corrected using TextBlob, a text processing library, and inspected manually for any incorrect corrections. Special characters were removed using regular expressions. All operations were performed in Python 3.

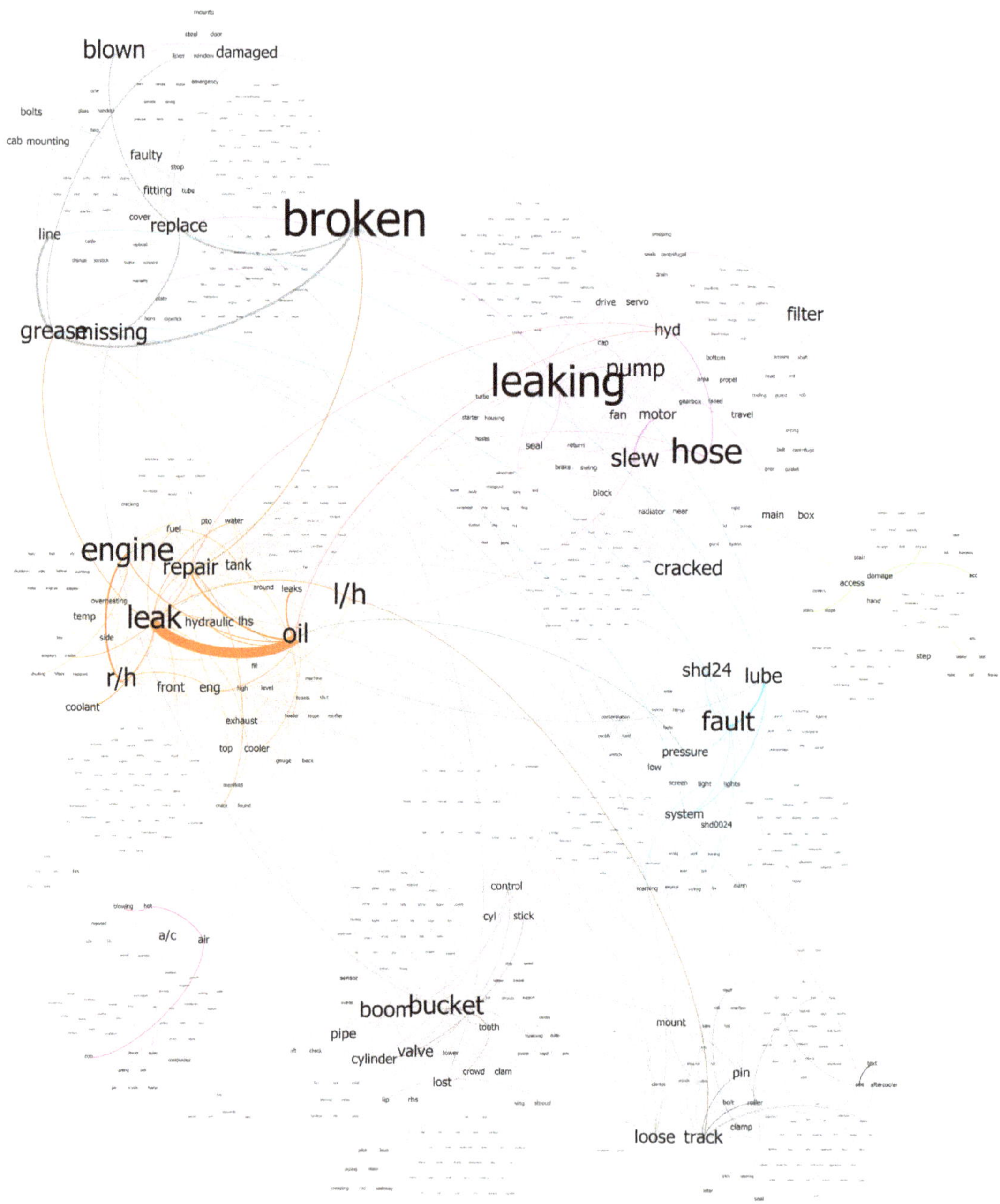

Figure 7. Complete co-occurrence network of corpus.

6. Results and Discussion

6.1. Topic Modeling

There are several approaches to inferring the failure modes using an unsupervised learning approach. The most popular approach is LDA. The biggest challenge when attempting to model LDA is deciding the number of topics to use as a hyperparameter. A coherence score is typically used to assess the quality of the topics, and a higher coherence score is better. The LDA was modeled using genism [62] in Python, making it easier to calculate the model coherence score. Unfortunately, for short texts, the number of topics to pick is not obvious, as shown in Figure 8. Increasing the number of topics continues to increase the coherence score, thus providing little guidance to select the number of topics. Another possibility is to select the number of topics as the first peak (topic = 10) according to Figure 8. This would result in losing information on certain failure modes or mis-categorizing them. Furthermore, this makes it difficult to ascertain failure modes when selecting the sensors to detect failure modes.

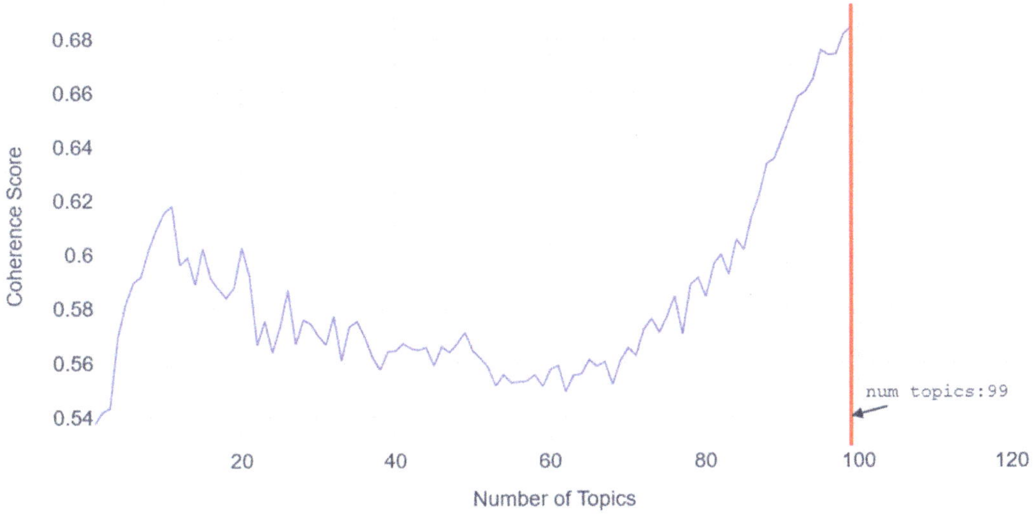

Figure 8. LDA coherence score as a function of the number of topics.

Although HDP had a lower coherence score, the number of topics discovered were closer to the ground truth. However, if the contents of the topics are plotted as a word cloud, as shown in Figure 9, it becomes evident that the topics have several of the same words that make up the bulk of the topic. This makes it difficult to identify the failure mode in each topic.

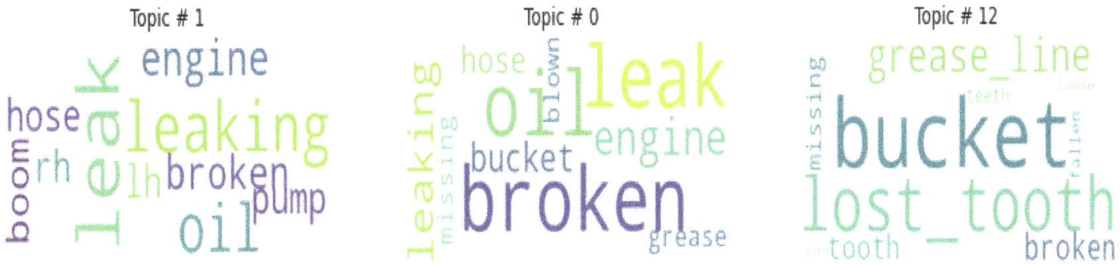

Figure 9. Word Cloud of Topics from HDP.

GSDMM provided the same result as HDP. Although it had a lower coherence score, the number of topics discovered was close to the ground truth. The summarized result of all three algorithms is provided in Table 3. The caveat here is that the coherence score alone is not a good statistic to indicate the performance of the model. The words that make up the topic should always be investigated before drawing a conclusion.

Table 3. Results of Topic Modeling Algorithms.

Model	Number of Topics Discovered	Coherence Score
HDP	22	0.35
LDA	99	0.68
GSDMM	25	0.33

While some of the topics discovered made sense, many of the topics were hard to interpret or to establish a one-to-one relationship between the failure mode and the topic. This is due to the issues discussed in the section Overview of Maintenance Text Issues. Clustering with sentence embeddings is another option to extract meaningful information from short texts.

6.2. Clustering Texts with Sentence Embeddings

To cluster text data, it is necessary to represent them numerically. This is achieved using either embeddings or vector representations. There are several sentence embedding models that have been made available for use as pre-trained models. These models are known to perform well for sentences with semantical similarities. Given the size of the dataset, it is best to use a pre-trained model. Three different BERT models were used (all-mpnet-base-v2, all-MiniLM-L6-v2, and all-distilroberta-v1).

The embedded sentences have more than seven hundred dimensions, making them computationally expensive. Hence, dimensionality reduction was performed using Uniform Manifold Approximation and Projection for Dimension Reduction (UMAP) [63]. Another issue with high dimensional data is the meaning of measures such as Euclidean distance and Manhattan distance in high dimensional space.

HDBSCAN was the clustering algorithm used due to its robustness on variable density clusters. The model itself does not require the number of clusters as the hyperparameters. The clustering algorithm was used on all three sentence embedding models. However, the results were far from satisfactory. All three embedding models suggested more than fifty clusters, which is far from the ground truth. The comparison of clustering using the three different embedding models is shown in Table 4.

Table 4. Performance of clustering sentence embeddings.

Model	ARI	NMI	Loss	Clusters
all-mpnet-base-v2	0.036	0.344	0.100	82
all-MiniLM-L6-v2	0.041	0.373	0.122	93
all-distilroberta-v1	0.040	0.326	0.125	67

There can be issues with the performance of embedding models, which are known to perform well on short texts, due to the data used for training. For example, sentence embedding models use everyday language for training. This is not suitable for this case as most of the words used in the maintenance records are not everyday words. This suggests a need to train a custom embedding model using only maintenance data. While this has been accomplished [64], unfortunately, the model is not available to the public. This is most likely due to the confidentiality associated with the data.

6.3. Active Learning with Ensemble Modeling

In this approach, multiple models are combined in the prediction process. This is a typical solution when the following challenges are encountered during modeling:

- High Variance: The model is very sensitive to the inputs provided.
- Low Accuracy: One model fit of the entire training data may not provide accurate results.
- Noise and Bias: The model relies heavily on one or a few features for prediction.

It is challenging to create a model that would produce highly accurate results given the limitations of machine learning. Multiple models are combined to boost the overall accuracy. Reducing model error and maintaining generalization objectives can be achieved with aggregating the output from the models. There are several ensemble models such as Adaptive Boosting (AdaBoost), Random Forest, and XGBoost, to name a few.

A random forest classifier is chosen to perform the classification task. Random Forest uses a subset of training samples as well as subset of features to build multiple trees. The goal is not to select and compare the performance of different ensemble models but to demonstrate the effectiveness of active learning.

Different sampling methods were used to test the active learning approach using the modAL library [64] in Python. Uncertainty sampling provided the highest accuracy. The results were achieved with default parameters without any hyperparameter optimization. The model was trained on 185 out of 2303 observations. This approach improved the accuracy by 10%. It took 52 h to manually identify all failure modes present within the maintenance records. However, the active learning model took less than 10 min to identify all failure modes. This saves substantial time and cost in terms of labor hours required to identify the failure modes.

6.4. Model Comparison

This section compares the active learning model with both unsupervised and supervised learning models that have been introduced earlier in this paper. The comparison with unsupervised learning models is shown in Figure 10. While the standard algorithm for topic modeling LDA overestimates the number of topics present in the document, both HDP and GSDMM, which are meant for short texts (<180 characters), are much closer in their estimation of the actual number of failure modes present. However, the pre-trained embedding models perform similarly to LDA, indicating that these models are not meant for short texts, especially technical texts. It is important to note that short texts are typically defined in terms of characters present in a tweet, which is not the same for technical texts (<50 characters).

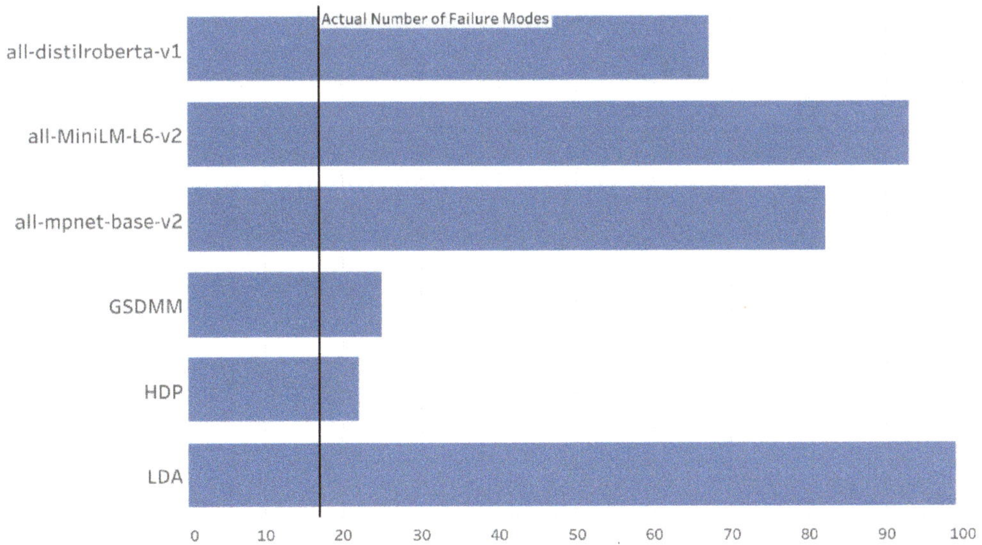

Figure 10. Number of Topics Discovered.

An ensemble learning model, namely random forest, was the core of the active learning model. Instead of accuracy, the F-1 score is a more accurate metric for classification models with imbalanced classes. The F-1 score for the random forest model was found to be 0.86. The F-1 scores of different supervised learning models and active learning models is presented in Table 5.

Table 5. F-1 scores of supervised and active learning models.

Model	F-1 Score
Random Forest	0.86
Stochastic Gradient Descent	0.83
K-nearest neighbors	0.80
Decision Trees	0.67
ADA Boost	0.70
Support Vector Machines	0.82
Multi-Layer Perceptron	0.83
Entropy Sampling	0.85
Uncertainty Sampling	0.88
Margin Sampling	0.83

It is evident from the comparisons with different models that active learning models perform better when compared to supervised or unsupervised learning models.

7. Conclusions and Future Work

In this paper, a framework to extract failure modes from maintenance records using online active learning is proposed. This work illustrates the challenges faced in identifying failure modes from maintenance records. The limitations of current unsupervised learning models to perform a cluster then label approach are illustrated with a case study. These are primarily due to the limitations of the NLP tools, such as vectorization and sentence embeddings, being geared towards non-technical terms. The case study is also used to demonstrate the effectiveness of using a human-in-the-loop approach with the same NLP tools.

While contributions made by this work are significant, there remain research avenues that can be explored to further improve the performance of both unsupervised and semi-supervised methods. An example would be training and embedding a model specifically with maintenance data. However, this requires large amounts of data, most of which are confidential, making partnerships with industry necessary. An additional area for further development using the output from this framework is to map the sensors to failure modes. In turn, this would serve as an input for sensor selection and sensor specification selection models.

Author Contributions: Conceptualization, A.K.; methodology, A.K.; software, A.K.; case study, A.K.; investigation, A.K.; resources, A.K.; data curation, A.K.; writing—original draft preparation, A.K.; writing—review and editing, V.P., J.T.; visualization, A.K.; supervision, V.P., J.T. All authors have read and agreed to the published version of the manuscript.

Funding: This research received no external funding.

Institutional Review Board Statement: Not applicable.

Informed Consent Statement: Not applicable.

Data Availability Statement: https://prognosticsdl.systemhealthlab.com/dataset/excavator-maintenance-work-orders (accessed on 17 January 2023).

Acknowledgments: This material is based upon work while Terpenny has been serving at the National Science Foundation. Any opinion, findings, and conclusions or recommendations expressed in this material are those of the author(s) and do not necessarily reflect the views of the National Science Foundation.

Conflicts of Interest: The authors declare no conflict of interest.

Appendix A

Figure A1. A marginal distribution plot of the cost of each asset over time.

References

1. Haldar, A.; Mahadevan, S. *Reliability Assessment Using Stochastic Finite Element Analysis*; John Wiley & Sons: Hoboken, NJ, USA, 2000.
2. Trivedi, N.; Singh, R. Prediction of impact induced failure modes in reinforced concrete slabs through nonlinear transient dynamic finite element simulation. *Ann. Nucl. Energy* **2013**, *56*, 109–121. [CrossRef]
3. Hodzic, A.; Kalyanasundaram, S.; Lowe, A.; Stachurski, Z. The microdroplet test: Experimental and finite element analysis of the dependance of failure mode on droplet shape. *Compos. Interfaces* **1998**, *6*, 375–389. [CrossRef]
4. Cormier, Y.; Dupuis, P.; Jodoin, B.; Ghaei, A. Finite Element Analysis and Failure Mode Characterization of Pyramidal Fin Arrays Produced by Masked Cold Gas Dynamic Spray. *J. Therm. Spray Technol.* **2015**, *24*, 1549–1565. [CrossRef]
5. Xu, Z.; Dang, Y.; Munro, P.; Wang, Y. A data-driven approach for constructing the component-failure mode matrix for FMEA. *J. Intell. Manuf.* **2019**, *31*, 249–265. [CrossRef]
6. GlobalData. Mining capital expenditure to rise by 22% across leading miners in 2022. *Min. Technol.* **2022**. Available online: https://www.mining-technology.com/comment/mining-capital-expenditure/ (accessed on 23 July 2022).
7. Hodkiewicz, M.R.; Batsioudis, Z.; Radomiljac, T.; Ho, M.T. Why autonomous assets are good for reliability—The impact of 'operator-related component' failures on heavy mobile equipment reliability. *Annu. Conf. PHM Soc.* **2017**, *9*, 1–7. [CrossRef]
8. Yun, H.; Carlone, M.; Liu, Z. Topic modeling of maintenance logs for linac failure modes and trends identification. *J. Appl. Clin. Med. Phys.* **2021**, *23*, e13477. [CrossRef]
9. Tanguy, L.; Tulechki, N.; Urieli, A.; Hermann, E.; Raynal, C. Natural language processing for aviation safety reports: From classification to interactive analysis. *Comput. Ind.* **2016**, *78*, 80–95. [CrossRef]
10. Kuhn, K.D. Using structural topic modeling to identify latent topics and trends in aviation incident reports. *Transp. Res. Part C Emerg. Technol.* **2018**, *87*, 105–122. [CrossRef]
11. James, A.T.; Gandhi, O.; Deshmukh, S. Knowledge management of automobile system failures through development of failure knowledge ontology from maintenance experience. *J. Adv. Manag. Res.* **2017**, *14*, 425–445. [CrossRef]
12. Collins, J.A.; Hagan, B.T.; Bratt, H.M. The Failure-Experience Matrix—A Useful Design Tool. *J. Eng. Ind.* **1976**, *98*, 1074–1079. [CrossRef]
13. Arunajadai, S.G.; Stone, R.B.; Tumer, I.Y. A Framework for Creating a Function-Based Design Tool for Failure Mode Identification. In *Volume 4: 14th International Conference on Design Theory and Methodology, Integrated Systems Design, and Engineering Design and Culture*; ASME: New York, NY, USA, 2002; pp. 195–206. [CrossRef]
14. Wu, Y.; Denny, J.C.; Rosenbloom, S.T.; Miller, R.A.; Giuse, D.A.; Song, M.; Xu, H. A Preliminary Study of Clinical Abbreviation Disambiguation in Real Time. *Appl. Clin. Inform.* **2015**, *6*, 364–374. [CrossRef]
15. Wani, M.F.; Jan, M. Failure Mode Analysis of Mechanical Systems at Conceptual Design Stage. In *Volume 1: Advanced Energy Systems, Advanced Materials, Aerospace, Automation and Robotics, Noise Control and Acoustics, and Systems Engineering*; ASME: New York, NY, USA, 2006; pp. 985–995. [CrossRef]
16. Meng, L.; Wang, H.; Xue, Y.; Ma, L. Research on automatic generation of software failure modes. *J. Front. Comput. Sci. Technol.* **2018**, *12*, 63–71.
17. Chen, L.; Nayak, R. A case study of failure mode analysis with text mining methods. In Proceedings of the 2nd International Workshop on Integrating Artificial Intelligence and Data Mining (AIDM 2007), Gold Coast, Australia, 1 December 2007; pp. 49–60.
18. Chang, W.L.; Tay, K.M.; Lim, C.P. Clustering and visualization of failure modes using an evolving tree. *Expert Syst. Appl.* **2015**, *42*, 7235–7244. [CrossRef]
19. Rajpathak, D.; De, S. A data- and ontology-driven text mining-based construction of reliability model to analyze and predict component failures. *Knowl. Inf. Syst.* **2015**, *46*, 87–113. [CrossRef]

20. Salton, G.; Wong, A.; Yang, C.-S. A vector space model for automatic indexing. *Commun. ACM* **1975**, *18*, 613–620. [CrossRef]
21. Sabbagh, R.; Ameri, F. A Framework Based on K-Means Clustering and Topic Modeling for Analyzing Unstructured Manufacturing Capability Data. *J. Comput. Inf. Sci. Eng.* **2019**, *20*, 011005. [CrossRef]
22. Blei, D.M.; Lafferty, J.D. A correlated topic model of Science. *Ann. Appl. Stat.* **2007**, *1*, 17–35. [CrossRef]
23. Alharthi, H.; Inkpen, D.; Szpakowicz, S. Unsupervised Topic Modelling in a Book Recommender System for New Users. 2017. Available online: http://ceur-ws.org (accessed on 17 January 2023).
24. Choi, H.S.; Lee, W.S.; Sohn, S.Y. Analyzing research trends in personal information privacy using topic modeling. *Comput. Secur.* **2017**, *67*, 244–253. [CrossRef]
25. Blei, D.M.; Ng, A.; Jordan, M.I. Latent dirichlet allocation. *J. Mach. Learn. Res.* **2003**, *3*, 993–1022.
26. Hajjem, M.; Latiri, C. Combining IR and LDA Topic Modeling for Filtering Microblogs. *Procedia Comput. Sci.* **2017**, *112*, 761–770. [CrossRef]
27. Onan, A.; Korukoğlu, S.; Bulut, H. Ensemble of keyword extraction methods and classifiers in text classification. *Expert Syst. Appl.* **2016**, *57*, 232–247. [CrossRef]
28. Tong, Z.; Zhang, H. A Text Mining Research Based on LDA Topic Modelling. *Comput. Sci. Inf. Technol.* **2016**, 201–210. [CrossRef]
29. Chen, P.; Zhang, N.L.; Liu, T.; Poon, L.K.; Chen, Z.; Khawar, F. Latent tree models for hierarchical topic detection. *Artif. Intell.* **2017**, *250*, 105–124. [CrossRef]
30. Jiang, B.; Li, Z.; Chen, H.; Cohn, A.G. Latent Topic Text Representation Learning on Statistical Manifolds. *IEEE Trans. Neural Netw. Learn. Syst.* **2018**, *29*, 5643–5654. [CrossRef] [PubMed]
31. Cigarrán, J.; Castellanos, Á.; García-Serrano, A. A step forward for Topic Detection in Twitter: An FCA-based approach. *Expert Syst. Appl.* **2016**, *57*, 21–36. [CrossRef]
32. Waheeb, S.A.; Khan, N.A.; Shang, X. Topic Modeling and Sentiment Analysis of Online Education in the COVID-19 Era Using Social Networks Based Datasets. *Electronics* **2022**, *11*, 715. [CrossRef]
33. Bao, T.; Ren, N.; Luo, R.; Wang, B.; Shen, G.; Guo, T. A BERT-Based Hybrid Short Text Classification Model Incorporating CNN and Attention-Based BiGRU. *J. Organ. End User Comput.* **2021**, *33*, 1–21. [CrossRef]
34. Settles, B. *Active Learning*; Springer: Cham, Switzerland, 2012. [CrossRef]
35. Angluin, D. Queries and Concept Learning. *Mach. Learn.* **1988**, *2*, 319–342. [CrossRef]
36. Angluin, D. *Queries Revisited*; Springer: Berlin/Heidelberg, Germany, 2001; pp. 12–31. [CrossRef]
37. Cohn, D.; Atlas, L.; Ladner, R. Improving generalization with active learning. *Mach. Learn.* **1994**, *15*, 201–221. [CrossRef]
38. Baum, E.B.; Lang, K. Query learning can work poorly when a human oracle is used. *Int. Jt. Conf. Neural Netw.* **1992**, *8*, 8.
39. Hanneke, S. A bound on the label complexity of agnostic active learning. In Proceedings of the 24th International Conference on Machine Learning, Corvalis, OR, USA, 20–24 June 2007; pp. 353–360. [CrossRef]
40. Lewis, D.D. A Sequential Algorithm for Training Text Classifiers. In *SIGIR '94*; Springer: London, UK, 1994; pp. 3–12.
41. Tong, S.; Koller, D. Support vector machine active learning with applications to text classification. *J. Mach. Learn. Res.* **2001**, *2*, 45–66.
42. Hoi, S.C.H.; Jin, R.; Lyu, M.R. Large-scale text categorization by batch mode active learning. In Proceedings of the 15th International Conference on World Wide Web, Edinburgh, Scotland, 23–26 May 2006; p. 633. [CrossRef]
43. Nigam, K.; McCallum, A.K.; Thrun, S.; Mitchell, T. Text Classification from Labeled and Unlabeled Documents using EM. *Mach. Learn.* **2000**, *39*, 103–134. [CrossRef]
44. Dagan, I.; Engelson, S.P. Committee-Based Sampling For Training Probabilistic Classifiers. In *Machine Learning Proceedings 1995*; Elsevier: Tahoe City, CA, USA, 1995; pp. 150–157. [CrossRef]
45. Kim, D.; Lee, S.; Kim, D. An Applicable Predictive Maintenance Framework for the Absence of Run-to-Failure Data. *Appl. Sci.* **2021**, *11*, 5180. [CrossRef]
46. Chen, J.; Zhou, D.; Guo, Z.; Lin, J.; Lyu, C.; Lu, C. An Active Learning Method Based on Uncertainty and Complexity for Gearbox Fault Diagnosis. *IEEE Access* **2019**, *7*, 9022–9031. [CrossRef]
47. Bull, L.; Rogers, T.; Wickramarachchi, C.; Cross, E.; Worden, K.; Dervilis, N. Probabilistic active learning: An online framework for structural health monitoring. *Mech. Syst. Signal Process.* **2019**, *134*, 106294. [CrossRef]
48. Brundage, M.P.; Sexton, T.; Hodkiewicz, M.; Dima, A.; Lukens, S. Technical language processing: Unlocking maintenance knowledge. *Manuf. Lett.* **2020**, *27*, 42–46. [CrossRef]
49. Cohn, D.A.; Ghahramani, Z.; Jordan, M.I. Active Learning with Statistical Models. *J. Artif. Intell. Res.* **1996**, *4*, 129–145. [CrossRef]
50. Cai, H.; Zheng, V.; Chang, K.C.-C. Active Learning for Graph Embedding. *arXiv* **2017**, arXiv:1705.05085.
51. Aodha, O.; Campbell, N.; Kautz, J.; Brostow, G.J. Hierarchical Subquery Evaluation for Active Learning on a Graph. In Proceedings of the 2014 IEEE Conference on Computer Vision and Pattern Recognition, Columbus, OH, USA, 23–28 June 2014; pp. 564–571. [CrossRef]
52. Banerjee, S.; Ramanathan, K.; Gupta, A. Clustering short texts using wikipedia. In Proceedings of the 30th Annual International ACM SIGIR Conference on Research and Development in Information Retrieval, Amsterdam, The Netherlands, 23–27 July 2007; pp. 787–788.
53. Yin, J.; Wang, J. A model-based approach for text clustering with outlier detection. In Proceedings of the 2016 IEEE 32nd International Conference on Data Engineering (ICDE), Helsinki, Finland, 16–20 May 2016; pp. 625–636. [CrossRef]

54. Hadifar, A.; Sterckx, L.; Demeester, T.; Develder, C. A Self-Training Approach for Short Text Clustering. In Proceedings of the 4th Workshop on Representation Learning for NLP (RepL4NLP-2019), Florence, Italy, 2 August 2019; pp. 194–199. [CrossRef]
55. Cheng, X.; Yan, X.; Lan, Y.; Guo, J. BTM: Topic Modeling over Short Texts. *IEEE Trans. Knowl. Data Eng.* **2014**, *26*, 2928–2941. [CrossRef]
56. Zheng, C.T.; Liu, C.; Wong, H.S. Corpus-based topic diffusion for short text clustering. *Neurocomputing* **2018**, *275*, 2444–2458. [CrossRef]
57. Zhang, H.; Zhong, G. Improving short text classification by learning vector representations of both words and hidden topics. *Knowl.-Based Syst.* **2016**, *102*, 76–86. [CrossRef]
58. Chen, Y.; Zhang, H.; Liu, R.; Ye, Z.; Lin, J. Experimental explorations on short text topic mining between LDA and NMF based Schemes. *Knowl.-Based Syst.* **2018**, *163*, 1–13. [CrossRef]
59. Yin, J.; Wang, J. A dirichlet multinomial mixture model-based approach for short text clustering. In Proceedings of the 20th ACM SIGKDD International Conference on Knowledge Discovery and Data Mining, New York, NY, USA, 24–27 August 2014; pp. 233–242. [CrossRef]
60. Yang, N.; Zhang, Y. Railway Fault Text Clustering Method Using an Improved Dirichlet Multinomial Mixture Model. *Math. Probl. Eng.* **2022**, *2022*, 7882396. [CrossRef]
61. Ho, M.T. *A Shared Reliability Database for Mobile Mining Equipment*; University of Western Australia: Perth, Australia, 2015.
62. Prognostics Data Library. Excavator Maintenance Work Orders. 21 May 2022. Available online: https://prognosticsdl.systemhealthlab.com/dataset/excavator-maintenance-work-orders (accessed on 26 July 2022).
63. Hodkiewicz, M.; Ho, M.T.-W. Cleaning historical maintenance work order data for reliability analysis. *J. Qual. Maint. Eng.* **2016**, *22*, 146–163. [CrossRef]
64. Danka, T.; Horvath, P. modAL: A modular active learning framework for Python. *arXiv* **2018**, arXiv:1805.00979.

sensors MDPI

Article

Maintenance Cost Minimization for an Agricultural Harvesting Gripper

Florina Maria Șerdean [1,*], Mihai Dan Șerdean [2,*] and Silviu-Dan Mândru [2]

[1] Faculty of Industrial Engineering, Robotics and Management Production, Technical University of Cluj-Napoca, 103-105 Muncii Blvd., 400641 Cluj-Napoca, Romania
[2] Faculty of Automotive, Mechatronics and Mechanical Engineering, Technical University of Cluj-Napoca, 103-105 Muncii Blvd., 400641 Cluj-Napoca, Romania
* Correspondence: florina.rusu@omt.utcluj.ro (F.M.Ș.); mihai.serdean@mdm.utcluj.ro (M.D.Ș.)

Abstract: A crucial aspect that has to be considered in all fields and, especially, in smart farming, a rapidly developing industry, is maintenance. Due to the costs generated by both under-maintaining and over-maintaining the components of a system, a balance has to be achieved. The paper is focused on presenting an optimal maintenance policy used to ensure cost minimization by determining the optimal time to make a preventive replacement of the actuators of a harvesting robotic system. First, a brief presentation of the gripper with Festo fluidic muscles used in a novel way instead of fingers is given. Then, the nature-inspired optimization algorithm, as well as the maintenance policy are described. The paper also includes the steps and the obtained results of the developed optimal maintenance policy applied for the Festo fluidic muscles. The outcome of the optimization shows that a significant reduction in the costs is obtained if one performs a preventive replacement of the actuators a few days before the lifetime provided by the manufacturer and the lifetime estimated using a Weibull distribution.

Keywords: optimization; evolutionary algorithm; maintenance cost minimization; smart farming; fluidic muscles; preventive maintenance

1. Introduction

In the last decades, a key role in increasing productivity and reducing costs in terms of industrial agricultural production has been played by robotics and automation. However, high implementation costs and failed results have slowed down the research activities in the field of robotics for agriculture. Recently, more projects and more useful applications have emerged in this field due to the rising cost of labor and new regulations on occupational safety.

Researchers are highly interested in mechanization and automatic harvesting as they can effectively reduce the overall costs of agricultural production. Nevertheless, due to the variations in fruit and legume sizes and shapes, as well as different harvesting techniques, researchers must develop customized robotic harvesting systems for different crop categories [1].

When designing independent robotic systems for harvesting crops such as fruits, several issues should be addressed. The main ones are enumerated in [2]: "high precision in fruit identification, real-time decision making about harvesting and storing, handling without deteriorating both the plant and the fruit". Additionally, the cost of designing and maintaining such a robotic system must be considered because there are studies whose outcome indicates that to be successful on the market, the harvesting costs must be reduced by half [1].

Research regarding harvesting grippers shows that different manipulator structures were developed considering the crop, such as a gripper with soft tips fingers [3], soft protection [4], or a soft structure [5], a gripper with spoon-shaped gripping elements [6], a vacuum-grasping dual arm robot [7], and so on. An overview of the agricultural grippers used to control weeds and harvesting is presented in [1] in an easy-to-read graphical list. Another comprehensive overview in the form of a table summary of the techno-functional properties of the different types of agricultural grippers is encompassed in [2].

In the context of the accelerated development of technology, maintenance has become an increasingly topical subject due to the fact that when it is lacking, or when it is overlooked, the consequences can be tragic in most industries. This is also true for the smart farming industry. However, a balance must be kept between under-maintaining and over-maintaining assets. According to [8], "30 to 40% of preventive maintenance costs are spent on assets with negligible failure impact". Considering this existing situation an optimal maintenance policy is required.

Over time, numerous maintenance policies have been developed, with extensive research dedicated to the subject as indicated by various sources [9–13]. However, the conventional maintenance method of replacing a component only when it fails can prove to be too costly in terms of financial and time resources. An alternative maintenance technique is scheduled preventive maintenance, where preventive replacements are carried out at fixed intervals or when a component reaches a specified age. The optimal replacement interval or the optimal preventive replacement age needs to be determined to optimize this policy [14]. Nonetheless, this policy can significantly raise costs. To minimize costs, it is necessary to consider approaches that reduce over-maintenance, such as condition-based maintenance (CBM). This approach uses condition-monitoring techniques, such as vibration monitoring or thermography, to identify specific defects [15].

This paper is focused on the maintenance cost minimization for the actuators of a proposed harvesting gripper. Section 2 presents the gripper-type harvesting solution using Festo fluidic muscles. The proposed gripper can be mounted on an independent mobile platform that has the capability to detect the crop and harvest it in the fields without human assistance. Due to the lack of a human operator, a certain malfunction can be detected only at the end of the workday when the platform with the gripper returns. Therefore, a preventive maintenance policy is recommended for the gripper actuators in order to minimize the risk of losing a workday and even damaging the crop. The Festo actuators are one of the elements more likely to fail and directly affect the crop and the harvesting process, hence a maintenance cost minimization is conducted for them.

The third part of the paper is focused on briefly presenting the Cuckoo Search (CS) optimization algorithm and the CBM policy used to solve the proposed optimization problem. Section 4 of the paper is devoted to the actual maintenance cost minimization performed in MATLAB and to presenting the obtained results.

2. Proposed Harvesting Gripper

The proposed harvesting system consists of seven main elements. Its 3D model was designed with SolidWorks 2019 SP03 software, and it is shown in Figure 1a in the deactivated state, and in Figure 1b in the activated state.

The elements are attached to base 1, which is a circular plate with the role of supporting the entire structure. Another role of the base plate 1 is to allow the gripper to be mounted on a manipulator arm by means of which the necessary movements can be made for good positioning in relation to the fruit for a correct harvesting process. Clamping couplings 2 are two rotary couplings clamped to base plate 1. They are mounted diametrically opposite each other by means of two countersunk head screws with an M5 hex socket. Another possibility of mounting them is by welding.

The structural support element 3 is a fixed component mounted along the diameter of base plate 1 so that it is perpendicular to the line joining the 2 rotation couples. The main function of this element 3 is to fix the Festo fluidic muscles, which are attached to its tip with the help of fixing components that are an integrated part of the Festo actuators and the gripper. The structural element supporting the actuators also has another function, namely, that of controlling the gripping direction of the gripper. Due to the particular way of assembly, the Festo fluidic muscles cannot contract in the grip direction, but their activation leads to the approach of the 2 mobile elements 4.

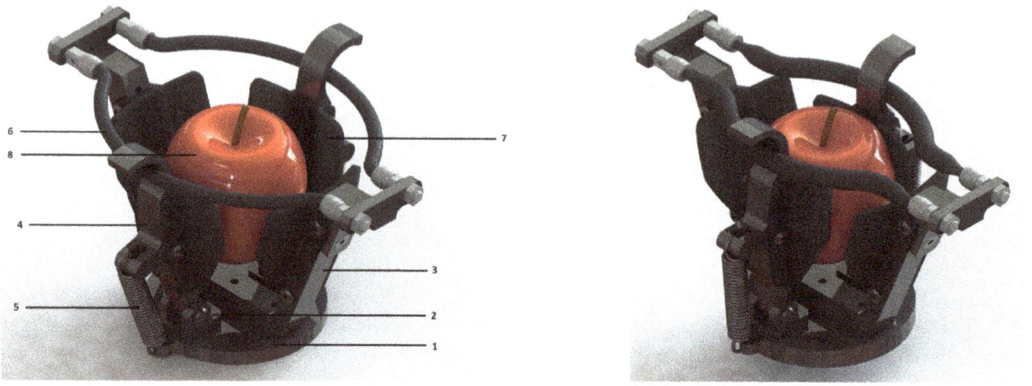

(a) (b)

Figure 1. A 3D model for the proposed harvesting gripper in: (**a**) deactivated state, where: 1—base, 2—joint, 3—structural element for sustaining the actuators, 4—mobile connecting element, 5—elastic element (spring) in a compressed state, 6—Festo actuator DMSP-5-150N-RM-CM, 7—rubber part, 8—agricultural product to be harvested; (**b**) activated state.

The mobile clamping element 4 has a hook-like feature which contains the artificial muscles in the upper part of the gripper and can make a rotational movement thanks to the coupling 2 and the clamping bolt in the plane perpendicular to that of the structural support element 3. Thus, at the moment in which the Festo fluidic muscles are activated, the agricultural product to be harvested is contained according to Figure 1b. Moreover, the hook area, where the muscle directly gets in contact with the mobile connecting element 4, has been designed with a chamfer in order to reduce the sharp edges and prevent the muscle from prematurely bursting. A top view of the 3D model for the proposed gripper is presented in Figure 2 both in activated and deactivated state. The 2 elastic elements 5 are clamped by both the base plate 1 and the mobile clamping elements 4, contributing to the opening of the latter which contain the actuators in the hook-like feature. With the deactivation of the artificial muscles, the elastic elements 5 ensure the opening of the gripper and its preparation in order to grip a new agricultural product.

The last main element of the proposed harvesting system is the 2 Festo fluidic muscles which are actuators activated by air pressure. They tighten the gripper around the agricultural product to be harvested so that the agricultural product 8 can be detached from the stem of the fruit tree (see Figure 1b). To protect the product 8, the gripper has four rubber-like parts 7 which are mounted both on the structural element and on the mobile connecting elements. Because the gripper can be attached to a manipulator arm, it can be programmed to perform complex orienting movements depending on the position of the fruit both for the intended harvest and for directing it to the collection system. The choice of the actuators that have similar operating principle as McKibben artificial muscles, such as the Festo fluidic muscles, is sustained by the McKibben artificial muscles popularity and interest of researchers [16,17]. Of course, McKibben can be used for prototyping and laboratory solutions, whereas Festo offers industrial pneumatic artificial muscles solutions.

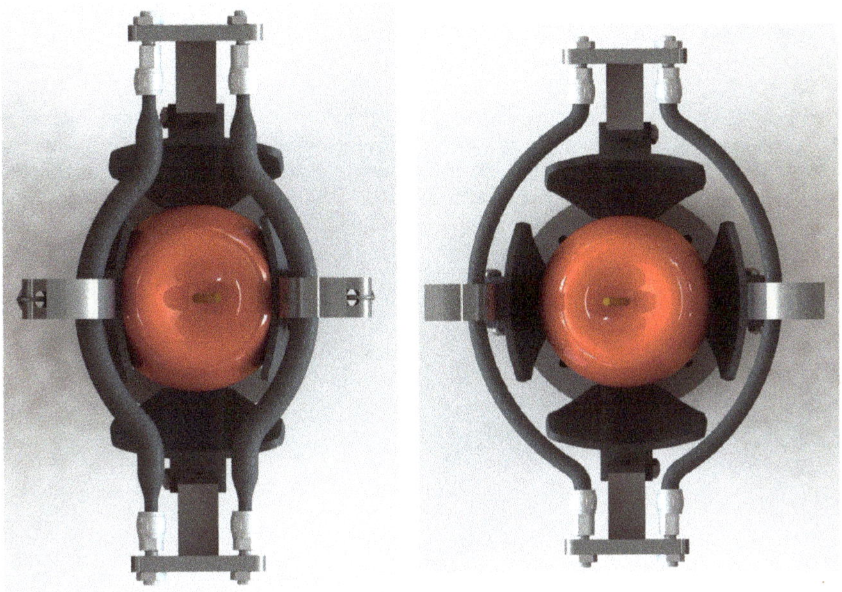

(a) (b)

Figure 2. Top view of the 3D model for the proposed harvesting gripper in: (**a**) activated state; (**b**) deactivated state.

Actuation of the artificial muscle and control of the contraction is achieved by adjusting the pressure applied to the muscle. This can be achieved, for example, using a proportional pressure regulator Festo model VPPE-3-1/8-6-010. The proportional pressure regulator controls the pressure in proportion to an analog electrical quantity (voltage or current) reference. The valve operates in a closed loop, the pressure sensor being integrated into the valve.

This pneumatic proportional regulator provides at the output a minimum pressure Pl at the command voltage of 0 V or 4 mA and a maximum pressure Ph at the command voltage of 10 V or 20 mA. The maximum supply pressure is 8 bar and the regulated pressure is 0–6 bar. The supply of the valve is conducted with a continuous voltage of 24 V and the control of the reference is conducted with a voltage of 0–10 V or 4–20 mA.

Because in most harvesting situations a control of the muscle deformation or a limitation of the closing stroke of the gripper is not necessarily desired, the Festo fluidic muscles can be activated by a simple ON/OFF system.

The detection of the actuator state can be performed using a stereo-vision system mounted on the same platform as the gripper, a system also used to detect agricultural products to be harvested. This solution was chosen to reduce the number of electronic components and therefore, to minimize the risk of failure.

More details regarding the structure and operation principle of these types of actuators are described in [18], where the McKibben artificial muscles are considered. In the same paper, the CAD design of a robotic system that incorporates McKibben actuators is included. Moreover, its control system and a developed prototype are also presented in the same paper.

3. Materials and Methods

3.1. Cuckoo Search Algorithm

The CS optimization algorithm is an evolutionary algorithm inspired by nature, specifically by the breeding behavior of cuckoo birds. These birds are known for being a migratory species that search for areas with more abundant food and better nests for laying their eggs. Interestingly, they do not build their own nests but lay their eggs in communal nests. In 2009, Xin-She Yang and Suash Deb developed the CS algorithm based on this behavior [19]. The migratory flight pattern is based on a type of movement called Lévy flights. Researchers have observed that animals in their natural habitat take a seemingly chaotic path when searching for food, but through the use of probability functions, the search direction can be predicted. Lévy flights are a type of random walk that mimics the search for food of certain animals and insects and have proven successful in natural habitats, leading scientists to incorporate them into some optimization algorithms. A Lévy flight random walk involves a random step length drawn from a Lévy distribution [20].

$$Lvy \sim u = t^{-\lambda}, 1 < \lambda \leq 3 \tag{1}$$

which has infinite mean with an infinite variance. When implementing a few steps, it can be seen that a flight characterized by a batch of small steps is interrupted by significant jumps (Figure 3). A new solution generated based on the random walk using Lévy flights for a given cuckoo x_i is obtained as follows:

$$x_i^{(k+1)} = x_i^{(k)} + \alpha \cdot Lévy(\lambda) \tag{2}$$

where $\alpha > 0$ is the scaling factor of the step size and k is the current generation of cuckoos.

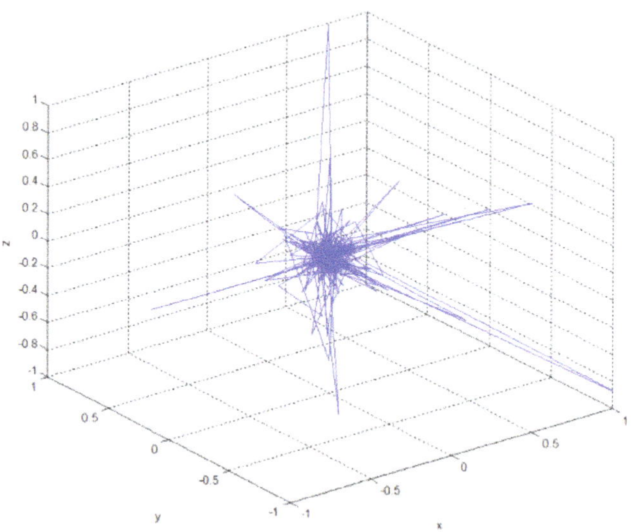

Figure 3. A 3D Lévy flight of 500 steps with (0, 0, 0) as starting point.

The standard version of CS mimicking cuckoo migration using Lévy flights has been customized so that it manages to achieve a balance between exploration and exploitation. A new and improved version of the algorithm was developed starting from the idea that with more knowledge accumulated, not so much exploration is needed. This was achieved by measuring the evolution of knowledge collected by the cuckoo population during migration, measurement made based on the Knowledge Gradient (KG) policy. This improved algorithm was named KGCS [20].

The algorithm was enhanced by introducing two different phases. The first one is dedicated to the search space exploration. In the second one, the intensity of the exploration is diminished while the intensity of the exploitation is boosted.

To achieve maximum exploration of the search space, KGCS uses three populations of cuckoos in its first phase. Initially, the cuckoos within each population are assigned random values within the boundaries of the search space, and their fitness is evaluated based on the value of the objective function. The exploration is carried out independently by all three populations, with more generations obtained by laying eggs in new nests whose positions are determined using Lévy flights. At the end of each generation, the best cuckoo in each population is identified and their values are saved in the corresponding archive. This process continues until a specified stopping criterion is met, ensuring that the algorithm has explored the search space thoroughly.

After 10% of the maximum number of generations has passed, the first phase ends. At this moment, based on the KG policy applied to the existing archives with the best cuckoos, the knowledge gradient value is calculated for each of the three populations. In the next phase, the exploitation is conducted only by the population with the highest knowledge gradient value because, according to the KG policy, it has the largest estimated advancement. The exploitation phase ends when the maximum allowed number of generations is reached. The enhanced KGCS algorithm has been tested on several benchmark functions, as well as used to solve some optimal design problems [20].

3.2. CBM Policy

In this subsection, a CBM maintenance policy that was developed and tested in [21] is briefly presented. It was derived from the maintenance policy introduced in [22] where the model does not consider the improvement of the accuracy of predictions. The developed CBM policy is then used to solve a maintenance cost minimization problem for the gripper presented in Section 2, namely when the optimal time to replace the Festo fluidic muscles in order to minimize the maintenance cost of the harvesting system is.

This maintenance policy uses a probability of failure threshold to prevent a premature replacement of a particular component under analysis. More components of the same type are observed over time, inspections are performed at a constant time interval T, and data are collected. When conducting an inspection, the estimated probability of failure in the next time interval until the next inspection Pr_{est} can be calculated. By expressing the total maintenance cost as a function of the probability of failure and performing a minimization of the maintenance cost, the optimal probability of failure threshold Pr_o can be obtained. Based on this optimal threshold, the proposed CBM maintenance policy can be formulated using the following rules [21]:

1. If the component failed during the last inspection interval, a failure replacement will be performed.
2. If the Pr_{est} value obtained for the next time interval until the next inspection is greater than Pr_o, the decision maker will perform a preventive replacement of the monitored component. Otherwise, the scheduled inspections continue.

The proposed CBM maintenance policy has four steps (Figure 4). The first step consists in collecting significant data about the operating conditions and lifetime of the monitored components. In the next step, the collected data is processed. If there is no software or mathematical model to predict the lifetime of the monitored component, then an approximative model can be built using some of the collected data, for example using the Kriging interpolation method. The data that are not used to develop the interpolation model can be used to test it. A mean value μ and a standard deviation σ of the lifetime prediction errors is obtained from those tests. Moreover, using the gathered information, a Weibull distribution can be modeled, which is a commonly used life distribution to predict the life characteristics of a component.

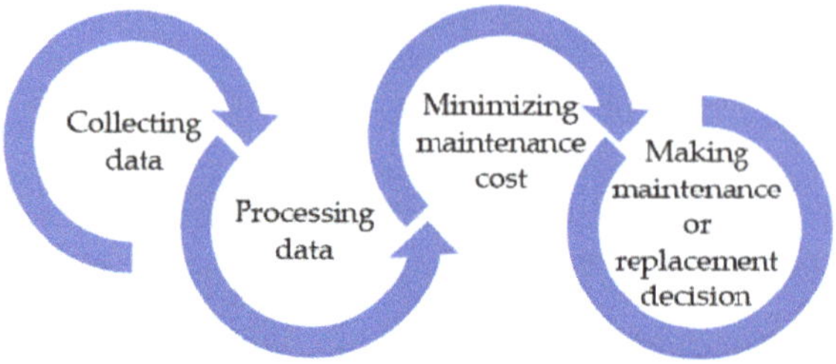

Figure 4. CBM policy's main steps.

The third step is to minimize the maintenance costs. This can be achieved by combining the KGCS optimization algorithm presented in the previous subsection with a stochastic approach in which the expected value is calculated based on small Monte Carlo simulations. In this third step, assuming a normal distribution for the lifetime prediction error, the failure probability estimated in the time interval until the next inspection is computed as follows [21]:

$$Pr_{est}(t, t_P) = \frac{\int_{t}^{t+T} \frac{1}{\sigma\sqrt{2\pi}} e^{-\frac{1}{2}\left(\frac{x-t_P}{\sigma}\right)^2} dx}{\int_{t}^{\infty} \frac{1}{\sigma\sqrt{2\pi}} e^{-\frac{1}{2}\left(\frac{x-t_P}{\sigma}\right)^2} dx} \tag{3}$$

where t_P is the estimated lifetime considering uncertainty and t is the current time of inspection.

The predicted failure time with respect to the actual lifetime of the component, t_W, according to the Weibull distribution, is assumed to have a normal distribution, namely $T_{pW} \sim N(t_W, \sigma^2)$. The same assumption is made for the predicted failure time with respect to t_P, namely, $T_{pP} \sim N(t_P, \sigma^2)$, where t_P is computed by subtracting the mean μ from the predicted lifetime.

Given a certain estimated lifetime and a threshold for the failure probability, the preventive replacement time denoted t_{PR} is determined as the inspection time when the failure probability Pr_{est} given by Equation (3) becomes greater than the threshold of the failure probability.

Denoting the preventive replacement cost of a component by C_P and the replacement cost in case of failure by C_F, the total expected replacement cost is obtained by computing the sum between the total expected preventive cost C_{TP} and the total expected replacement cost in case of failure C_{TF}, where [21]:

$$C_{TP}, C_{TF} : [0;1] \times [t_W - 3\sigma; t_W + 3\sigma] \to (0;\infty) \tag{4}$$

$$C_{TP}(Pr, t_w) = \sum_{i=1}^{nS} C_P \cdot I(t_{PR}(Pr, T_{pP}(i)) < t_w) \tag{5}$$

$$C_{TF}(Pr, t_w) = \sum_{i=1}^{nS} C_F \cdot I(t_{PR}(Pr, T_{pP}(i)) \geq t_w) \tag{6}$$

$$I(t_{PR}(Pr, T_{pP}(i)) < t_w) = \begin{cases} 1, & if\ t_{PR}(Pr, T_{pP}(i)) < t_w \\ 0, & otherwise \end{cases} \tag{7}$$

where nS is the sample size of the Monte Carlo simulations used to estimate the costs. This sample contains the values $T_{pP}(i)$, $i = 1, \ldots, nS$ for the predicted failure time taking into account elements of uncertainty, which were generated according to the probability distribution of the random variable T_{pP}.

Correspondingly, the expected replacement time is obtained by computing the sum between the expected preventive time T_{TP} and the expected replacement time in case of failure T_{TF}, where [21]:

$$T_{TP}, T_{TF} : [0;1] \times [t_W - 3\sigma; t_W + 3\sigma] \rightarrow (0;\infty) \tag{8}$$

$$T_{TP}(Pr, t_w) = \sum_{i=1}^{nS} t_{PR}(Pr, T_{pP}(i)) \cdot I(t_{PR}(Pr, T_{pP}(i)) < t_w) \tag{9}$$

$$T_{TF}(Pr, t_w) = \sum_{i=1}^{nS} t_w \cdot I(t_{PR}(Pr, T_{pP}(i)) \geq t_w) \tag{10}$$

Assuming the α și β are the parameters for the Weibull distribution, the function used to model the total expected cost of the replacement per unit of time with respect to the threshold of the failure probability Pr is given by [21]:

$$C_{\exp} : [0;1] \rightarrow (0;\infty); C_{\exp}(Pr) = \frac{C_T(Pr)}{T_T(Pr)} \tag{11}$$

where

$$C_T, T_T : [0;1] \rightarrow (0;\infty) \tag{12}$$

$$C_T(Pr) = \sum_{j=1}^{nS} (C_{TP}(Pr, T_{pW}(j)) + C_{TF}(Pr, T_{pW}(j))) \tag{13}$$

$$T_T(Pr) = \sum_{j=1}^{nS} (T_{TP}(Pr, T_{pW}(j)) + T_{TF}(Pr, T_{pW}(j))) \tag{14}$$

The Monte Carlo simulations used to estimate the expected cost per unit of time have the same sample size nS. This time, the sample contains the values $T_{pW}(j)$, $j = 1, \ldots, nS$ for the predicted failure time with respect to t_W, which were generated according to the probability distribution of the random variable T_{pW}.

The optimization is carried out with the aim of determining the optimal Pr_o value of the failure probability threshold, a value that minimizes the expected maintenance cost. The optimization problem is a minimization one and can be formulated as follows:

$$\min_{Pr \in [0,1]} C_{\exp}(Pr) \tag{15}$$

In the fourth and last step of the proposed maintenance policy, the decision to make a preventive replacement is taken only if the calculated failure probability for the next time interval is larger than Pr_o, the optimal probability threshold, and if, of course, the component has not failed in the meantime. A summary of the CBM policy as it was implemented in MATLAB can be found in Table 1.

Table 1. Procedure of the proposed CBM policy.

Step 1	Collect historical data relevant to the lifetime of the component
Step 2	Build the surrogate model based on available historical data Obtain the prediction error based on tests using the surrogate model Determine the distribution of the component lifetime
Step 3	Model the expected cost per unit of time in relation to a given failure probability Determine the optimal failure probability, Pr_o, that minimizes the expected cost

Step 4	For each inspection point:
	    If the component has failed
	            Make failure replacement
	    Else
	          Calculate failure probability during next inspection interval Pr_{est}
	          If $Pr_{est} > Pr_o$
	                  Make preventive replacement
	          End if
	    End if
	End for

4. Results and Discussion

The problem proposed and solved in this paper consists of determining the best replacement time for the Festo fluidic muscles that minimizes the maintenance cost for the gripper-type harvesting system proposed in Section 2.

For the first step of the CBM maintenance policy, which is to collect historical data on the lifetime of the Festo fluidic muscles, there was no possibility to collect enough data related to the actual lifetime of several such artificial muscles. Therefore, 100 values were randomly generated representing lifetimes for Festo fluidic muscles, which, according to the data provided by the manufacturer Festo, have an average duration of 1,000,000 contractions for a theoretical force of 140 N on the pneumatic muscle at the maximum functioning pressure of 6 bar. The actual mean value of the 100 randomly generated from a Gaussian distribution is 1,001,038.90 contractions, and the standard deviation is 7077.29.

In the second step of the proposed CBM policy, the parameters of the Weibull distribution followed by the generated data were determined. The obtained values for the parameters are α = 1,004,527, β = 150.4681, and therefore the actual lifetime of the component according to the Weibull distribution is t_W = 994,482. Another 100 values were then generated also from a Gaussian distribution to simulate the estimated lifetimes for the Festo fluidic muscles. The average lifetime for the estimated value is 998,443.77 contractions. At this step, the estimated errors were also calculated, and it was determined that they follow the Gaussian distribution with a mean of μ = 2595.13 and a standard deviation of σ = 10,643.31. Thus, T_{pW}~N(994,482; 10,643.312) and T_{pP}~N(100,1038.90; 10,643.312). The graphical representation of the probability density functions corresponding to the Weibull distribution and the distributions of T_{pW} and T_{pP} are shown in Figure 5.

In the third step of the CBM maintenance policy, the minimization of the maintenance cost for the proposed harvesting system was carried out using the improved evolutionary algorithm KGCS presented in Section 3.1. The expected cost was estimated using Monte Carlo simulations of 100 values for both the estimated failure time with respect to the actual Festo fluidic muscles lifetime given by the Weibull distribution (T_{pW}) and the estimated failure time with respect to the duration of their estimated lifetime taking into account elements of uncertainty (T_{pP}). As a rough estimate, the replacement cost of a set of Festo fluidic muscles is USD 50. If such an actuator fails and the harvesting process doesn't take place until the failure is discovered at the end of the workday, the total replacement cost is around USD 800 for one gripper, but the total cost is significantly increased when several grippers are attached to different manipulator arms within the same autonomous mobile platform. The optimization was repeated three times using these values and the obtained results are included in Table 2, whereas the statistics regarding these results are included in Table 3. As it can be observed in these tables, the average value for the optimal failure probability threshold is 0.0542 and the approach is stable providing remarkably similar results even if only 100 values were generated for the estimated lifetime of the considered Festo actuators.

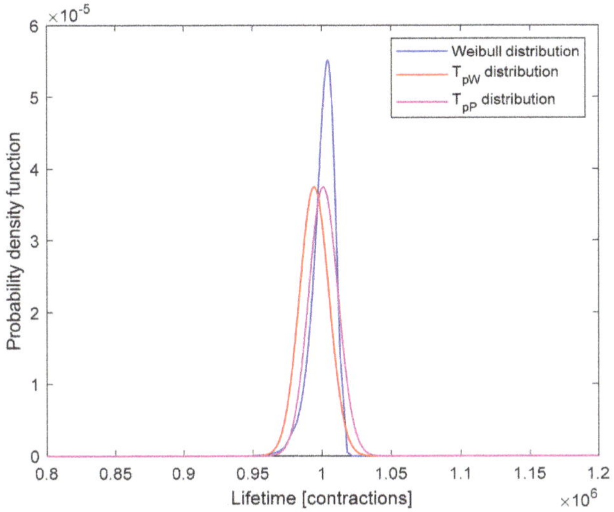

Figure 5. Probability density functions.

Table 2. Results obtained in the maintenance cost optimization.

No.	Optimal Failure Probability	Total Expected Cost per Time Unit [$/day]
1	0.0546	0.3049
2	0.0538	0.3004
3	0.0543	0.3032

Table 3. Statistics for the results obtained in the maintenance cost optimization.

Statistics	Optimal Failure Probability	Total Expected Cost per Time Unit [$/day]
Average	0.0542	0.3028
Variance	1.6333×10^{-7}	5.1633×10^{-6}
Standard deviation	4.0415×10^{-4}	0.023

In the last step, considering the optimal threshold of the probability of failure given by the average of the three results obtained in the optimization step, namely 0.0542, the conditional probability of failure can be determined for each inspection interval. We consider that an inspection takes place at the end of each harvesting day and the Festo fluidic muscles perform daily ~6000 contractions. The values for the conditional failure probability were computed for each inspection interval and the results are illustrated in Figure 6. As it can be observed, the value of the conditional failure probability exceeds the optimal threshold for the interval [984,000; 990,000] contractions. Thus, preventive replacement of the muscles is recommended after 984,000 contractions, a value that corresponds to day 164 of the 166.6 days of operation given by the manufacturer.

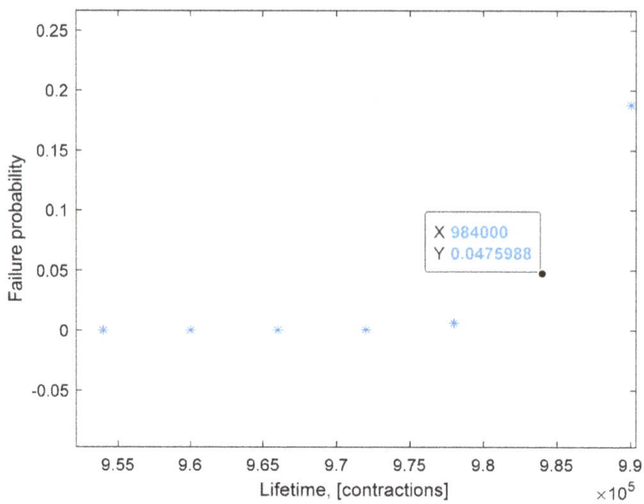

Figure 6. Failure probability at the beginning of each inspection interval.

For example, if the decision to replace the Festo fluidic muscles is delayed until 990,000 contractions (a value much closer to the value of t_W, according to the Weibull distribution) have been performed, the failure probability increases dramatically and causes a significant increase in the average of the total expected cost per time unit from USD 0.30 per day to USD 5.15 per day. This increase of approximately 17 times when considering the estimated lifetime given by the Weibull distribution becomes quite worrying when not just one gripper is considered but several harvesting platforms with multiple grippers.

5. Conclusions

Smart farming is one of the rapidly developing engineering fields because it can substantially cut down the total costs of agricultural production so crucial in a world with high population growth. Harvesting crops is one of the time-consuming and cost-increasing aspects of farming, therefore mechanization and automation are essential, especially for large-scale farming.

The gripper-type harvesting unit proposed in this paper uses Festo fluidic muscles in a novel way, providing a solution easy to manufacture with low maintenance and which can be used for more types of crops due to the actuator choice. Moreover, the gripper can be attached to an independent mobile platform equipped with a vision system that allows crop detection, and therefore, large-scale harvesting can be conducted without human assistance.

The harvesting process can be hindered and the crop can be damaged if one of the Festo fluidic muscles of the proposed harvesting system fails. Hence, an optimal maintenance policy is developed in this paper for the gripper actuators in order to minimize the costs. Due to a lack of available historical data for Festo fluidic muscles, a surrogate model is developed using randomly generated artificial muscle lifetime values in order to illustrate the use and importance of the developed maintenance policy. Using the model and an enhanced evolutive optimization algorithm, the optimal threshold of the failure probability is determined. Based on this threshold, a preventive replacement can be performed in order to minimize maintenance costs.

The results obtained after applying the optimal maintenance policy developed in this paper for the Festo fluidic muscles indicate that a considerable decrement in the costs can be achieved if one performs a preventive replacement of the actuators when the estimated probability of failure in the time interval until the next inspection becomes larger than the determined optimal threshold failure probability. Specifically, a replacement conducted a few days before the lifetime provided by the manufacturer and the lifetime estimated using the well-known Weibull distribution could decrease the maintenance costs approximately 17 times for the Festo fluidic muscles.

The optimal maintenance policy proposed in this paper can be easily customized for other engineering systems and the notable results obtained using this method are proof of its high usefulness in industry.

Author Contributions: Conceptualization, F.M.Ș., M.D.Ș. and S.-D.M.; methodology and supervision, S.-D.M.; software, F.M.Ș. and M.D.Ș.; validation and data curation, F.M.Ș. and M.D.Ș.; formal analysis and writing—original draft preparation, F.M.Ș. and M.D.Ș.; writing—review and editing, F.M.Ș., M.D.Ș. and S.-D.M.; visualization, F.M.Ș. and M.D.Ș. All authors have read and agreed to the published version of the manuscript.

Funding: This research received no external funding.

Institutional Review Board Statement: Not applicable.

Informed Consent Statement: Not applicable.

Data Availability Statement: The data presented in this study are available on request from the corresponding author. The data are not publicly available due to privacy.

Acknowledgments: The authors would like to thank the Technical University of Cluj-Napoca, Romania, for providing technical support in conducting the research.

Conflicts of Interest: The authors declare no conflict of interest.

References

1. Vu, Q.; Kuzov, M.; Ronzhin, A. Hierarchical classification of robotic grippers applied for agricultural object manipulations. *MATEC Web Conf.* **2018**, *161*, 03015. [CrossRef]
2. Șerdean, M.; Șerdean, F.; Mândru, D. An Overview of Grippers in Agriculture Robotic Systems. In *New Advances in Mechanisms, Mechanical Transmissions and Robotics*; Mechanisms and Machine Science; Lovasz, E.C., Maniu, I., Doroftei, I., Ivanescu, M., Gruescu, C.M., Eds.; Springer: Cham, Switzerland, 2020; Volume 88, pp. 212–225.
3. Ceccarelli, M.; Figliolini, G.; Ottaviano, E.; Mata, A.S.; Criado, E.J. Designing a robotic gripper for harvesting horticulture products. *Robotica* **2000**, *18*, 105–111. [CrossRef]
4. Font, D.; Pallejà, T.; Tresanchez, M.; Runcan, D.; Moreno, J.J.; Martínez, D.; Teixidó, M.; Palacín, J. A Proposal for Automatic Fruit Harvesting by Combining a Low Cost Stereovision Camera and a Robotic Arm. *Sensors* **2014**, *14*, 11557–11579. [CrossRef] [PubMed]
5. Hemming, J.; Bac, C.W.; van Tuijl, B.A.J.; Ruud, B.; Bontsema, J.; Pekkeriet, E. A robot for harvesting sweet-pepper in greenhouses. In Proceedings of the International Conference of Agricultural Engineering, Zurich, Switzerland, 6–10 July 2014.
6. Zhao, D.-A.; Lv, J.D.; Ji, W.; Zhang, Y.; Chen, Y. Design and control of an apple harvesting robot. *Biosyst. Eng.* **2011**, *110*, 112–122. [CrossRef]
7. Zhao, Y.; Gong, L.; Liu, C.; Huang, Y. Dual-arm Robot Design and Testing for Harvesting Tomato in Greenhouse. *IFAC-PapersOnLine* **2016**, *49*, 161–165. [CrossRef]
8. Anderson, D. Oniqua Enterprise Analytics. Available online: http://www.plant-maintenance.com/articles/PMCostReduction.pdf (accessed on 14 March 2023).
9. Durán, O.; Vergara, B. Maintenance Strategies Definition Based on Systemic Resilience Assessment: A Fuzzy Approach. *Mathematics* **2022**, *10*, 1677. [CrossRef]
10. Gashi, M.; Gursch, H.; Hinterbichler, H.; Pichler, S.; Lindstaedt, S.; Thalmann, S. MEDEP: Maintenance Event Detection for Multivariate Time Series Based on the PELT Approach. *Sensors* **2022**, *22*, 2837. [CrossRef] [PubMed]
11. Rykov, V.; Kochueva, O.; Rykov, Y. Preventive Maintenance of the k-out-of-n System with Respect to Cost-Type Criterion. *Mathematics* **2021**, *9*, 2798. [CrossRef]
12. Kang, J.; Wang, Z.; Guedes Soares, C. Condition-Based Maintenance for Offshore Wind Turbines Based on Support Vector Machine. *Energies* **2020**, *13*, 3518. [CrossRef]
13. Bucci, G.; Ciancetta, F.; Fioravanti, A.; Fiorucci, E.; Mari, S.; Silvestri, A. Online SFRA for Reliability of Power Systems: Characterization of a Batch of Healthy and Damaged Induction Motors for Predictive Maintenance. *Sensors* **2023**, *23*, 2583. [CrossRef] [PubMed]
14. Tian, Z.; Wu, B.; Chen, M. Condition-based maintenance optimization considering improving prediction accuracy. *J. Oper. Res. Soc.* **2014**, *65*, 1412–1422. [CrossRef]
15. Eti, M.; Ogaji, S.; Probert, S. Development and implementation of preventive-maintenance practices in nigerian industries. *Appl. Energy* **2006**, *83*, 1163–1179. [CrossRef]

16. Zhong, S.; Gai, Z.; Yang, Y.; Zhao, Y.; Qi, Y.; Yang, Y.; Peng, Y. A contraction length feedback method for the McKibben pneumatic artificial muscle. *Sens. Actuators A Phys.* **2022**, *334*, 113321. [CrossRef]
17. Soleymani, R.; Khajehsaeid, H. A mechanical model for McKibben pneumatic artificial muscles based on limiting chain extensibility and 3D application of the network alteration theories. *Int. J. Solids Struct.* **2020**, *202*, 620–630. [CrossRef]
18. Serdean, M.; Besoiu, S.; Lungu, I.; Mandru, D. Robotic System Actuated by McKibben Artificial Muscles. *Robot. Manag.* **2011**, *16*, 37–42.
19. Yang, X.-S.; Deb, S. Cuckoo Search via Lévy flights. In Proceedings of the World Congress on Nature & Biologically Inspired Computing (NaBIC), Coimbatore, India, 9–11 December 2009; pp. 210–214.
20. Tudose, L.; Rusu, F.; Tudose, C. Optimal design under uncertainty of bearing arrangements. *Mech. Mach. Theory* **2016**, *98*, 164–179. [CrossRef]
21. Rusu, F. Optimizations under Uncertainty with Applications in Rolling Bearing Industry. Ph.D. Thesis, Technical University of Cluj-Napoca, Cluj-Napoca, Romania, 2014.
22. Wu, B. Condition-Based Maintenance Optimization Using Data Driven Methods. Ph.D. Thesis, Concordia University, Montreal, QC, Canada, 2013.

sensors

MDPI

Article

Explainable Remaining Tool Life Prediction for Individualized Production Using Automated Machine Learning

Lukas Krupp [1,*], Christian Wiede [1], Joachim Friedhoff [2] and Anton Grabmaier [1,3]

1 Fraunhofer Institute for Microelectronic Circuits and Systems,47057 Duisburg, Germany
2 CAX Technologies, University of Applied Sciences Ruhr West, 45407 Mülheim an der Ruhr, Germany
3 Department of Electronic Components and Circuits, University of Duisburg-Essen, 47057 Duisburg, Germany
* Correspondence: lukas.krupp@ims.fraunhofer.de

Abstract: The increasing demand for customized products is a core driver of novel automation concepts in Industry 4.0. For the case of machining complex free-form workpieces, e.g., in die making and mold making, individualized manufacturing is already the industrial practice. The varying process conditions and demanding machining processes lead to a high relevance of machining domain experts and a low degree of manufacturing flow automation. In order to increase the degree of automation, online process monitoring and the prediction of the quality-related remaining cutting tool life is indispensable. However, the varying process conditions complicate this as the correlation between the sensor signals and tool condition is not directly apparent. Furthermore, machine learning (ML) knowledge is limited on the shop floor, preventing a manual adaption of the models to changing conditions. Therefore, this paper introduces a new method for remaining tool life prediction in individualized production using automated machine learning (AutoML). The method enables the incorporation of machining expert knowledge via the model inputs and outputs. It automatically creates end-to-end ML pipelines based on optimized ensembles of regression and forecasting models. An explainability algorithm visualizes the relevance of the model inputs for the decision making. The method is analyzed and compared to a manual state-of-the-art approach for series production in a comprehensive evaluation using a new milling dataset. The dataset represents gradual tool wear under changing workpieces and process parameters. Our AutoML method outperforms the state-of-the-art approach and the evaluation indicates that a transfer of methods designed for series production to variable process conditions is not easily possible. Overall, the new method optimizes individualized production economically and in terms of resources. Machining experts with limited ML knowledge can leverage their domain knowledge to develop, validate and adapt tool life models.

Keywords: automated machine learning; explainable artificial intelligence; free-form machining; individualized production; remaining useful lifetime; tool condition monitoring

1. Introduction

With the advancement of Industry 4.0, the demand for highly customized products is increasing. A growing proportion of single-part and small-batch production manifests this in the manufacturing industry. The resulting frequent machine and process reconfigurations increase the susceptibility to process errors, which is unacceptable for applications requiring a high product quality and reliability. Die making and mold making combine high product quality requirements and a dominant share of individualized production [1,2]. The core technology in die making and mold making is machining, particularly multi-axis milling, for manufacturing complex free-form workpieces [3]. The decisive quality parameters are their dimensional accuracy and surface roughness, which are significantly influenced by the wear of the cutting tools [4–6]. Even minor deviations from the specification can lead to defective end products, e.g., in injection molding or die-casting processes. Therefore, monitoring the machining processes and tool wear is essential to avoid scrap and rework [7].

In recent years, remaining tool life prediction based on sensor-driven process monitoring has been increasingly investigated in this context [8]. The remaining tool life prediction enables a joint estimation of the current tool condition based on the monitoring data and the duration until quality-related specifications of the process are violated. Thus, the quality and productivity of machining processes become controllable and the job-shop scheduling is simplified due to increased plannability [9]. Simultaneously, process-integrated sensors allow for reductions in time-consuming measurements using manufacturing metrology.

However, single-part and small-batch production conditions have made the development of remaining tool life prediction methodologies considerably difficult [10]. In particular, frequent changes in the workpiece geometries and process parameters do not allow for the direct inference of the tool condition from the sensor data due to a lack of comparability. Furthermore, the prediction of the remaining tool life is affected by the increased uncertainty regarding future process conditions. Therefore, previous approaches mainly focus on series production under constant process conditions, implying that the used prediction models are not adaptable.

This paper aims to develop and investigate a methodology enabling the remaining tool life prediction for individualized production, i.e., single-part and small-batch production. The prediction should be based on process-integrated sensors to ensure permanent process monitoring under constantly changing workpieces and cutting parameters.

The contributions of our work include the following:

1. A remaining tool life prediction methodology adaptable to new process conditions without manual intervention through automated machine learning (AutoML) while jointly explaining the predictions for model validation and optimization;
2. A new dataset and the methodology for its generation, representing gradual tool wear and its influence on the workpiece surface in individualized production under continuous variation in workpieces and cutting parameters;
3. A detailed evaluation of the methodology based on the new dataset, comparing it with a state-of-the-art approach for series production and investigating its explanation and generalization capabilities as well as its potential to increase the prediction robustness.

This paper is structured as follows. Section 2 provides an overview of the background and current state-of-the-art methods in the field of remaining tool life prediction, focusing on approaches based on sensor-driven process monitoring. Section 3 introduces the new remaining tool life prediction methodology for individualized production and highlights its three major components. Section 4 outlines the implementation details of the methodology and describes the dataset generation to evaluate it. In Section 5, the evaluation procedure is presented, and the results are discussed. Section 6 summarizes and concludes the paper.

2. Related Work

The state-of-the-art methods in sensor-driven remaining tool life prediction comprise two main approaches: direct [11] and criterion-based [12]. Direct methods use models generating a temporal output from the process-describing sensor data. The tool state and the decision threshold regarding the end of the tool life are, therefore, only implicitly part of the model and cannot be extracted or set separately. Criterion-based methods integrate an intermediate step via a tool life criterion to indicate the tool condition. The subsequent extrapolation of the tool life criterion allows for the setting of arbitrary decision thresholds. In addition, the tool condition is directly available for further applications, e.g., for integration into a simulation. Due to its comprehensive significance for the process quality, the tool condition and the end of tool life are usually determined based on tool wear [13]. Alternatively, quality parameters, such as the workpiece surface roughness, can be used as tool life criteria.

The data basis for the remaining tool life models is generated using the state-of-the-art sensor types in process monitoring [8]. The monitoring variables are the cutting force [14], vibration [15], drive current and power [16] or machine tool controller signals [17]. Since the sensor data provide the input for the predictions, either purely data-based [18] or hybrid physics- and data-based models [19] are used. In the area of data-based remaining tool life prediction, machine learning (ML) models and particularly neural network architectures, like convolutional neural networks (CNNs) [20], temporal convolutional networks (TCNs) [21] or long short-term memory (LSTM) networks [22], are current research topics due to their high adaptivity, accuracy and suitability for temporal predictions.

Their underlying production scenarios and datasets are the most significant distinguishing characteristics of the prediction models. A remaining tool life dataset comprises the sensor and target data over the life cycle of multiple tools. Possible variants of datasets are shown in Figure 1, depending on the respective degree of process condition variations during single and multiple tool life cycles. Process conditions refer to influencing factors, i.e., the tool shape and material; workpiece shape and material; cutting parameters; machine tool design and its condition or tool path; and process kinematics. While the dataset variants I, e.g., Ref. [23], and II, e.g., Ref. [24], mainly represent series production, the combinations III [25] and IV describe the individualized production scenario.

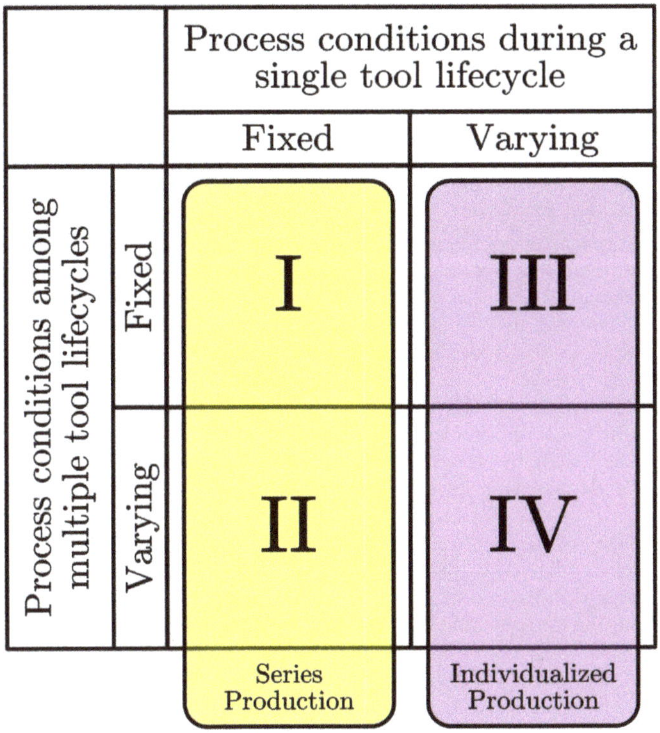

Figure 1. Possible dataset variants in the area of remaining tool life prediction depending on the degree of process condition variations during single and multiple tool life cycles.

Previous work on sensor-driven remaining tool life prediction mainly investigates series production scenarios [8]. Individualized production, i.e., the variation in process conditions during a tool life cycle, is hardly considered. A single approach analyzes varying cutting parameters during the tool life cycle [25]. However, the same workpiece is manufactured repeatedly. In [9], a methodology for small-batch production is developed using a dataset containing several identical cutting operation sequences. The approach of Matsumura et al. [26], while considering varying workpieces during the tool life cycle, requires direct wear measurements and is thus outside the scope of sensor-driven predictions.

Overall, sensor-driven remaining tool life prediction has not yet been analyzed under variable cutting parameters and workpiece geometries during single and multiple tool life cycles. Datasets according to variant III, like [25], do not include the implied degree of process condition variation, and datasets according to variant IV do not exist. Therefore, it is still unknown whether the feature extraction methodologies for sensor data under fixed process conditions are applicable. Furthermore, previous remaining tool life prediction models do not mitigate the uncertainty due to variable future process conditions.

Several papers investigate the use of AutoML methods to make the benefits of ML-based models even easier to apply to tool condition monitoring [6,27–29]. AutoML leverages the autonomous adaptation of models to changing process conditions, especially in individualized production. However, with increased autonomy in the generation of models, their validity must be ensured. Although the first approaches to explaining ML models in the context of machining process monitoring exist [30–33], methods combining AutoML and model explainability are missing so far.

3. Methodology

Our methodology extends the state-of-the-art methods described in the following three points to enable and overcome the challenges of the remaining tool life prediction for individualized production: an extended and non-invasively acquired feature set as input to the prediction model (Section 3.1); a new AutoML-based prediction model incorporating future feature estimates (Section 3.2); and a feature importance ranking algorithm that exploits the AutoML-based nature of our prediction model to explain its output (Section 3.3).

3.1. Sensors Signals and Feature Extraction

The model inputs are features obtained from the data of process-integrated sensors. The characteristics of individualized production must be considered when selecting the sensors. There are frequent changes in process-related components, such as the tool, the tool holder or the clamping device. In addition, different machine types are used due to the variety of workpieces. Accordingly, permanent and comprehensive process monitoring is only possible if sensors that are neither directly dependent on the process-related components nor the machine tool are used.

Vibration sensors fulfill this requirement and allow a trade-off between process proximity and independence. Due to the property of vibration propagation via the machine structure, sensors can be placed on machine components that are only indirectly involved in the process, such as the spindle. If the sensor sensitivity is sufficient, it is still possible to detect process emissions with a high resolution. Figure 2 shows the sensor system and signal processing approach followed in this paper based on these considerations.

To ensure a high degree of process information, both the mass-dependent vibration of the machine component due to the cutting forces and the process-related structure-borne sound, marked as ① and ② in Figure 2, are acquired. State-of-the-art micro-electro-mechanical system (MEMS) accelerometers for industrial condition monitoring enable this, featuring low noise, a high sensitivity and a high bandwidth. In addition, their small size and high energy efficiency make them suitable for use in embedded systems on dynamic machine components. The signals from the acceleration sensors on the three spatial axes form the basis for feature extraction.

In total, three types of features are distinguished: instantaneous, parameter and cumulative features. Instantaneous features are time- or frequency-domain features describing the current milling process at the signal level and are therefore derived directly from the sensor data. They contain information about the tool state at the current time but are also influenced by the process conditions, i.e., in particular, cutting parameters and workpiece geometry or the resulting engagement conditions. In previous work, for the most part, only instantaneous features were used due to the assumed constant process conditions per tool life cycle. Since process conditions are constantly changing in individualized production, it becomes difficult to distinguish the tool condition based on instantaneous features only.

Therefore, in this paper, we investigate parameter features as model inputs in addition to the instantaneous features. The parameter features describe the process parameters set during process planning, i.e., in particular, cutting parameters, such as the feed per tooth f_z, the cutting speed v_c and the axial depth of cut a_p or the geometry parameters of the workpieces. We assume that the parameter features can be extracted from the numerical control (NC) code, which represents a second machine- and process-independent data source in addition to the acceleration sensors. The NC code is transferred to the sensor system from the computer-aided manufacturing (CAM) software running on a dedicated computer or in the cloud.

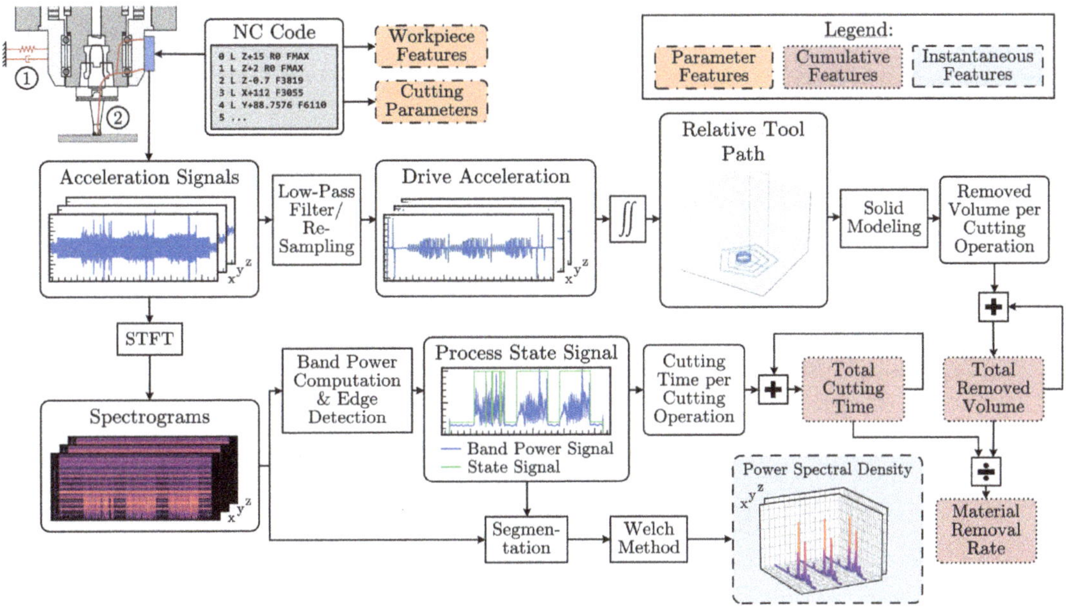

Figure 2. Sensor system architecture showing the signal processing and feature extraction steps based on three spindle-mounted accelerometers covering the spatial axis. The accelerometers acquire the mass-dependent vibration of the spindle due to the cutting forces ① and the process-related structure-borne sound ②.

The cumulative features are process-describing variables summed over time, such as the tool's cutting time or the volume of material already removed. Due to their resulting monotonic behavior, the cumulative characteristics correlate very well with potential target variables that are also monotonic, such as gradual tool wear.

The basis for feature extraction is the short-time Fourier transform (STFT), generating time-frequency spectrograms from the acceleration signals. The spectrograms are converted into a process state signal by determining the band power in the dominant frequency range of the machining process. Then, an edge detection algorithm is applied to the band power signal to receive the state signal. The process state signal enables the logical segmentation of continuous cutting operations. For each cutting operation, using the Welch method, the instantaneous features in the form of the power density spectra of the triaxial signals are determined from the spectrograms. It is known that the power density spectra as instantaneous features in series production scenarios show good correlations with tool wear [30,34]. Furthermore, the cutting time per cutting operation is calculated based on the process state signal. The cumulative feature of the total cutting time of the tool is then derived by accumulating the cutting time per cutting operation.

The low-frequency drive acceleration signals are extracted from the broadband acceleration signals via low-pass filtering and resampling in a second signal processing path. The drive acceleration signals can be converted into relative position information of the axes via double integration with intermediate error correction. If the tool geometry is known, solid modeling allows for an estimation of the removed material volume per cutting operation via Boolean operations [35]. The cumulative feature of the total removed volume is derived by accumulating the removed volume per cutting operation. The quotient of the total removed volume and the total cutting time of the tool represents the material removal rate. The material removal rate is a measure of the process productivity. Since the feature is derived from two cumulative features, it is also attributed to the cumulative features.

3.2. Prediction Model

In this paper, the criterion-based approach shown in Figure 3 is followed for the remaining tool life prediction. Since the workpiece surface roughness and dimensional accuracy are mainly influenced by the flank wear of the milling tool, the wear mark width *VB* is considered the primary tool life criterion in the following. In this paper, the *VB* is defined as the average of the maximum wear mark widths per cutting edge. The procedure to determine the *VB* in practice is described in Appendix A. The criterion-based approach allows for an arbitrary choice of the end-of-life threshold, i.e., the tool wear threshold VB_t. The end-of-life threshold is critical in individualized production since it may change depending on the required workpiece tolerances.

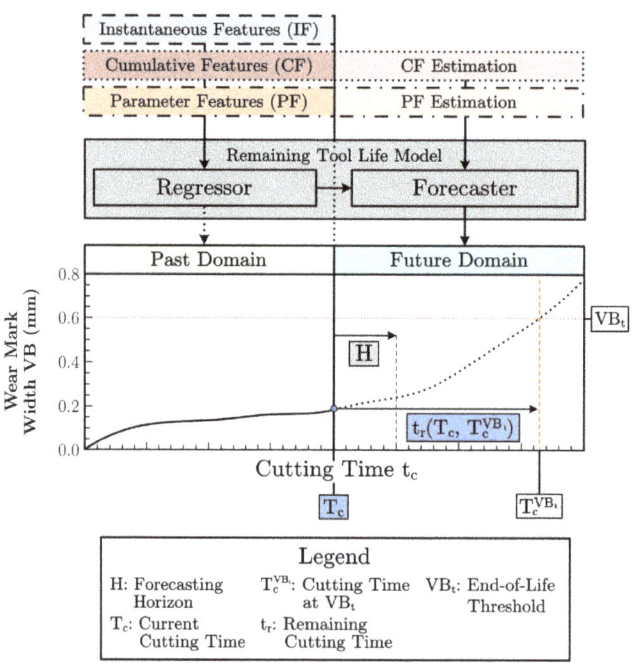

Figure 3. Remaining tool life prediction methodology based on an extended feature set. The feature set contains not only instantaneous features with information on the current tool state but also cumulative and parameter features with context information on current and past processes. Additionally, the model allows for estimates of future features as inputs to include a priori knowledge.

The goal is to predict at the current cutting time step T_c the future time step $T_c^{VB_t}$ where the wear threshold value VB_t is reached. The remaining tool life t_r can then be derived according to Equation (1):

$$t_r\left(T_c, T_c^{VB_t}\right) = T_c^{VB_t} - T_c \quad (1)$$

The criterion-based prediction approach represents a multi-series forecasting problem, with the particularity that a regression of the tool state from the sensor data features must first be performed to perform the forecast. Accordingly, the model shown in Figure 3 includes a regression and a forecasting component. The dataset $\mathcal{D}$ for training and testing the model contains N tool instance series and is thus defined as $\mathcal{D} = \{\mathcal{D}_i\}_{i=1}^{N}$. The samples of $\mathcal{D}_i$ are defined as $\mathcal{D}_i = \left\{\mathbf{X}^{(p)}_{\mathbf{i},1:T_c}, \mathbf{X}^{(f)}_{\mathbf{i},T_c+1:T_c+H}, \mathbf{y}_{\mathbf{i},1:T_c}, \mathbf{y}_{\mathbf{i},T_c+1:T_c+H}\right\}$with T_c being the instantaneous cutting time step beyond which the wear curve f_{VB} is to be forecasted and H being the forecasting horizon, i.e., the cutting time steps over which a model predicts the f_{VB}. The tensors $\mathbf{X}^{(p)}_{\mathbf{i},1:T_c}$ and $\mathbf{X}^{(f)}_{\mathbf{i},T_c+1:T_c+H}$ describe the past features and estimated future features. The vectors $\mathbf{y}_{\mathbf{i},1:T_c}$ and $\mathbf{y}_{\mathbf{i},T_c+1:T_c+H}$ describe the past and future targets. For simplicity, the index of the tool instance i is omitted in the following.

An abstracted time step in the model refers to a continuous cutting operation j. Hence, each time step can be assigned a sample of the form $\{\mathbf{x}_j, y_j\}$, with $\mathbf{x}_j$ being the feature vector and the scalar y_j being a wear value. In contrast to the state-of-the-art methods, the prediction approach shown in Figure 3 is based on inputs from past cutting operations and allows for the integration of estimates of future features. Thus, the constantly changing conditions of individualized production can be considered. According to Equations (2a) and (2b), the feature tensors of past cutting operations $\mathbf{X}^{(p)}_{1:T_c}$ are composed of the instantaneous features $\mathbf{f_i}$, the parameter features $\mathbf{f_p}$ and the cumulative features $\mathbf{f_c}$. The features are obtained based on the methodology introduced in Section 3.1.

$$\mathbf{X}^{(p)}_{1:T_c} = [\mathbf{x}_1, \dots, \mathbf{x}_{T_c}] \quad (2a)$$

$$\mathbf{x}_j = \left[\mathbf{f}^T_{\mathbf{i},j}, \mathbf{f}^T_{\mathbf{p},j}, \mathbf{f}^T_{\mathbf{c},j}\right]^T \quad (2b)$$

The future feature tensors $\mathbf{X}^{(f)}_{T_c+1:T_c+H}$ are based on estimates and prior knowledge about future manufactured workpieces and cutting parameters. Since the instantaneous features can only be generated based on a machining process that has taken place, only the parameter features $\hat{\mathbf{f}}_\mathbf{p}$ and cumulative features $\hat{\mathbf{f}}_\mathbf{c}$ according to Equations (3a) and (3b) can be considered for the future feature tensors.

$$\mathbf{X}^{(f)}_{T_c+1:T_c+H} = [\hat{\mathbf{x}}_{T_c+1}, \dots, \hat{\mathbf{x}}_{T_c+H}] \quad (3a)$$

$$\hat{\mathbf{x}}_j = \left[\hat{\mathbf{f}}^T_{\mathbf{p},j}, \hat{\mathbf{f}}^T_{\mathbf{c},j}\right]^T \quad (3b)$$

Based on the feature tensors, the prediction is performed according to Figure 3. The presented model consisting of the regression component f_R and the forecasting component f_F can be formally described according to Equations (4a) and (4b).

$$\hat{\mathbf{y}}_{1:T_c} = f_R\left(\mathbf{X}^{(p)}_{1:T_c}, \boldsymbol{\theta}^R\right) \quad (4a)$$

$$\hat{\mathbf{y}}_{T_c+1:T_c+H} = f_F\left(f_R, \mathbf{X}^{(f)}_{T_c+1:T_c+H}, \boldsymbol{\theta}^F\right) \quad (4b)$$

The tensors $\boldsymbol{\theta}^R$ and $\boldsymbol{\theta}^F$ represent the parameters of the prediction model, which are adjusted during training.

3.3. Explainablity Methodology

Given the prediction model and the dataset, the goal is to identify the features relevant to the model output to ensure model explainability. At the same time, the best possible model performance should be achieved, and the manual configuration effort during training, e.g., due to hyperparameter searches, should be minimized. Therefore, we pursue model explainability based on AutoML-driven training in this paper.

It is assumed in the following that the prediction model is generally represented by an ensemble of ML pipelines $\mathcal{M}_E$. An individual ML pipeline $\mathcal{M}_\lambda$ with the model parameters $\boldsymbol{\theta}$ and hyperparameters λ describes the transfer from the input features to the final prediction. The pipelines are trained based on a training dataset $\mathcal{D}_{train}$ generated from the previously introduced dataset $\mathcal{D}$. Besides the adaptation of the model parameters $\boldsymbol{\theta}$, the training includes the combined architecture search and hyperparameter (CASH) optimization of the pipelines and a subsequent selection of several pipelines resulting in the ensemble $\mathcal{M}_E$ [36]. The ensemble combines the predictions of the included pipelines to increase the prediction accuracy. It is assumed that all pipelines contain an importance score that they use for their internal feature selection to obtain an overall measure of feature importance. Algorithm 1 formally describes the procedure for generating a global feature importance measure to explain the model decision.

Algorithm 1 AutoML-based combined remaining tool life prediction model generation and feature importance ranking

Input: Pipelines $\mathcal{M}_\lambda$ hyperparemeterized by $\lambda \in \boldsymbol{\Lambda}$ including feature selection based on importance score vectors $\mathbf{s}$, Empirical generalization error function $\hat{E}_G$, Training dataset $\mathcal{D}_{train}$, Training time budget T, Number N of pipelines to include in a final ensemble
Output: Best-performing ensemble $\mathcal{M}_E$ of pipelines, Global feature importance vector $\mathbf{s_G}$
1: Solve $\mathcal{M}_{\lambda^*} \in \underset{\lambda \in \boldsymbol{\Lambda}}{\operatorname{argmin}}\, \hat{E}_G(\mathcal{M}_\lambda, \mathcal{D}_{train})$ s.t. $\sum t_{\lambda_i} < T$
▷ Standard time-bounded CASH optimization
2: $\mathcal{M}_E$ = **EnsembleSelection**$(\mathcal{M}^*_\lambda, \mathcal{D}_{train}, N)$
▷ Standard weighted average ensemble method
3: $\boldsymbol{\mathcal{S}}$ = **CollectFeatureImportance**$(\mathcal{M}_E)$
▷ Returns a tensor containing the importance vectors of all pipelines in the ensemble
4: $\mathbf{s_G} = \frac{1}{N} \sum_{j=1}^{N} \mathbf{s}_j$ with $\boldsymbol{\mathcal{S}} = [\mathbf{s}_1, \dots, \mathbf{s}_N]$

Algorithm 1 is a global methodology, meaning that the feature importance output $\mathbf{s_G}$ relates to the complete training dataset. However, the procedure can be extended without a loss of generality using state-of-the-art local algorithms, as evaluated in [37], to obtain importance scores for individual feature vectors.

4. Implementation

4.1. Sensor System

The sensor system implementation for analyzing the previously introduced remaining tool life prediction methodology requires the use of sensitive MEMS accelerometers as a core component, according to Section 3.1. Therefore, the CN-0549 platform from Analog Devices is used [38]. The platform consists of the ADXL1002 acceleration sensor with a 3 dB bandwidth of 11 kHz [39], the CN-0540 signal acquisition board [40] and the Cora-Z7 field-programmable gate array (FPGA) system-on-chip (SoC). The CN-0540 signal acquisition board features a 24-bit Σ-Δ analog-to-digital converter allowing sampling rates up to 256 kHz. A high sampling rate is required to support all milling applications, including high-speed scenarios with common spindle rotation speeds up to 60,000 rpm. With the maximum sampling frequency, a single spindle rotation is then still devoted to 256 sensor signal samples.

Furthermore, according to Section 3.1, the methodology requires the sensors to be mounted to the machine tool's spindle. We assume that the spindle moves only along the three spatial axes. The principle is, therefore, only partially applicable in machines whose spindles can be tilted in addition to the translational movements. However, extending the sensor system with additional inclination sensors can restore the unrestricted applicability. A Hermle C30 U five-axis indexed milling center is used for further investigations. The machine tool's spindle performs only translational movements and the machine table can be tilted and rotated. Thus, the machine tool fulfills the previous assumption. The ADXL1002 is a single-channel acceleration sensor. Therefore, three sensors are installed on the spindle to cover the spatial axes according to the setup shown in Figure 2.

4.2. Experimental Setup

To date, no dataset exists that represents tool wear in the context of individualized production with continuous variation in workpieces and cutting parameters. Therefore, in this paper, the methodology shown in Figure 4 is proposed for the dataset generation based on the pocket milling process.

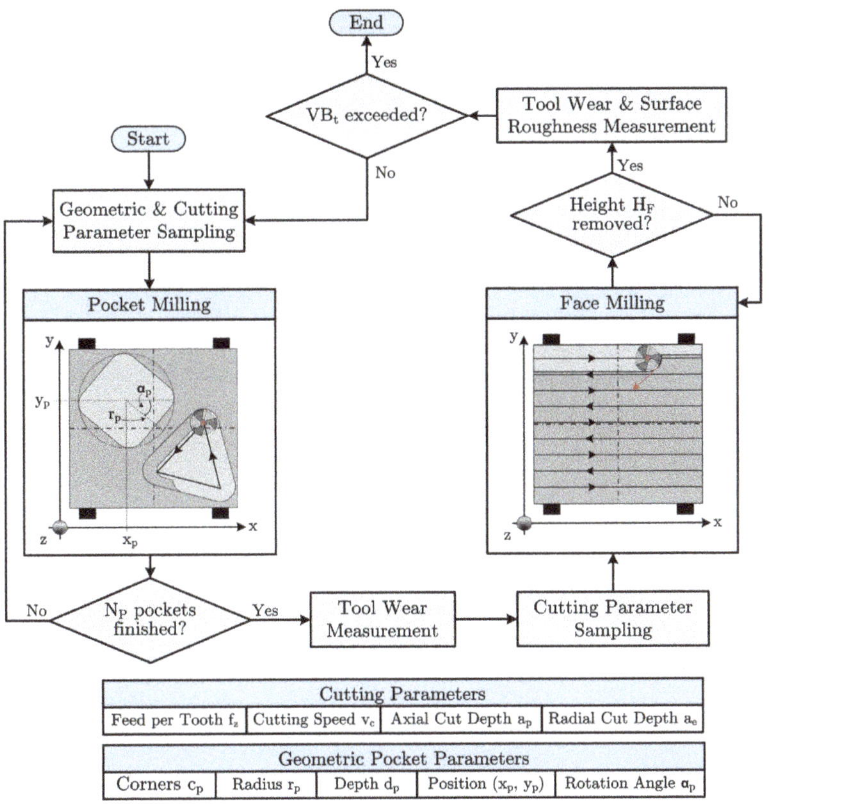

Figure 4. Methodology for the dataset generation based on a parameterizable pocket milling process. The dataset represents tool wear until exceeding an end-of-life threshold VB_t in individualized production scenarios under continuous variation in workpieces and cutting parameters.

The pocket milling process is selected due to the easily parameterizable workpiece geometries. The workpiece geometry is a polygon with a variable number of corners c_p. The variation in the number of corners allows for controllable variation in the engagement conditions of the cutter as its engagement angle changes along the tool path. The number of corners defines the maximum engagement angle. Furthermore, the radius of the polygon r_p, its depth d_p, its position as 2D coordinates (x_p, y_p) and its rotation angle α_p are varied. A computer-aided design (CAD) workpiece generator based on the geometry parameters is developed to sample the workpieces randomly. In addition to the workpiece geometries, the cutting parameters, i.e., the feed per tooth f_z, the cutting speed v_c and the axial depth of cut a_p, are actively varied. The sampling of the cutting parameters is based on latin hypercube sampling (LHS) to cover the parameter space with the limited number of milling operations that can be performed during a tool's life. The axial depth of cut a_e varies automatically along the tool path due to the variable engagement angle.

A fixed number N_p of pocket geometries are manufactured in a sequence. Subsequently, face milling removes the remaining material at a height H_f. The H_f must be greater than the maximum pocket depth. The face milling process applies a zig-zag strategy and LHS-derived cutting parameters. After completing a sequence of pocket and face milling processes, the maximum wear mark widths of the cutting edges are measured and the VB is derived. In addition, the surface roughness of the workpiece is measured for reference after each face milling process. The combined pocket and face milling are repeated until the wear mark width exceeds a threshold VB_t.

Since individualized production is dominantly represented in die making and mold making, the experimental requirements are based on industrial practice in this field. To manufacture the workpieces on an X155CrVMo12-1 steel cube with an edge length of 200 mm, a toroidal milling tool with three circular inserts (⌀= 8 mm) is used. The processing is performed on the Hermle C30 U five-axis indexed milling center described in the previous section. The tool wear is measured using a Garant MM1 video measuring microscope and the workpiece surface roughness is measured using a MarSurf PS 10 device. Based on the sensor data obtained during the process execution and the NC code, the input features of the dataset are generated for each tool according to Section 3.1. Table 1 shows the components of the final input feature set. The maximum wear mark width VB averaged over the three inserts is used as the target. The detailed methodology to derive the VB is described in Appendix A. An end-of-life threshold of VB_t = 0.8 mm is selected based on the tool manufacturer's recommendation. The milling process specifications, including the detailed cutting and workpiece parameter intervals, are summarized in Appendix B.

Table 1. Components of the feature set used as input to the prediction methodology, divided into instantaneous, cumulative and parameter features.

Feature Subset	Parameter	Description
Instantaneous	$\mathbf{P}_{\mathbf{xx}}^{(0-6\ \mathbf{kHz})}$	Tensor, including the bins of the power spectra between 0 and 6 kHz for the three accelerometer channels (x, y, z)
Cumulative	T_c	Total cutting time
	V	Total removed material volume
	Q	Material removal rate
Cutting Parameters	f_z	Feed per tooth
	v_c	Cutting speed
	a_p	Axial depth of cut
Workpiece Parameters	c_p	No. of pocket corners
	r_p	Pocket radius
	d_p	Pocket depth
	(x_p, y_p)	Pocket center point
	α_p	Rotation angle

4.3. Model Implementation

In the following, a model for the remaining tool life prediction can be trained using the dataset generated in the previous section. Figure 5 shows the model implementation according to the methodology described in Sections 3.2 and 3.3.

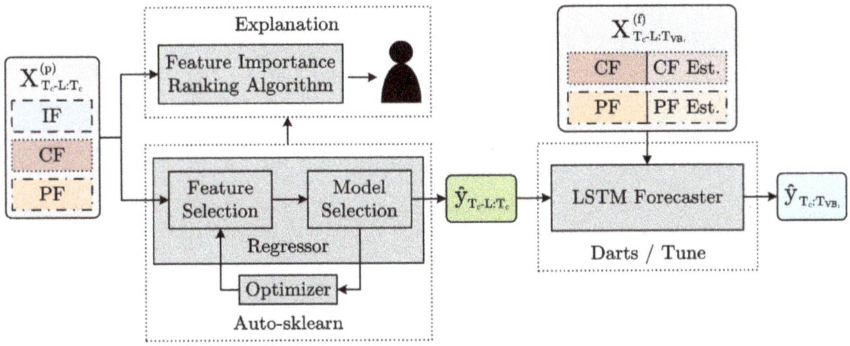

Figure 5. Framework implementing the automated and explainable remaining tool life prediction. The feature tensors $\mathbf{X}^{(p)}_{T_c-L:T_c}$ and $\mathbf{X}^{(f)}_{T_c-L:T_c^{VB_t}}$, including the instantaneous features (IFs), cumulative features (CFs) and parameter features (PFs), are the model inputs. The tool wear vectors $\hat{\mathbf{y}}_{T_c-L:T_c}$ and $\hat{\mathbf{y}}_{T_c+1:T_c^{VB_t}}$ are the model outputs.

The regression component is based on the Auto-sklearn library [41], enabling the implementation of the feature importance ranking algorithm (Algorithm 1). Using Auto-sklearn, CASH optimization can be performed based on ML pipelines with dedicated feature selection, restriction to importance-based feature selection methods and ensembling multiple ML pipelines. Auto-sklearn is selected due to its large model architecture and hyperparameter search space and especially its high number of included feature selection methods. A comparison between Auto-sklearn and two other state-of-the-art AutoML frameworks (LightAutoML [42] and FLAML [43])is described in Appendix C. The forecasting component is implemented using the Darts library [44]. An LSTM neural network is selected as the underlying model since LSTMs enable future features to be taken into account in time-series forecasting. Therefore, the remaining tool life prediction can be extended with prior knowledge about future machining operations as described in Section 3.2. To ensure a holistic AutoML approach, an additional wrapper is implemented around the forecaster model using the Tune library [45]. The wrapper combines the asynchronous successive halving algorithm (ASHA) [46] as a search algorithm with the tree-structured Parzen estimator [47] as a scheduler to enable the joint neural architecture and hyperparameter search.

In practical implementation, LSTMs base their predictions on the features in a limited window of length L from the past and do not include the entire history, as this quickly leads to intractability in the calculations. Therefore, the feature tensor $\mathbf{X}^{(p)}_{T_c-L:T_c}$ between $T_c - L$ and T_c is the input of the prediction model, as shown in Figure 5. Furthermore, the forecast must extend to the time point T_{VB_t} where the wear mark threshold VB_t is reached to enable the estimation of the remaining tool life. However, forecasting models have a fixed forecasting horizon H. Thus, the LSTM must be autoregressive to allow for an estimate of the remaining tool life at any time. The LSTM receives as input the tensor $\mathbf{X}^{(f)}_{T_c-L:T_c^{VB_t}}$ between $T_c - L$ and $T_c^{VB_t}$ as well as the predictions of the regressor component $\hat{\mathbf{y}}_{T_c-L:T_c}$ between the $T_c - L$ and T_c and predicts the future tool wear $\hat{\mathbf{y}}_{T_c+1:T_c^{VB_t}}$ between the $T_c + 1$ and T_{VB_t}.

5. Results and Discussion

5.1. Dataset and Evaluation Approach

Following the methodology introduced in Section 4.2, the dataset is generated as a foundation for the evaluations performed in the following. In total, the data of nine tools are acquired over their lifetime. Figure 6 shows the measured wear curves of the tools. In addition, the average material removal rate $\bar{Q}$ according to Equation (5) is shown per tool. A discrete number N_{T_c} of cutting time steps T_c represents the tool life. The total removed material volume per time step is denoted as V_{T_c}.

$$\bar{Q} = \frac{1}{N_{T_c}} \sum_{T_c} \frac{V_{T_c}}{T_c} \tag{5}$$

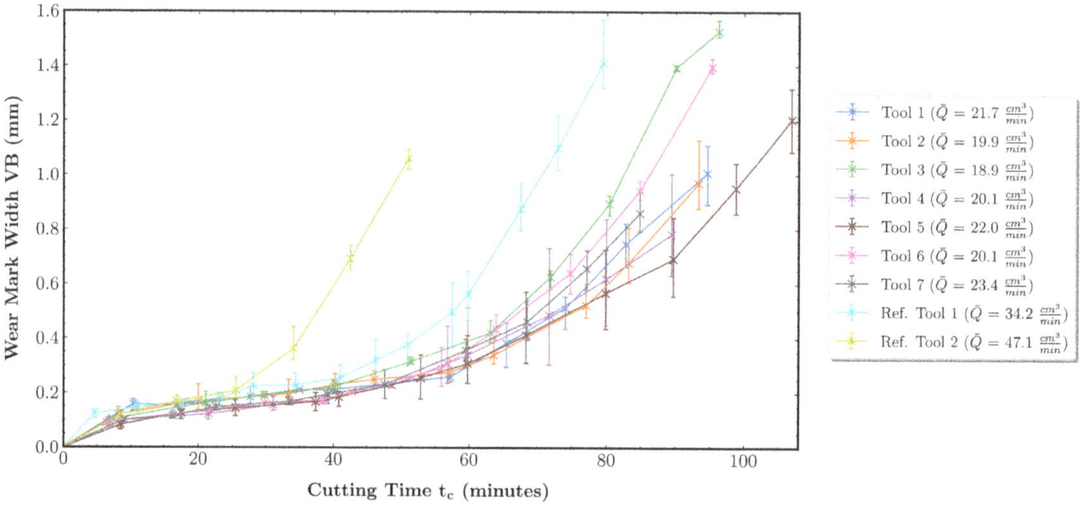

Figure 6. Tool wear mark width measurements and average material removal rates for the nine tools of the dataset. A cross marks the average maximum wear mark width over all cutting inserts $i \in \{1, 2, 3\}$. The vertical bars denote the maximum and minimum individual wear mark width values VB_i among the inserts.

Tools 1 to 7 manufacture variable pocket geometries with variable cutting parameters. This results in material removal rates between 18.9 $\frac{\text{cm}^3}{\text{min}}$ and 23.4 $\frac{\text{cm}^3}{\text{min}}$. Furthermore, the data during the lifetimes of two reference tools are acquired. Reference tool 1 manufactures variable pocket geometries under fixed cutting parameters. The cutting parameters are set to the maximum values of the intervals specified for tools 1 to 7. The maximized cutting parameters lead to an increase in $\bar{Q}$ to 34.2 $\frac{\text{cm}^3}{\text{min}}$ since the machining time of the pockets decreases. Thus, higher productivity is achieved. Reference tool 2 is applied in pure face milling based on the zig-zag strategy with fixed, maximum cutting parameters. Since only face milling is performed, the workpiece geometry can also be considered fixed. In this case, the maximum material removal rate of 47.1 $\frac{\text{cm}^3}{\text{min}}$ is achieved as no pocket milling is performed. The data from the reference tools are used to evaluate the generalization performance of the prediction models. For reference tools 1 and 2, the wear progress increases due to the increased productivity, while this is not the case for tools 1 to 7. An explanation for this is the influence of the varying workpiece geometries and, thus, loads on the tool cutting edges. Not only do the cutting parameters and the resulting machining speed affect the wear progress, but the combination with the workpiece geometry must always be considered.

To achieve an optimal test coverage of the prediction models with a limited number of available tools and ensure their robustness, the validation and test strategy shown in Figure 7 based on the leave-one-group-out methodology is used. The strategy is denoted as leave-one-tool-out cross-validation and testing (LOTO-CVT).

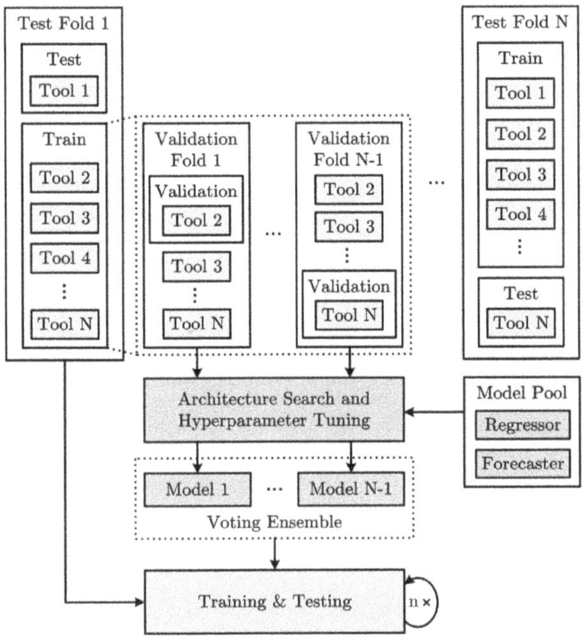

Figure 7. Leave-one-tool-out cross-validation and testing (LOTO-CVT) strategy for the remaining tool life prediction methodology based on a dataset with a limited number of tools N.

The data from N tools are divided into training and test sets to generate the regression and forecasting models. The data of a particular test tool are excluded from model training. Each tool is used once for testing to ensure that the prediction methodology is functional for arbitrary permutations and that its performance is not just based on the random selection of individual test tools. The model architecture and hyperparameter search are then performed based on the training set containing the data of $N-1$ tools. Model architecture and hyperparameter configurations are sampled from a model pool. When searching for the best configuration, a search criterion is required, enabling the evaluation of the configurations and their optimization. As with training, the prediction error can be used for this purpose. However, an additional validation tool has to be kept out of the training set. The evaluation of a model using the data of the validation tool guides the search.

It is problematic that selecting a single random validation tool can overfit the models, thus misleading the architecture and hyperparameter search. Hence, each tool is used once for validation to generate a model robust to the test tool data. The resulting models of the $N-1$ validation folds are combined into a voting ensemble. The outputs of the models are averaged to compensate for overfitted models. After the model architecture and hyperparameter search is complete, the voting ensemble models are trained using the data from all training tools. Subsequently, the evaluation is performed based on the data of the test tool. The stochastic nature of the parameter initialization and optimization of machine learning models may lead to different model outputs for multiple training runs. Training and testing are repeated n times to enable reliable model quality assessment.

5.2. Prediction Model Evaluation

The ability of the model approach introduced in Section 3.2 to predict tool wear and remaining tool life under variable process conditions is investigated using tools 1 to 7 in the following. First, the regression component and then the overall model extended by the forecasting component are investigated. The regression component quantifies the tool condition based on a tool wear prediction up to the current time point T_c. Previous approaches rely primarily on instantaneous features derived from sensor data as an input to tool wear prediction models. The reason for this is the fixed process conditions during a tool life cycle assumed in previous work, resulting in comparable cutting processes and a direct correlation with gradual tool wear. A common approach based on vibration data, as in [30] or [34], is to perform spectral analysis of the cutting operations, with frequency bins of power or amplitude spectra representing the features. Our methodology also incorporates spectral analysis in the form of the power spectral density. However, it goes beyond that by using the cumulative features, workpiece and cutting parameters as model inputs.

We evaluate the explainable state-of-the-art approach given in [30] based on our dataset described in Section 5.1 and the LOTO-CVT strategy. In [30], the wear prediction is a classification problem based on a random forest model, which receives the frequency spectra from structure-borne sound signals acquired during cutting operations as input. The approach is transferable to our regression component since random forest models can also be used for regression problems. First, only the power spectra of the accelerometer signals are used as input to the regression model. Before the evaluation is performed using tools 1 to 7, the correlation of the power spectra with tool wear is ensured under fixed process conditions using the data from reference tool 2. We then compare the results based on the methodology described in [30] with the wear predictions of our AutoML-based regression model using the extended feature set proposed in this paper.

For the regression model training, the Auto-sklearn environment is configured. Both the meta-learning and ensembling capabilities of Auto-sklearn are enabled. The maximum time budgets are set to 10 min for the entire CASH optimization and 30 s for training a single pipeline configuration with a memory limit of 20 GB per pipeline. The R2 score function is used as a metric for training. The training and testing steps are repeated five times according to the LOTO-CVT strategy. Since the regression is only required up to the end-of-life criterion VB_t, the range for prediction and evaluation is limited to 0.8 mm. Figure 8 shows the comparison of the regression results. For a comprehensive error analysis, the prediction errors in terms of root-mean-square error (RMSE) and mean absolute error (MAE) over the dataset are summarized in Table 2.

Figure 8 and Table 2 show that the state-of-the-art method for tool wear prediction described in [30], which is purely based on the instantaneous spectral features, is not easily transferable to the case of variable process conditions during the tool life cycle. Estimating the wear measurement curve is only partially possible to a limited extent, as seen in Figure 8a, e.g., for tools 2 and 4. In comparison, the predictions based on our method with the extended feature set achieve a reduction in the RMSE of between 43.4 and 80.2% and in the MAE of between 54.8 and 78.8%. As seen in Figure 8b, the prediction is possible for all tools and is mainly within the measured wear value intervals of the tool cutting edges. For tool 3 only, the prediction lies outside the wear value interval starting from a cutting time of 40 min. An explanation for this could be that tool 3 has the highest material removal rate of tools 1 to 7. Thus, the wear curve represents an extreme case of the dataset and the regression model has to perform an extrapolation during inference, which is much more error-prone than an interpolation. Overall, the better performance of our method compared to purely spectral feature-based prediction can be explained by the additional features. Under variable process conditions, their influence on the signals dominates, reducing the correlation between the instantaneous features and the tool wear. Particularly, the new cumulative features allow our method to restore the comparability of the cutting operations. The feature importance is investigated in Section 5.3 to confirm this hypothesis using Algorithm 1.

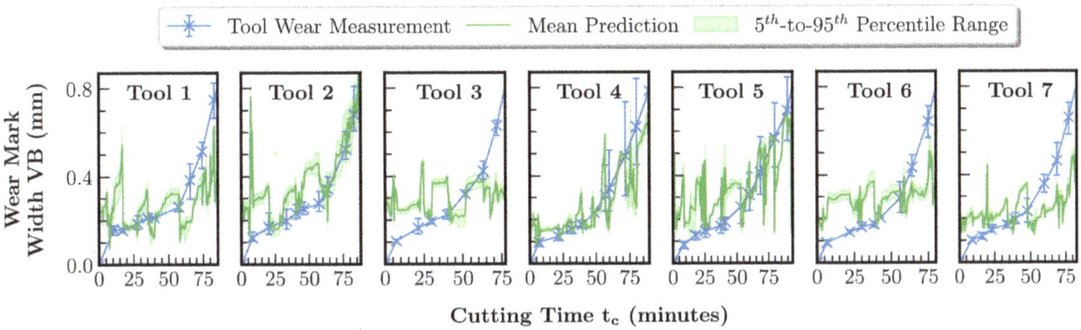

(**a**) Prediction results using only instantaneous features and the state-of-the-art hand-crafted model architecture.

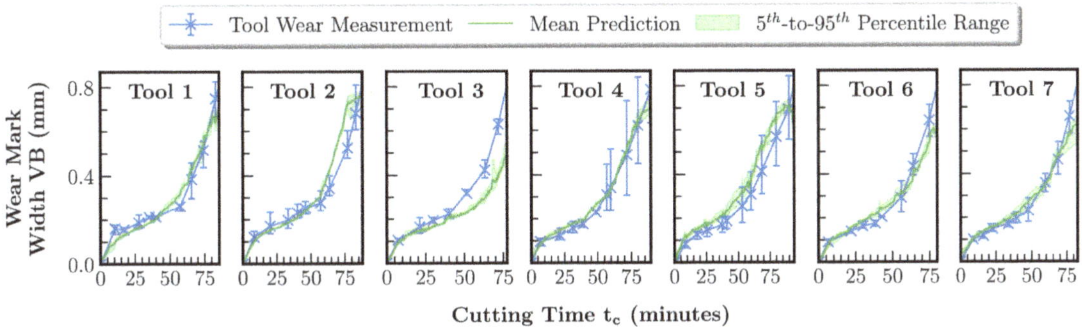

(**b**) Prediction results using our AutoML-based approach and the extended feature set.

Figure 8. Comparison of tool wear regression results using only the instantaneous features in combination with a state-of-the-art hand-crafted model architecture and the AutoML-based tool wear regression using the extended feature set proposed in this paper. The tool's data whose prediction results are displayed have been excluded from the training set.

Table 2. Comparison of tool wear regression errors using only the instantaneous features in combination with a state-of-the-art hand-crafted model architecture and the AutoML-based tool wear regression using the extended feature set proposed in this paper.

Method	Metric	Prediction Errors per Test Tool (mm)							
		Tool 1	Tool 2	Tool 3	Tool 4	Tool 5	Tool 6	Tool 7	Mean
Hand-crafted	RMSE	0.169	0.140	0.189	0.131	0.199	0.182	0.162	0.167
	MAE	0.133	0.102	0.155	0.097	0.147	0.143	0.132	0.130
Ours	RMSE	**0.038**	**0.068**	**0.107**	**0.026**	**0.066**	**0.065**	**0.041**	**0.059**
	MAE	**0.031**	**0.046**	**0.070**	**0.021**	**0.055**	**0.045**	**0.028**	**0.042**

In advance, the evaluation of the remaining tool life prediction based on the previously trained regression component is performed. The goal is to analyze how the extension of the remaining tool life prediction model compared to the state-of-the-art method through the possibility of entering future feature estimates affects the predictions. For this purpose, the LSTM-based forecasting component, according to Section 4.3, is trained and tested using the data from tools 1 to 7. Based on the LSTM output, the remaining tool life is calculated using Equation (1). The forecasting component is first tuned and trained based on the LOTO-CVT strategy. The LSTM model has a single layer and a hidden dimension of 25. The length of the model input sequences between 12 and 60 samples and the output sequences between 1 and 36 samples is subject to the model tuning. Furthermore, the hyperparameters of the batch size in the range of $[4, 32]$ and learning rate in the range of $[10^{-5}, 10^{-2}]$ are tuned, guided by the MAE. An LSTM instance can train for a maximum of 30 epochs while early stopping is employed. In total, the training and testing of the models are repeated five times. In the testing phase, the outputs of the regression component shown in Figure 8b are input to the LSTM. Additionally, an exploration of non-spectral feature combinations as future feature inputs is performed. Figure 9 and Table 3 show the remaining tool life prediction results for the LSTM without future features and the best-performing LSTM with future features.

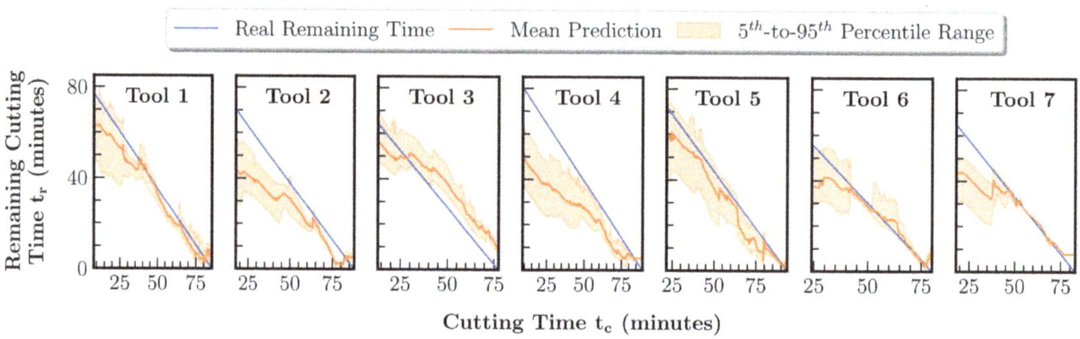

(**a**) Prediction results of the LSTM without future feature inputs.

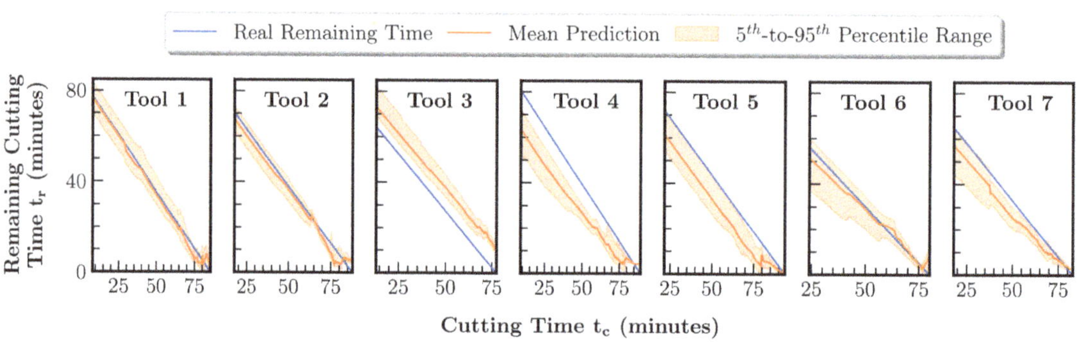

(**b**) Prediction results of the LSTM with the cutting time T_c as future feature inputs.

Figure 9. Comparison of the remaining tool life prediction results using the LSTM without future feature inputs and the LSTM with the cutting time T_c as future feature inputs. The tool's data whose prediction results are displayed have been excluded from the training set.

Table 3. Comparison of the remaining tool life prediction errors using the LSTM with no future feature inputs and the LSTM with the cutting time T_c as future feature inputs.

Future Features	Metric	Prediction Errors per Test Tool (Minutes)							
		Tool 1	Tool 2	Tool 3	Tool 4	Tool 5	Tool 6	Tool 7	Mean
No	RMSE	5.7	12.9	**7.9**	16.5	**6.9**	7.1	9.6	9.5
	MAE	4.7	10.8	**7.2**	14.0	**6.0**	4.9	6.5	7.8
Yes	RMSE	**2.0**	**2.7**	9.9	**12.2**	9.8	**3.0**	**6.1**	**6.5**
	MAE	**1.6**	**2.4**	9.9	**11.4**	9.4	**2.6**	**5.6**	**6.1**

The remaining tool life prediction without future features has an average RMSE of 9.5 min and an MAE of 7.8 min. With future features, the RMSE is reduced by 32% to 6.5 min and the MAE by 22% to 6.1 min. The results are achieved using the total cutting time T_c as a single future feature input. In Figure 9, the difference in prediction accuracy between the two model instances becomes evident. Without future features, the prediction is primarily inaccurate in the early stages of tool life, as shown in Figure 9a. Including the future features allows for a mostly accurate estimation of the remaining tool life at arbitrary time points. Only the predictions for tools 3, 4 and 5 in Figure 9b are characterized by a constant offset error. However, for tools 4 and 5, the predictions converge toward the real remaining time in the last 10 min of their respective lifetime. For tool 3 only, the offset remains constant until the end of its life. In this case, the offset is because the wear regression lies outside the measured wear values, as already described in the context of Figure 8b. Therefore, the regression error is propagated to the forecast and prevents the correct estimation of the future course of the wear curve. For tool 2, the maximum optimization of the prediction is achieved by an error reduction of 79% in RMSE and 78% in MAE. In addition, the dispersion of the predicted values over the entire tool life represented by the 5th-to-95th percentile range can be reduced using future features. The decreased dispersion indicates a reduction in the model uncertainty regarding the future. Overall, it can be confirmed that the remaining tool life prediction is possible under variable process conditions. In addition, an increase in accuracy and higher robustness of the prediction can be achieved by including process-describing information about future machining operations.

5.3. Feature Importance Analysis

Based on the evaluation of the remaining tool life prediction method in the previous section, the feature importance analysis is performed in the following. The aim is to demonstrate and evaluate the feature importance ranking method according to Algorithm 1. Furthermore, it should be investigated why state-of-the-art prediction methods for fixed process conditions based on instantaneous features, such as [30], are not directly applicable to variable process conditions. Therefore, Figure 10 shows the feature importance scores derived according to Algorithm 1 for all input features of the regression model whose predictions are depicted in Figure 8b. The feature importance scores are averaged over the complete dataset, i.e., over all tools, and split by spectral and non-spectral features. The mean feature importance scores and the standard deviations are displayed.

The maximum feature importance scores of the cumulative features with mean values of 0.8, 0.7 and 0.3 for the total cutting time T_c, the total removed volume V and the material removal rate Q, respectively, indicate that they contribute more to the model decisions than the parameters or instantaneous features. The low weighting of the instantaneous spectral features supports the hypothesis that the correlation between the values of the spectra per frequency bin and the target, i.e., the tool wear, decreases due to the influence of the variable process conditions on the sensor signals. Due to their higher level of abstraction and inherent memory capability, cumulative features can maintain correlation with the target despite variable process conditions. The memory capability also distinguishes them significantly from the parameter features, which have low feature importance scores, similar to those of the instantaneous features. A detailed influence analysis of the separate feature subsets and individual high-importance features on the tool wear prediction performance can be found in Appendix D.

It has to be noted that the methods for tool wear prediction studied in this paper, i.e., the state-of-the-art method from [30] and our AutoML-based approach, rely purely on classical ML models. However, the influences of variable process parameters may be filtered out from the spectra using deep learning models, such as CNN-LSTMs, which are particularly good at representing spatio-temporal relationships. Thus, the correlation with the target could also be recovered for variable process parameters. Furthermore, the dataset used in this paper represents gradual tool wear. Abrupt tool wear, e.g., the breakage of the cutting edges due to excessive cutting forces caused by critical engagement conditions, is not included. Although the manufacturing of pocket geometries causes a variation in the engagement conditions, their influence on the sensor signals is only moderate. The influence increases for critical engagement conditions. Thus, the instantaneous features gain importance again for detecting abrupt tool wear. Moreover, the instantaneous features provide a simple wear indicator at fixed process conditions. Overall, despite their low feature importance for the dataset used in this paper, the instantaneous features are a necessary component of tool wear and tool life prediction models.

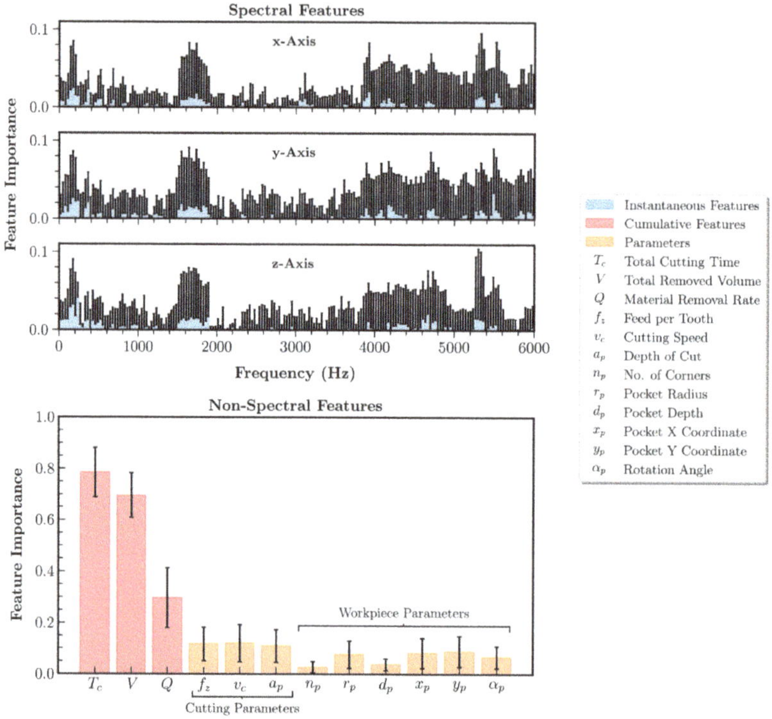

Figure 10. Feature importance analysis of the spectral instantaneous, cumulative and parameter features derived from the AutoML-based feature importance ranking algorithm introduced in this paper. The mean feature importance scores and their standard deviation over all tools are shown.

5.4. Generalization Performance

In the final evaluation step, the remaining tool life prediction method proposed in this paper is investigated in terms of its generalization performance. The aim is to evaluate whether a prediction of tool wear and remaining tool life is possible with increased productivity of the machining process by reducing the machining time without explicit training. Productivity is determined by the material removal rate derived from the cutting parameters feed per tooth f_z and cutting speed v_c defining the feed rate and the axial depth of cut a_p. Increasing the cutting parameters introduces uncertainty regarding the changing tool wear development and its impact on the workpiece quality. If the approach proposed in this paper allows transferability to increased yet unknown cutting parameter configurations, process reliability can be ensured nonetheless.

The evaluation is performed using two reference tools, reference tools 1 and 2. The data from reference tool 1 represents the test set. It is based on pocket manufacturing at fixed, maximum cutting parameters, resulting in an increase in the material removal rate and thus productivity of between 32 and 45% compared to tools 1 to 7. Two training set scenarios are distinguished to study the transferability to the variable pocket manufacturing with increased productivity:

1. **Tools 1–7 and reference tool 2**: Knowledge of the target wear curve for variable pocket manufacturing using variable cutting parameters and of the wear curve for face milling using fixed, maximum cutting parameters.
2. **Tools 1–7**: Knowledge of the target wear curve for variable pocket manufacturing using variable cutting parameters only.

The regression and forecasting components are trained with both training sets. Figure 11 and Table 4 show the tool wear prediction results of the regression component.

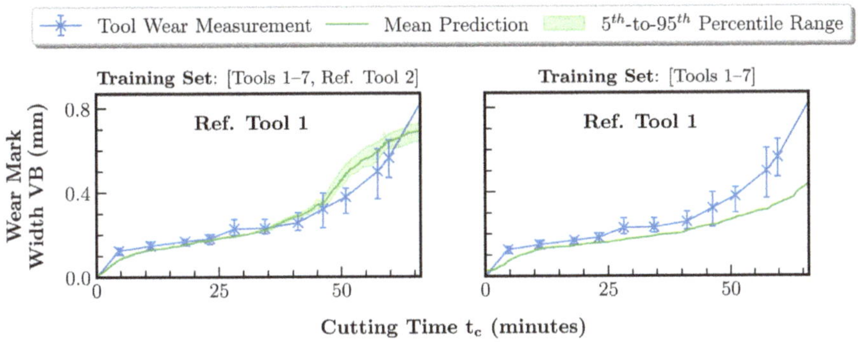

Figure 11. Comparison of the tool wear regression results using the two training sets and the data of reference tool 1 as the test set. The data of reference tool 1 have been excluded from the training set.

Table 4. Comparison of the tool wear regression errors using the two training sets and the data of reference tool 1 as the test set.

Metric	Prediction Errors for Ref. Tool 1 per Training Set (mm)	
	Training Set 1: Tools 1–7 and Ref. Tool 2	Training Set 2: Tools 1–7
RMSE	**0.054**	0.108
MAE	**0.041**	0.078

Based on training set 1, the prediction results lie mostly within the wear measurement intervals, leading to an RMSE of 0.054 mm and an MAE of 0.041 mm. In the case of training set 2, excluding reference tool 2, the same behavior as for tool 3 in Figure 8b is obtained. Reference tool 1 represents the tool life cycle with the maximum material removal rate and the fastest wear progress. This leads to a significant underestimation of the wear curve with an RMSE of 0.108 mm and an MAE of 0.078 mm. The result supports the hypothesis that the regression component of our approach is not able to extrapolate the wear curve. With additional knowledge of the wear curve for face milling (training set 1) with a material removal rate of 47.1 $\frac{\text{cm}^3}{\text{min}}$ exhibiting faster wear progress than reference tool 1, the regression model performs an interpolation, leading to a feasible prediction. Furthermore, the investigation can also verify the high feature importance scores of the three cumulative features T_c, V and Q noted in Section 5.3. The material removal rate and thus the two parameters T_c and V define the wear progress in the considered scenario of gradual tool wear and are thus crucial for the regression. Overall, the transferability of the regression component to pocket milling at increased productivity is given, provided that the task represents an interpolation.

For evaluating the forecasting component, the output of the regression model based on training set 1 is used as the LSTM input in the testing phase. Furthermore, an exploration of non-spectral feature combinations as future feature inputs is performed, as in Section 5.2.

The results of the remaining tool life prediction for reference tool 2 are shown in Figure 12 and Table 5.

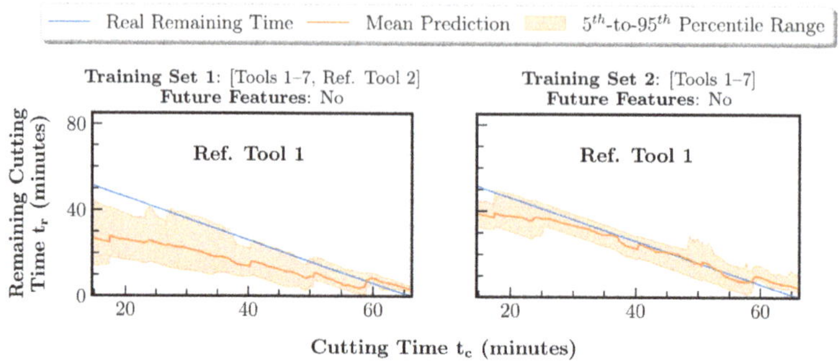

(**a**) Prediction results of the LSTM without future feature inputs.

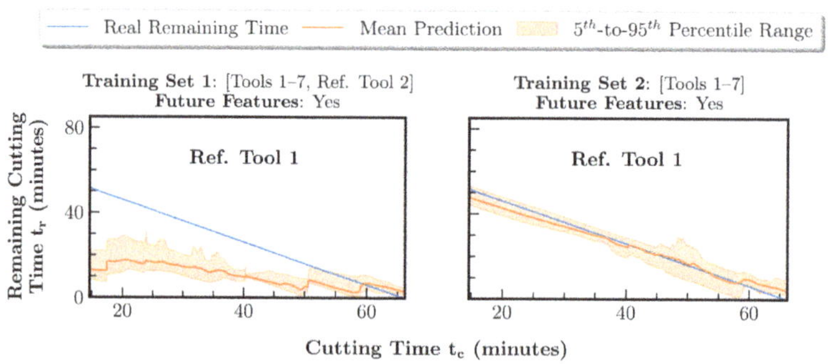

(**b**) Prediction results of the LSTM with the cutting time T_c as future feature inputs.

Figure 12. Comparison of the remaining tool life prediction results using the two training sets combined with the LSTM without future feature inputs and the LSTM with the cutting time T_c as future feature inputs. The data of reference tool 1 have been excluded from the training set.

Table 5. Comparison of the remaining tool life prediction errors using the two training sets combined with the LSTM with no future feature inputs and the LSTM with the cutting time T_c as future feature inputs.

Future Features	Metric	Prediction Errors for Ref. Tool 1 per Training Set (Minutes)	
		Training Set 1: Tools 1–7 and Ref. Tool 2	Training Set 2: Tools 1–7
No	RMSE	14.9	4.9
	MAE	12.7	3.5
Yes	RMSE	22.8	<u>2.2</u>
	MAE	19.3	<u>1.8</u>

Figure 12 shows that, based on training set 2, the prediction is feasible with an RMSE of 4.9 min and an MAE of 3.5 min. In contrast to the regression component, which provides a feasible prediction based on training set 1, the forecasting component trained with training set 1 significantly underestimates the remaining tool life with an RMSE of 14.9 min and an MAE of 12.7 min. An explanation for this behavior is the sensitivity of the LSTM to the characteristic temporal wear curve progression during pocket manufacturing. In this context, face milling represents a modified workpiece geometry and, as part of the training set, mitigates the transferability of the LSTM to pocket manufacturing with different cutting parameter configurations. When the LSTM input is extended to include the future features, as shown in Figure 12b, the prediction for the model based on training set 1 deteriorates with an RMSE of 22.8 min and an MAE of 19.3 min. For the feasible LSTM based on training set 2, the behavior already observed in Figure 9b for tools 1 to 7 repeats. In the early stage of the tool life up to a cutting time of 40 min, the prediction accuracy can be increased and the uncertainty can be reduced. This results in minimum values of the RMSE of 2.2 min and the MAE of 1.8 min.

Overall, the remaining tool life prediction approach introduced in this paper allows transferability to pocket manufacturing at increased, previously unknown parameter configurations, thus ensuring process reliability at increased productivity. The evaluation in this paper refers to a specific combination of workpiece material and tool type within the milling process. Due to its general architecture, the remaining tool life prediction methodology applies to other combinations and machining processes without a loss of generality. However, the extent to which the model generated in this paper needs to be re-trained depends on the distance of the resulting data distributions.

6. Conclusions and Future Work

In this paper, a new method for predicting remaining tool life under the variable process conditions of individualized machining production was presented. The method is criterion-based, i.e., it inherently uses a dedicated variable to represent the tool condition based on which the remaining tool life is determined. In contrast to the traditional approach, where the remaining tool life models are created manually, the introduced method is based on AutoML. The model decisions depending on feature importance scores are then extracted and visualized. Domain experts from the field of machining are thus enabled to develop, validate and optimize remaining tool life models without extensive ML knowledge. The AutoML-based modeling procedure is complemented by a feature set optimized for prediction in individualized production scenarios, obtained purely using non-invasive vibration-based process monitoring. A prototypical vibration sensor system was implemented using an FPGA-SoC-based hardware platform on a five-axis indexed milling center for evaluation. The setup was used to generate a dataset representing gradual tool wear under continuous variation in workpiece geometries and cutting parameters. Based on the dataset, several experiments were conducted to evaluate the method for predicting the remaining tool life.

The experiments reveal that a transfer of manually created remaining tool life prediction models for the case of series production, i.e., constant process conditions, fails due to the input features used in previous work. In order to enable a feasible prediction of the remaining tool life for individualized production, methods extracting features invariant to variable process conditions and preserving the correlation to the tool condition are necessary. Our methodology improves the prediction accuracy over manually created state-of-the-art models by up to 80% with an average MAE of 6.1 min. This corresponds to an accuracy of 7% of the average tool life throughout the dataset. Furthermore, it can be shown that integrating prior knowledge about future machining operations improves the accuracy of the remaining tool life prediction by up to 22% and increases its robustness. The consideration of prior knowledge in the models can also be exploited to perform process optimization, e.g., targeting productivity through virtual exploration of the effect of yet-unknown cutting parameters on the tool condition without affecting process reliability. Thus, the remaining tool life prediction enables the full utilization of the tool life and increased plannability at the job-shop level in individualized production. Overall, the remaining tool life prediction leverages Pareto optimization in manufacturing, targeting product quality, tool costs and productivity.

The AutoML-based modeling approach presented in this paper is, at its core, universally applicable to data- and criterion-based remaining life predictions of technical systems. Nevertheless, the overall methodology is highly specialized due to the domain-specific feature extraction. In order to achieve transferability to other applications, the methodology will be extended in the future using a general feature extraction mechanism, e.g., using CNNs. Furthermore, the dataset generated in the paper represents only gradual tool wear. However, a fully comprehensive tool life model must also handle abrupt tool wear, e.g., due to critical tool engagement conditions. An extension of the dataset will be necessary in the future to investigate abrupt tool wear.

Author Contributions: Conceptualization, L.K.; methodology, L.K.; software, L.K.; validation, L.K.; formal analysis, L.K.; investigation, L.K.; resources, L.K.; data curation, L.K.; writing—original draft preparation, L.K.; writing—review and editing, C.W. and J.F.; visualization, L.K.; supervision, C.W., J.F. and A.G.; project administration, L.K.; funding acquisition, A.G. All authors have read and agreed to the published version of the manuscript.

Funding: This research was part of the project KI-MUSIK 4.0 funded by the German Federal Ministry of Education and Research under grant number 16ME0076.

Institutional Review Board Statement: Not applicable.

Informed Consent Statement: Not applicable.

Data Availability Statement: The data presented in this study are available on request from the corresponding author. The data are not publicly available due to privacy reasons.

Acknowledgments: The authors thank Markus Künkler and Niclas Georg from Aura Tools GmbH and Romas Auder and Frederik Simon from the University of Applied Sciences Ruhr West for supporting this research.

Conflicts of Interest: The authors declare no conflict of interest.

Appendix A. Wear Mark Width Measurement

The target parameter of the dataset introduced in this paper is the wear mark width VB. The VB is acquired using a video measuring microscope according to the procedure shown in Figure A1.

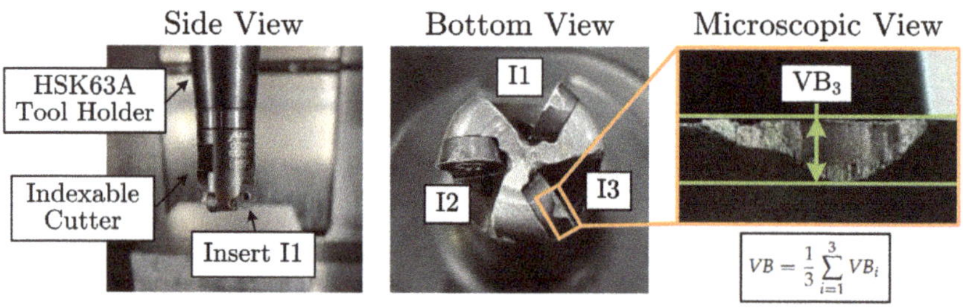

Figure A1. Measurement procedure of the individual wear mark width VB_i per cutting insert i.

First, the maximum wear mark width VB_i per insert $i \in \{1, ..., z_c\}$ is measured. z_c is the number of inserts of the cutting tool. Then, the average over the maximum wear mark widths is calculated according to Equation (A1) to obtain the target value VB.

$$VB = \frac{1}{z_c} \sum_{i=1}^{z_c} VB_i \tag{A1}$$

Appendix B. Milling Process Details

Table A1 summarizes the specifications of the reference milling process developed in this paper for the dataset generation. In addition to the specifications, Figure A1 shows the used tool holder, cutter and cutting inserts. A tool instance of the dataset refers to a set of three cutting inserts.

Table A1. Fixed and variable parameters of the reference milling process for the dataset generation.

Category	Parameter	Unit	Range/Value
Cutting Process	Feed per Tooth (f_z)	mm	$[0.4, 0.7]$
	Cutting Speed (v_c)	$\frac{mm}{s}$	$[170, 200]$
	Axial Cut Depth (a_p)	mm	$[0.5, 0.8]$
	Radial Cut Depth (a_e)	mm	R
Pocket Geometry	Corners (c_p)	-	$[3, 9]$
	Radius (r_p)	mm	$[30, 100]$
	Depth (d_p)	mm	$[1, 10]$
	Position (x_p, y_p)	mm	$[50, 150]$, $[50, 150]$
	Rotation Angle (α_p)	°	$[0, 360]$
Milling Tool	Type	-	Indexable
	Tool Shape	-	Toroidal
	Edge Shape	-	Circular
	Number of Teeth (z_c)	-	3
	Cutter Radius (R)	mm	10
	Edge Radius (r)	mm	4
	End-of-life Threshold (VB_t)	mm	0.8
Workpiece	Dimensions (l, w, h)	mm	200, 200, 200
	Material	-	X155CrVMo12-1 (DIN 1.2379)
	Consecutive Pockets (N_p)	-	4

Appendix C. AutoML Framework Analysis

Before evaluating the remaining tool life prediction approach introduced in this paper, an AutoML framework must be selected as a core component of the methodology. For this purpose, a comparison between the three state-of-the-art AutoML systems, Auto-sklearn [41], LightAutoML [42] and FLAML [43], is performed. Table A2 compares the size of the search space of the systems based on the included feature selection methods and ML model types for regression tasks. LightAutoML has the smallest search space, directly followed by FLAML. Auto-sklearn offers the largest search space of the AutoML systems, with 13 feature selection methods and 12 model types. Thus, the space of not only the model architectures but also the tunable hyperparameters is larger compared to the other frameworks.

Table A2. Comparison of the search space size with respect to the number of feature selection methods and ML model types between the state-of-the-art frameworks Auto-sklearn [41], LightAutoML [42] and FLAML [43].

AutoML Framework	Search Space Size (Number of Included Methods)	
	Feature Selection	Model Types
LightAutoML	2	2
FLAML	3	6
Auto-sklearn	13	12

The frameworks are compared quantitatively based on the dataset generated in this paper in the context of tool wear regression. The entire feature set, as described in Section 5.2, is used as model input for the tool wear regression, and the model training is performed with a maximum time budget of 10 min. The prediction errors and search runtimes are shown in Tables A3 and A4, respectively.

Table A3. Comparison of tool wear regression errors using the full input feature set and the state-of-the-art frameworks Auto-sklearn [41], LightAutoML [42] and FLAML [43]. Every run is performed with a maximum search time budget of 10 min.

Method	Metric	Prediction Errors per Test Tool (mm)							
		Tool 1	Tool 2	Tool 3	Tool 4	Tool 5	Tool 6	Tool 7	Mean
Light AutoML	RMSE	**0.033**	0.050	**0.079**	**0.022**	**0.051**	**0.064**	0.036	**0.048**
	MAE	**0.027**	0.035	**0.054**	**0.016**	**0.042**	0.043	0.026	**0.035**
FLAML	RMSE	0.035	**0.044**	0.086	0.023	0.062	0.065	**0.034**	0.050
	MAE	0.028	**0.032**	0.057	0.017	0.047	**0.041**	**0.022**	**0.035**
Auto-sklearn	RMSE	0.038	0.068	0.107	0.026	0.066	0.065	0.041	0.059
	MAE	0.031	0.046	0.070	0.021	0.055	0.045	0.028	0.042

Table A4. Comparison of the final search runtimes required by the state-of-the-art frameworks Auto-sklearn [41], LightAutoML [42] and FLAML [43].

Method	Search Time (s)							
	Tool 1	Tool 2	Tool 3	Tool 4	Tool 5	Tool 6	Tool 7	Mean
Light AutoML	185.0	**185.1**	**186.6**	**186.2**	**185.7**	**186.3**	186.5	**185.9**
FLAML	**165.2**	308.2	280.8	214.6	238.6	214.8	**177.8**	228.6
Auto-sklearn	600.0	600.00	600.0	600.0	600.0	600.0	600.0	600.0

Table A3 shows that LightAutoML achieves the best regression result over the tools of the dataset with an RMSE of 0.048 mm and an MAE of 0.035 mm. At the same time, the mean search runtime is minimal at 3.1 min, according to Table A4. FLAML follows LightAutoML with an RMSE of 0.050 mm, an MAE of 0.035 mm and a mean search runtime of 3.8 min. Auto-sklearn has the maximum amount of errors with an RMSE of 0.059 mm and an MAE of 0.042 mm. However, the regression errors of the AutoML systems vary only slightly overall. Due to the size of the search space, Auto-sklearn uses the entire time budget of 10 min for the CASH optimization. LightAutoML and FLAML apply early stopping by default.

The evaluation results coincide with the size of the respective search spaces. A smaller number of possible model architectures and tunable hyperparameters allows feasible solutions to be found faster. However, to achieve a robust and reliable explainability of the models using the feature importance-based method according to Algorithm 1, a high degree of diversity of the model pipelines and the feature selection procedures is beneficial. Since the regression errors of the AutoML systems vary only slightly and further minimization can be achieved by a larger time budget of the CASH optimization, Auto-sklearn is therefore used in this paper due to the size of the search space.

Appendix D. Feature Subset Comparison

In addition to the feature importance analysis in Section 5.3, this section investigates the tool wear regression performance depending on the feature subsets (instantaneous, cumulative and parameters) introduced in this paper. Therefore, Auto-sklearn is configured as described in Section 5.2. Table A5 shows the prediction errors for the instantaneous, cumulative and parameter features. The prediction errors for the instantaneous features are copied from Table 2 for comparison.

Table A5. Comparison of tool wear regression errors using only the instantaneous, cumulative and parameter feature subsets.

Feature Subset	Metric	Prediction Errors per Test Tool (mm)							
		Tool 1	Tool 2	Tool 3	Tool 4	Tool 5	Tool 6	Tool 7	Mean
Instantaneous	RMSE	0.169	0.140	0.189	0.131	0.199	0.182	0.162	0.167
	MAE	0.133	0.102	0.155	0.097	0.147	0.143	0.132	0.130
Cumulative	RMSE	**0.036**	**0.072**	**0.061**	**0.032**	**0.054**	**0.066**	**0.025**	**0.049**
	MAE	**0.030**	**0.047**	**0.041**	**0.025**	**0.042**	**0.044**	**0.017**	**0.035**
Parameters	RMSE	0.244	0.153	0.171	0.202	0.235	0.277	0.190	0.210
	MAE	0.196	0.125	0.153	0.165	0.195	0.216	0.152	0.172

The regression model based on the cumulative features T_c, V and Q with an RMSE of 0.049 mm and an MAE of 0.035 mm performs best. The RMSE and MAE are both 17% lower than for the model based on the entire feature set, as shown in Table 2. Moreover, Auto-sklearn thus achieves approximately the same result as LightAutoML in Table A3. Overall, the particular importance of the cumulative features for the model performance, which can be seen in Figure 10, is confirmed. Based on the parameters, the regression model exhibits the worst performance with an RMSE of 0.210 mm and an MAE of 0.172 mm. This can be explained by the fact that the parameters do not correlate with the target. Therefore, their importance in Figure 10 is also consistently low. In the future, however, there is the potential to extract further wear-correlated features by combining the instantaneous features and the parameters. Since the instantaneous features significantly depend on the parameters, filtering these influences may allow the wear influence to be isolated.

As identified in Section 5.3, the total cutting time T_c, the removed material volume V and the material removal rate Q have the highest feature importance scores. Therefore, in addition to examining the feature subsets above, the prediction performance is evaluated based on the individual high-importance features. Table A6 summarizes the results.

Table A6. Comparison of tool wear regression errors using only individual high-importance features.

Feature	Metric	Prediction Errors per Test Tool (mm)							
		Tool 1	Tool 2	Tool 3	Tool 4	Tool 5	Tool 6	Tool 7	Mean
T_c	RMSE	**0.035**	0.046	**0.059**	0.035	**0.054**	**0.058**	**0.026**	**0.045**
	MAE	0.029	0.034	**0.041**	0.027	**0.042**	**0.037**	**0.018**	**0.033**
V	RMSE	0.042	**0.037**	0.117	**0.031**	0.080	0.077	0.057	0.063
	MAE	**0.026**	**0.029**	0.075	**0.024**	0.065	0.043	0.050	0.045
Q	RMSE	0.207	0.171	0.202	0.164	0.208	0.195	0.211	0.194
	MAE	0.185	0.129	0.137	0.134	0.181	0.152	0.180	0.157

With an RMSE of 0.045 mm and an MAE of 0.033 mm, the model trained with the feature T_c exhibits the best performance. The errors are 8% (RMSE) and 6% (MAE) lower than for the model based on the entire cumulative feature set shown in Table A5. This result suggests that only a single feature is needed. However, considering the generalization scenario according to Section 5.4 based on the individual features, it can be seen that this does not hold. Therefore, Figure A2 shows the wear prediction based on training dataset 1, as described in Section 5.4.

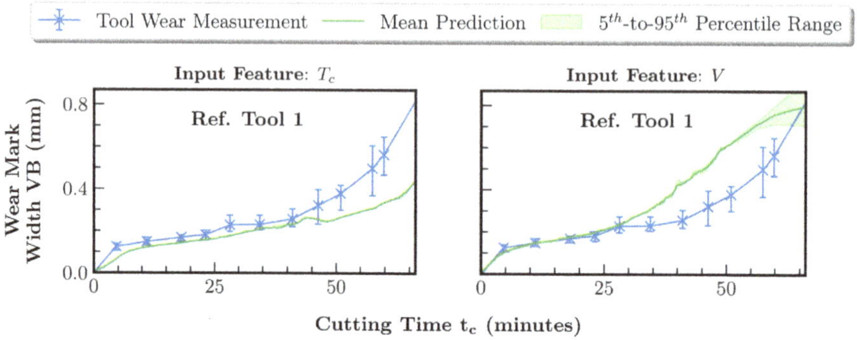

Figure A2. Comparison of the tool wear regression results using the individual high-importance features T_c and V based on the training set composed of tools 1 to 7 and reference tool 2. The data of reference tool 1 are the test set and are thus excluded from the training set.

The model trained using T_c underestimates the wear curve of reference tool 1 with an RMSE of 0.110 mm and an MAE of 0.076 mm. The model trained using V overestimates the wear curve with an RMSE of 0.142 mm and an MAE of 0.104 mm. As shown in Section 5.4, a robust and generalizable tool wear model as the foundation for the remaining tool life prediction can only be trained based on the combination of the high-importance features.

References

1. Boos, W.; Kelzenberg, C.; Prümmer, M.; Goertz, D.; Boshof, J.; Horstkotte, R.; Ochel, T.; Lürken, C. *Tooling in Germany 2020*; WZL of RWTH Aachen, Fraunhofer IPT: Aachen, Germany, 2020.
2. Boos, W.; Arntz, K.; Johannsen, L.; Prümmer, M.; Horstkotte, R.; Ganser, P.; Venek, T.; Gerretz, V. *Erfolgreich Fräsen im Werkzeugbau*; Fraunhofer IPT, WBA Aachener Werkzeubau Akademie: Aachen, Germany, 2018.
3. Norberto López de Lacalle, L.; Lamikiz, A. Sculptured Surface Machining. In *Machining—Fundamentals and Recent Advances*; Davim, J.P., Ed.; Springer: London, UK, 2008; pp. 225–249.
4. Möhring, H.-C.; Nguyen, Q.P.; Kuhlmann, A.; Lerez, C.; Nguyen, L.T.; Misch, S. Intelligent Tools for Predictive Process Control. *Procedia CIRP* **2016**, *57*, 539–544. [CrossRef]
5. Möhring, H.-C.; Eschelbacher, S.; Georgi, P. Fundamental investigation on the correlation between surface properties and acceleration data from a sensor integrated milling tool. *Procedia Manuf.* **2020**, *52*, 79–84. [CrossRef]
6. Denkena, B.; Dittrich, M.-A.; Lindauer, M.; Mainka, J.; Stürenburg, L. Using AutoML to Optimize Shape Error Prediction in Milling Processes. In Proceedings of the 2020 Machining Innovations Conference (2020), Online, 1–2 December 2020.
7. Mohamed, A.; Hassan, M.; M'Saoubi, R.; Attia, H. Tool Condition Monitoring for High-Performance Machining Systems-A Review. *Sensors* **2022**, *22*, 2206. [CrossRef] [PubMed]
8. Sayyad, S.; Kumar, S.; Bongale, A.; Kamat, P.; Patil, S.; Kotecha, K. Data-Driven Remaining Useful Life Estimation for Milling Process: Sensors, Algorithms, Datasets, and Future Directions. *Int. J. Adv. Manuf. Technol.* **2021**, *115*, 2683–2709. [CrossRef]
9. Denkena, B.; Krüger, M.; Schmidt, J. Condition-based tool management for small batch production. *Int. J. Adv. Manuf. Technol.* **2014**, *74*, 471–480. [CrossRef]
10. Arntz, C.; Brandstätter, T.C.; Dorißen, J.; Frye, M.; Krauß, J.; Krebs, L.; Holst, C.; Horstkotte, R.; Mende, H.; Schiller, S.; et al. *Künstliche Intelligenz in der Einzel- und Kleinserienfertigung*; Fraunhofer IPT: Aachen, Germany, 2021.
11. Wang, W.; Wang, B.; Li, N.; Lei, Y.; Yan, T. Remaining Useful Life Prediction Based on Multi-channel Attention Bidirectional Long Short-term Memory Network. In Proceedings of the 2021 IEEE International Conference on Sensing, Diagnostics, Prognostics, and Control (2021), Weihai, China, 13–15 August 2021; pp. 7–12.
12. Sun, H.; Zhang, J.; Mo, R.; Zhang, X. In-process tool condition forecasting based on a deep learning method. *Robot.-Comput.-Integr. Manuf.* **2020**, *64*, 101924. [CrossRef]
13. Astakhov, V.P.; Davim, J.P. Tools (Geometry and Material) and Tool Wear. In *Machining—Fundamentals and Recent Advances*; Davim, J.P., Ed.; Springer: London, UK, 2008; pp. 29–59.
14. Mebrahitom, A.G.; Seow, X.Y.; Azmir, A.; Tamiru, A.L. Remaining Tool Life Prediction Based on Force Sensors Signal During End Milling of Stavax ESR Steel. In Proceedings of the International Mechanical Engineering Congress and Exposition (2017), Tampa, FL, USA, 3–9 November 2017; pp. 1–7.
15. Zhang, J.; Zeng, Y.; Starly, B. Recurrent Neural Networks with Long Term Temporal Dependencies in Machine Tool Wear Diagnosis and Prognosis. *SN Appl. Sci.* **2021**, *3*, 442. [CrossRef]

16. Drouillet, C.; Karandikar, J.; Nath, C.; Journeaux, A.-C.; El Mansori, M.; Kurfess, T. Tool Life Predictions in Milling using Spindle Power with the Neural Network Technique. *J. Manuf. Process.* **2016**, *22*, 161–168. [CrossRef]
17. An, Q.; Tao, Z.; Xu, X.; El Mansori, M.; Chen, M. A Data-driven Model for Milling Tool Remaining Useful Life Prediction with Convolutional and Stacked LSTM Network. *Measurement* **2020**, *154*, 107461. [CrossRef]
18. Nasir, V.; Sassani, F. A Review on Deep Learning in Machining and Tool Monitoring: Methods, Opportunities, and Challenges. *IEEE Access* **2021**, *9*, 110255–110286. [CrossRef]
19. Li, Y.; Xiang, Y.; Pan, B.; Shi, L. A Hybrid Remaining Useful Life Prediction Method for Cutting Tool considering the Wear State. *Int. J. Adv. Manuf. Technol.* **2022**, *121*, 3583–3596. [CrossRef]
20. Guo, L.; Yu, Y.; Gao, H.; Feng, T.; Liu, Y. Online Remaining Useful Life Prediction of Milling Cutters Based on Multisource Data and Feature Learning. *IEEE Trans. Ind. Inform.* **2022**, *18*, 5199–5208. [CrossRef]
21. Jia, W.; Wang, W.; Li, Z.; Li, H. Prediction of Tool Wear in Sculpture Surface by a new Fusion Method of Temporal Convolutional Network and Self-Attention. *Int. J. Adv. Manuf. Technol.* **2022**, *121*, 2565–2583. [CrossRef]
22. Liu, Y.; Hu, X.; Jin, J. Remaining Useful Life Prediction of Cutting Tools based on Deep Adversarial Transfer Learning. In Proceedings of the 2019 8th International Conference on Computing and Pattern Recognition (2019), Beijing China, 23–25 October 2019 ; pp. 434–439.
23. Li, X.; Lim, B.S.; Zhou, J.H.; Huang, S.; Phua, S.J.; Shaw, K.C.; Er, M.J. Fuzzy Neural Network Modelling for Tool Wear Estimation in Dry Milling Operation. In Proceedings of the Annual Conference of the PHM Society (2009), San Diego, CA, USA, 27 September–1 October 2009.
24. Goebel, K. *Management of Uncertainty for Sensor Validation, Sensor Fusion and Diagnosis in Sensor Driven Mechanical Systems Using Soft Computing Techniques*; UC Berkeley: Berkeley, CA, USA, 1996.
25. Zhou, J.-T.; Zhao, X.; Gao, J. Tool Remaining Useful Life Prediction Method based on LSTM under Variable Working Conditions. *Int. J. Adv. Manuf. Technol.* **2019**, *104*, 4715–4726. [CrossRef]
26. Matsumura, R.; Nishida, I.; Shirase, K. Tool Life Prediction in End Milling using a Combination of Machining Simulation and Tool Wear Progress Data. *J. Adv. Mech. Des. Syst. Manuf.* **2023**, *17*, JAMDSM0025. [CrossRef]
27. Zegarra, F.C.; Vargas-Machuca, J.; Coronado, A.M. Tool Wear and Remaining Useful Life (RUL) Prediction based on Reduced Feature Set and Bayesian Hyperparameter Optimization. *Prod. Eng.* **2022**, *16*, 465–480. [CrossRef]
28. Lutz, B.; Reisch, R.; Kißkalt, D.; Avci, B.; Regulin, D.; Knoll, A.; Franke, J. Benchmark of Automated Machine Learning with State-of-the-Art Image Segmentation Algorithms for Tool Condition Monitoring. *Procedia Manuf.* **2020**, *51*, 215–221. [CrossRef]
29. Kißkalt, D.; Mayr, A.; Lutz, B.; Rögele, A.; Franke, J. Streamlining the development of data-driven industrial applications by automated machine learning. *Procedia CIRP* **2020**, *93*, 401–406. [CrossRef]
30. Schmetz, A.; Vahl, C.; Zhen, Z.; Reibert, D.; Mayer, S.; Zontar, D.; Garcke, J.; Brecher, C. Decision Support by Interpretable Machine Learning in Acoustic Emission Based Cutting Tool Wear Prediction. In Proceedings of the 2021 IEEE International Conference on Industrial Engineering and Engineering Management (2021), Singapore, 13–16 December 2021 ; pp. 629–633.
31. Sotubadi, S.V.; Liu, R.; Nguyen, V. Explainable AI for Tool Wear Prediction in Turning. In Proceedings of the ASME 2023 18th International Manufacturing Science and Engineering Conference (2023), New Brunswick, NJ, USA, 12–16 June 2023 .
32. Li, Y.; Wang, J.; Huang, Z.; Gao, R.X. Physics-informed Meta Learning for Machining Tool Wear Prediction. *J. Manuf. Syst.* **2022**, *62*, 17–27. [CrossRef]
33. Wang, J.; Li, Y.; Zhao, R.; Gao, R.X. Physics Guided Neural Network for Machining Tool Wear Prediction. *J. Manuf. Syst.* **2020**, *57*, 298–310. [CrossRef]
34. Schaefer, C. *Signaltechnische Voraussetzungen und Analyseverfahren zur Überwachung von Präzisions- und Ultrapräzisionsbearbeitungsverfahren*; WZL of RWTH Aachen: Aachen, Germany, 2013.
35. Benardos, P.; Vosniakos, G.-C. Removed Material Volume Calculations in CNC Milling by Exploiting CAD Functionality. *Int. J. Comput. Aided Eng. Technol.* **2017**, *10*, 491–503. [CrossRef]
36. Feurer, M.; Klein, A.; Eggensperger, K.; Springenberg, J.T.; Blum, M.; Hutter, F. Efficient and Robust Automated Machine Learning. In Proceedings of the 28th International Conference on Neural Information Processing Systems (2015), Montreal, QC, Canada, 7–12 December 2015; Volume 2, pp. 2755–2763.
37. Sharma, P.; Mirzan, S.R.; Bhandari, A.; Pimpley, A.; Eswaran, A.; Srinivasan, S.; Shao, L. Evaluating Tree Explanation Methods for Anomaly Reasoning: A Case Study of SHAP TreeExplainer and TreeInterpreter. In Proceedings of the International Conference on Conceptual Modeling (2020), Vienna, Austria, 3–6 November 2020 .
38. CN-0549: IEPE-Compliant, CbM Machine Learning Enablement Platform. Available online: https://www.analog.com/en/design-center/reference-designs/circuits-from-the-lab/cn0549.html#rd-overview (accessed on 13 August 2023).
39. CN-0532: IEPE-Compatible Interface for Wideband MEMS Accelerometer Sensors. Available online: https://www.analog.com/en/design-center/reference-designs/circuits-from-the-lab/cn0532.html (accessed on 13 August 2023).
40. CN-0540: 24-Bit Data Acquisition System for IEPE Sensors. Available online: https://www.analog.com/en/design-center/reference-designs/circuits-from-the-lab/cn0540.html (accessed on 13 August 2023).
41. Feurer, M.; Eggensperger, K.; Falkner, S.; Lindauer, M.; Hutter, F. Auto-Sklearn 2.0: Hands-free AutoML via Meta-Learning. *J. Mach. Learn. Res.* **2022**, *23*, 11936–11996.
42. Vakhrushev, A.; Ryzhkov, A.; Savchenko, M.; Simakov, D.; Damdinov, R.; Tuzhilin, T. LightAutoML: AutoML Solution for a Large Financial Services Ecosystem. *arXiv* **2021**, arXiv:2109.01528.

43. Wang, C.; Wu, Q.; Weimer, M.; Zhu, E. FLAML: A Fast and Lightweight AutoML Library. In Proceedings of the Fourth Conference on Machine Learning and Systems (2021), Online, 8–11 November 2022 .
44. Herzen, J.; Lässig, F.; Piazzetta, S.G.; Neuer, T.; Tafti, L.; Raille, G.; van Pottelbergh, T.; Pasieka, M.; Skrodzki, A.; Huguenin, N.; et al. Darts: User-Friendly Modern Machine Learning for Time Series. *J. Mach. Learn. Res.* **2022**, *23*, 5442–5447.
45. Liaw, R.; Liang, E.; Nishihara, R.; Moritz, P.; Gonzalez, J.E.; Stoica, I. Tune: A Research Platform for Distributed Model Selection and Training. *arXiv* **2018**, arXiv:1807.05118.
46. Li, L.; Jamieson, K.; Rostamizadeh, A.; Gonina, E.; Hardt, M.; Recht, B.; Talwalkar, A. A System for Massively Parallel Hyperparameter Tuning. In Proceedings of the 3rd Conference on Systems and Machine Learning (2020), Virtual, 24–26 November 2020.
47. Akiba, T.; Sano, S.; Yanase, T.; Ohta, T.; Koyama, M. Optuna: A Next-generation Hyperparameter Optimization Framework. In Proceedings of the 25th ACM SIGKDD International Conference on Knowledge Discovery & Data Mining (2019), Anchorage, AK, USA, 4–8 August 2019; pp. 2623–2631.

sensors

MDPI

Article

Monitoring Flow-Forming Processes Using Design of Experiments and a Machine Learning Approach Based on Randomized-Supervised Time Series Forest and Recursive Feature Elimination

Leroy Anozie [1,†], Bodo Fink [2], Christoph M. Friedrich [1,3,*] and Christoph Engels [1]

1 Department of Computer Science, University of Applied Sciences and Arts (FH Dortmund), 44227 Dortmund, Germany; leroy.anozie@mailbox.org (L.A.); christoph.engels@fh-dortmund.de (C.E.)
2 WF Maschinenbau und Blechformtechnik GmbH & Co.KG, 48324 Sendenhorst, Germany
3 Institute for Medical Informatics, Biometry and Epidemiology (IMIBE), University Hospital Essen, 45122 Essen, Germany
* Correspondence: christoph.friedrich@fh-dortmund.de
† Technical enquiries to: leroy.anozie@mailbox.org.

Abstract: The machines of WF Maschinenbau process metal blanks into various workpieces using so-called flow-forming processes. The quality of these workpieces depends largely on the quality of the blanks and the condition of the machine. This creates an urgent need for automated monitoring of the forming processes and the condition of the machine. Since the complexity of the flow-forming processes makes physical modeling impossible, the present work deals with data-driven modeling using machine learning algorithms. The main contributions of this work lie in showcasing the feasibility of utilizing machine learning and sensor data to monitor flow-forming processes, along with developing a practical approach for this purpose. The approach includes an experimental design capable of providing the necessary data, as well as a procedure for preprocessing the data and extracting features that capture the information needed by the machine learning models to detect defects in the blank and the machine. To make efficient use of the small number of experiments available, the experimental design is generated using Design of Experiments methods. They consist of two parts. In the first part, a pre-selection of influencing variables relevant to the forming process is performed. In the second part of the design, the selected variables are investigated in more detail. The preprocessing procedure consists of feature engineering, feature extraction and feature selection. In the feature engineering step, the data set is augmented with time series variables that are meaningful in the domain. For feature extraction, an algorithm was developed based on the mechanisms of the r-STSF, a state-of-the-art algorithm for time series classification, extending them for multivariate time series and metric target variables. This feature extraction algorithm itself can be seen as an additional contribution of this work, because it is not tied to the application domain of monitoring flow-forming processes, but can be used as a feature extraction algorithm for multivariate time series classification in general. For feature selection, a Recursive Feature Elimination is employed. With the resulting features, random forests are trained to detect several quality features of the blank and defects of the machine. The trained models achieve good prediction accuracy for most of the target variables. This shows that the application of machine learning is a promising approach for the monitoring of flow-forming processes, which requires further investigation for confirmation.

Keywords: flow forming; metal forming; multi-sensor; time series classification; design of experiments; recursive feature elimination; r-STSF; random forest

1. Introduction

Flow forming is a metal forming process for the production of rotationally symmetrical parts such as flanged components and pressure cylinders [1,2]. In a flow-forming process, a blank, i.e., a short and thick metal disk, is mounted on a rotating spindle and stretched out by mechanically guided rollers [3]. In the process, the workpiece becomes longer and its wall thickness thinner, while the inside diameter remains constant [3].

In the flow-forming process studied in the work at hand, a test machine, the VSTR 400-3S built by WF Maschinenbau, forms a circular metal blank (see Figure 1) in the gear component shown in Figures 2 and 3. First, the blank is placed on the spindle of the machine (see Figure 4, marking A). When the program is started from the control panel, the tailstock (see Figure 4, marking B) then moves in from above and fixes the blank on the spindle. The three rollers of the machine (see Figure 4, marking C1, C2 and C3) are responsible for the actual forming. During the process, they move along a programmatically defined path, while the blank is rotated by the spindle. The free-spinning rollers are pressed against the rotating workpiece along the axial feed axis, which also causes them to rotate and deform the workpiece. The shaping in the inner part of the workpiece with the toothing shown in Figure 2 is defined by the punch (see Figure 4, marking B). During the entire forming process, the workpiece is cooled with a cooling liquid, which is why this is also referred to as cold forming.

Figure 1. Ronde (Circular metal blank placed in the machine for processing).

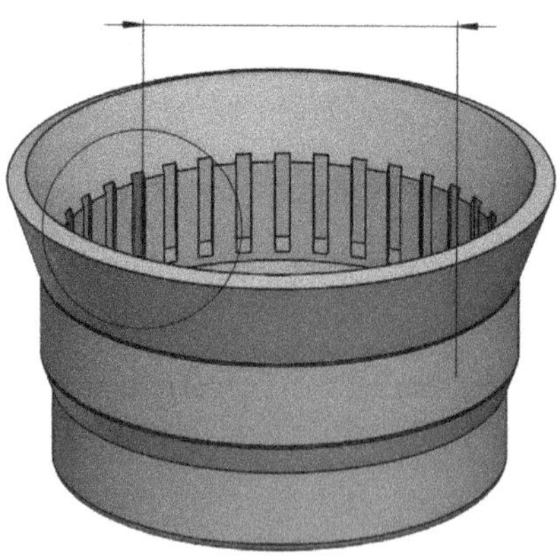

Figure 2. Technical drawing of the workpiece showing its toothed inner side.

Figure 3. Product of the studied forming process. The missing piece on the right edge was subsequently milled out in order to be able to perform measurements on the workpiece.

Figure 4. Interior of the VSTR 400-3S. Among other things, the fixed metal blank (marking A), the tailstock (marking B), the rollers (marking C1, C2 and C3) and the punch (marking D) are visible.

The quality of the produced workpieces is highly dependent on the conditions or input variables of the forming processes, which primarily include the quality of the metal blanks used in the process and the preconditions of the machine itself. Blanks with an inexpedient geometry, such as an irregular thickness or a hole that is displaced from the center, can lead to undesired properties in the resulting workpiece, such as being asymmetrical. The causes of blanks with such inexpedient geometry can be errors or misconfigurations in the manufacture of the blanks as well as the inadvertent use of incorrect blanks. Similarly, machine defects can also lead to undesirable process results, such as irregularities on the surface of the finished workpiece caused by a worn out roller. The resulting waste of time and material increases as long as these problems go unnoticed. This is particularly problematic if the feeding of the machine is automated, which is not the case for the test machine used in this study, but may be the case for other flow forming machines as well as extensions of the VSTR 400-3S. Therefore, automated process monitoring capable of detecting and responding to such defects by issuing warnings, recommending actions, or even stopping the machine can prevent significant financial loss.

However, the mechanics of flow-forming processes, are not fully understood [4], so it is not possible to model them completely physically in order to implement the required monitoring functions on this basis. Consequently, data-driven modeling using machine learning algorithms is used to automatically monitor the process and the machine.

Machine learning is a branch of artificial intelligence that is widely used in business, medicine, industry and other fields [5]. Machine learning methods can be used to identify complex relationships in existing data sets that cannot be captured by "handcrafted" rules [6]. The machine learning algorithm used in this work is random forest (see, e.g., [7]). In so-called supervised learning methods like random forest, a training data set is used to model the relationship between attributes (also termed "features" or "variables") of considered objects and a defined target variable. The resulting model can then be used to determine the target value for new or unseen objects [5]. Random forests can be used for both classification (i.e., discrete target variables) and for regression (i.e., continuous target variables) [7].

Flow forming has been the subject of various studies, many of which have focused on modeling the relationships between process input parameters and the geometry of the finished workpieces (see Section 2.4). To the author's knowledge, however, there is no literature describing approaches to data-driven monitoring of flow-forming processes to date. Furthermore, in the literature on the monitoring of other machine processing operations, prediction of process input variables is a rarely considered topic and the time series data used are mostly one-dimensional, which leaves possible potentials such as monitoring the machine using multiple sensors unaddressed.

The work at hand addresses these research gaps and proposes an approach for the acquisition and processing of sensor (and other) data as well as for the generation of machine learning models that predict process input variables using this data. The contributions of this work can be summarized as follows:

- Demonstration of the feasibility of utilizing machine learning and sensor data to monitor flow-forming processes.
- Development of a practical approach to monitor flow-forming processes. The approach includes an experimental design capable of providing the necessary data, as well as a procedure for preprocessing the data and extracting features that capture the information needed by the machine learning models to detect defects in the blank and the machine.
- Development of a feature extraction algorithm for multivariate time series classification, derived from the state-of-the-art algorithm r-STSF (see [8]), incorporating extensions for multivariate time series and metric target variables.

2. Related Work

2.1. Condition-Based Maintenance

Over time, machinery equipment suffers from wear and tear, which can lead to failures [9]. In many cases, e.g., because of safety risks or reduced productivity, such failures are unacceptable. Therefore, attempts are often made to prevent these failures through preventive maintenance measures [9]. In addition to the comparatively uneconomical approach of performing maintenance measures on the basis of a fixed schedule (calendar-based maintenance), there is condition-based maintenance or CBM for this purpose [9]. The aim of CBM is to monitor the condition of a machine by collecting data about it (condition monitoring or CM) and, based on this, to enable and carry out needs-oriented maintenance measures [9].

A variety of machine learning algorithms have been used for condition monitoring in manufacturing. Among others, commonly used are SVMs (see, e.g., [10–13]), artificial neural networks (see, e.g., [10,11,14]) and random forests (see, e.g., [10,13,15]). Comparisons of several of these algorithms in different condition monitoring applications are conducted in [14] (detection of faulty cycles of a cutting machine), [10] (classification of defects in ball bearings), [15] (detection of defects in robot arms) and [12] (classification of defects in vehicle transmissions). They show that, in this field, the results obtained with different learning algorithms usually differ only slightly, so that the choice of algorithm is not a high priority.

2.2. Data Acquisition

To obtain data for the training of machine learning models, data from production operations (see [15,16]) as well as from experiments conducted specifically for this purpose can be used. When conducting experiments, the Design of Experiments (DoE) methodology can be used to increase the effectiveness of individual experiments in order to reduce the number of experiments required, thus saving time and reducing costs [17]. Consequently, DoE is an integral part of the guideline for the development of monitoring systems for machining processes described in [17]. However, in practice, instead of using DoE, in practice, experiments are also often performed by testing the defects or unacceptable conditions one at a time in one or more experiments (see, e.g., [12,18]) or by running manufacturing processes until a defect occurs. The latter approach is used in cases such as [19,20], where the aim is to detect different stages of wear, and also in cases such as [21,22], where the run-to-failure data are used to predict the remaining time until a component breaks (Remaining Useful Life, RUL).

2.3. Sensors

The sensors used to monitor manufacturing processes can be of various types. These include, but are not limited to force, vibration, current and acoustic emission sensors [17]. Figure 5 shows an overview of the most common sensor technologies used in condition monitoring of manufacturing tools (Tool Condition Monitoring). Force sensors are particularly common [17,23]. The works [17] (machining), [24] (machining), and [23] (milling) deal with the different types of sensors, their advantages and disadvantages and their respective applications. In [13], a comparison is made of the accuracy achievable with different sensors individually, as well as with sensor fusion (feature set assembled from the signals of all sensors [13]), when classifying defects on the tools of an ultrasonic metal welder.

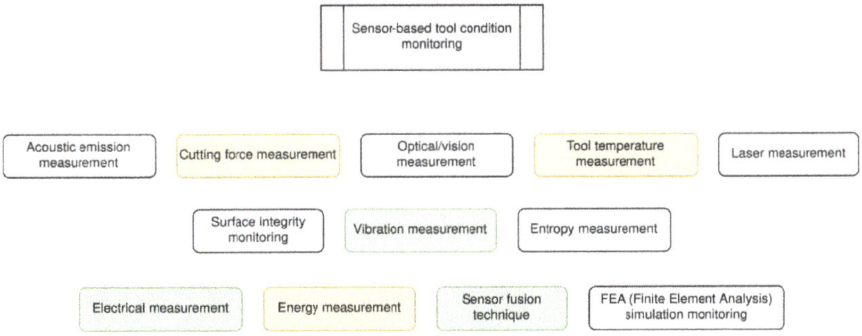

Figure 5. Commonly used sensor techniques in Tool Condition Monitoring. Sensor techniques used in the common work are marked in green; sensor techniques used indirectly or in a similar way are marked in yellow (Adapted from [24]).

2.4. Monitoring Flow forming Processes

While the monitoring of other manufacturing processes has been extensively studied in the literature, little work exists to date on the monitoring of flow-forming processes and, more generally, incremental rotary forming processes. This is also noted in [2,4]. To the author's knowledge, there is currently no literature on data-driven process or condition monitoring of flow-forming processes (including CBM, TCM and Predictive Maintenance), particularly using machine learning-based methods.

Papers [2,4] lay some groundwork in this area by investigating correlations between intentionally provoked process failures and changes in sensor data (acoustic, vibration, ultrasonic) recorded during the processes. The goal here is to evaluate the potential of different approaches to monitoring flow-forming processes and incremental rotary forming processes [4], but not to implement data-driven condition monitoring.

Furthermore, there are several works that deal with predicting the geometry of the finished workpiece based on the process input parameters such as the machine settings. For instance, in [3], linear and quadratic models are used to predict ovality based on roller feed, spindle speed, and radius of the roller. In [25,26], the researchers train artificial neural networks to predict various geometry variables based on process inputs and compare them with linear models, with the neural networks producing significantly better results in both cases. Both [25,26] also attempt to optimize the process input parameters to induce a desired geometry in the finished part. In [25], this is achieved using particle swarm optimization and composite desirability, while in [26], the authors trained neural networks to approximate the input parameters necessary for a desired geometry. An approach developed in [27] allows for manipulation of the forming process during its execution to achieve a desired geometry. For this purpose, physical models, a laser to measure wall thickness, and a sensor for indirect estimation of material properties are used. Analyses of the relationships between the process input parameters and the forces acting during flow-forming processes can be found in [28–30].

3. Materials and Methods

3.1. Experimental Setup

The data for the training and evaluation of the machine learning models are generated in experiments, which are carried out on the VSTR 400-3S (see Figure 6). The VSTR 400-3S is a machine that was built and specially modified for these experiments by WF Maschinenbau. It uses a rotating spindle and three mechanically guided rollers to form gear components with up to 400 mm in diameter. This process is called a flow-forming process [3] and is controlled by a CNC (Computerized Numerical Control) program, which contains the commands to operate the machine. The same CNC program is executed in all the experiments. Thus, the CNC program defines the example flow-forming process under study.

Multiple sensors installed in the machine collect several types of data during the forming process, such as pressure, vibration and temperature data. Figure 7 shows an example of a pressure sensor that measures the pressure on one side of a hydraulic oil cylinder. A complete list of all relevant sensors and signals can be found in Appendix A. Process monitoring using sensors is particularly challenging in flow-forming processes, mainly because both the workpiece and the processing tools, i.e., the rollers, rotate, making it impossible to place, for example, a vibration sensor directly at the site of deformation impossible. This is different to most other Tool Condition Monitoring applications, which typically involve a tool and a workpiece with one of them moving while the other is stationary and can be used to mount these sensors on. Additionally, in flow-forming processes, the area of contact between the tool and the workpiece is constantly moving and can change shape during the process, exacerbating the problem [4]. Because of these constraints, setups, like those, for example, in [31], who placed a vibration sensor directly on the tool to monitor turning operations, or in [32], who integrated a force sensor into a shear-cutting tool, cannot be implemented. The same applies to the approach of mounting the workpiece on the force sensor, as is carried out, for example, in [33] on a milling machine. Instead, to monitor flow-forming processes, sensors such as vibration sensors need to be placed further away from the source of the signal, which is a detrimental because this way, the system suffers more from noise introduced by vibrating components as well as loss of signal strength between components, both of which increase as the distance between the site of deformation and the sensor increases [4]. On the VSTR 400-3S, the tailstock traverse (common support beam for all three rollers) and the spindle bearing were chosen as locations for the vibration sensors with reasonable proximity to the deformation site, but without the need for significant modifications to the machine. The in-process forces applied to the workpieces are approximated using the pressures measured at the hydraulic oil cylinders and the physical model of the machine. A total of ten pressure sensors are used, one on each of the two sides of the hydraulic oil cylinders for the three mechanically guided rollers and the spindle, as well as one each for the hydraulic oil cylinders of the tailstock traverse and the ejector.

Figure 6. Front side of the VSTR 400-3S.

The sensor data are supplemented by the programmable logic controller (PLC), i.e., the control unit of the machine, with additional values such as position data of the rollers. The PLC, a Simatic S7 from Siemens, on the one hand, controls the movements of the machine by translating the CNC program into precise machine commands and sending them to the machine. On the other hand, it monitors the sensors and stores their signals at regular intervals. Since not all signals are recorded at the same sampling rate, interpolation of intermediate values is required for the signals with slower sampling rates. For this purpose, Akima splines are used, which is a mathematical method for fitting a "soft" and "natural" curve through a given set of data points [34].

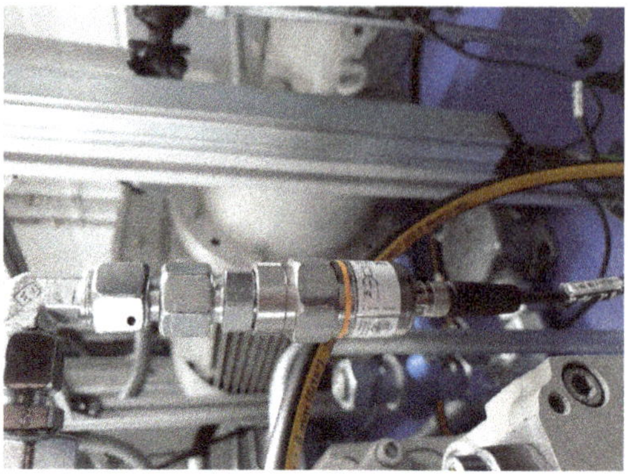

Figure 7. Sensor for measuring the pressure on one side of a hydraulic oil cylinder. A total of ten sensors of this type are installed in the machine. These are distributed on each of the two sides of the hydraulic oil cylinders of the three mechanically guided rollers and the spindle as well as one each for the hydraulic oil cylinders of the tailstock traverse and the ejector.

3.2. Experiments

Two series of experiments are carried out, both designed using DoE. These experimental designs determine the variation of influencing factors, such as a defect in the machine or a changed geometry of the blank for the individual experiments. In this way, changes in the forming process are provoked that the machine learning models will later be supposed to detect in the sensor data.

In the course of the experimental series, a total of ten factors (influencing variables) of the forming process, selected by the process experts at WF Maschinenbau, are varied. These include eight blank parameters, which are set by preparing the blanks, and two machine parameters, which can be set by making changes to the machine. The following parameters are investigated in the experiments:

1. Diameter of the blank;
2. Thickness of the blank;
3. Irregular diameter reduction in the blank by grinding (see Figure 8);
4. Local thickness reduction in the blank by grinding (see Figure 9);
5. Alloy from which the blank is made;
6. Deviation of the center of the hole from the center of the blank (see Figure 10);
7. A defined deformation of the blank where the blank is not completely flat (see Figure 11);
8. Straight piece milled from the blank (see Figure 12);
9. Flow rate of the cooling liquid;
10. Damage to one of the processing rollers by grinding (see Figure 13).

For all factors (except the alloy of the blank, where two different alloys were tested, which have different properties, while both being feasible for the process), there is an ideal or nominal value defined by the process experts. The deviations from these nominal values generated by preparation or configuration simulate causes for suboptimal process results occurring in practice.

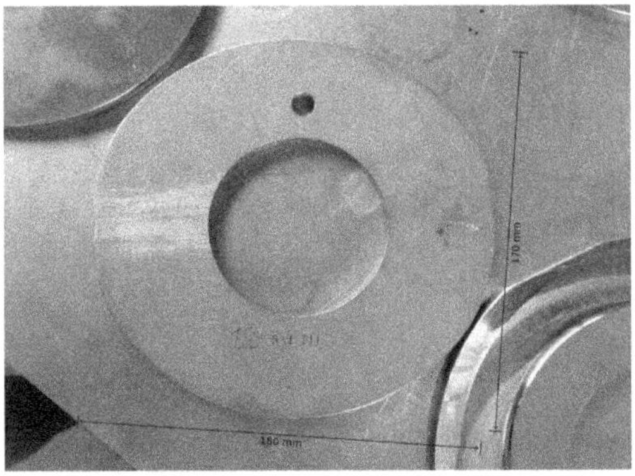

Figure 8. Metal blank with by grinding irregularly reduced diameter (change in influencing factor 3).

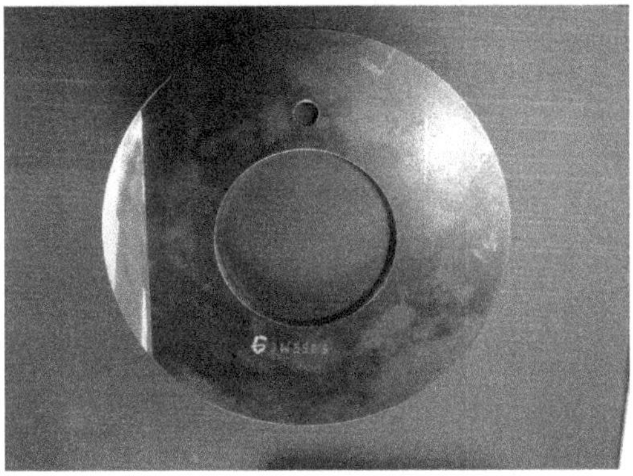

Figure 9. Metal blank with by grinding locally reduced thickness (change in influencing factor 4).

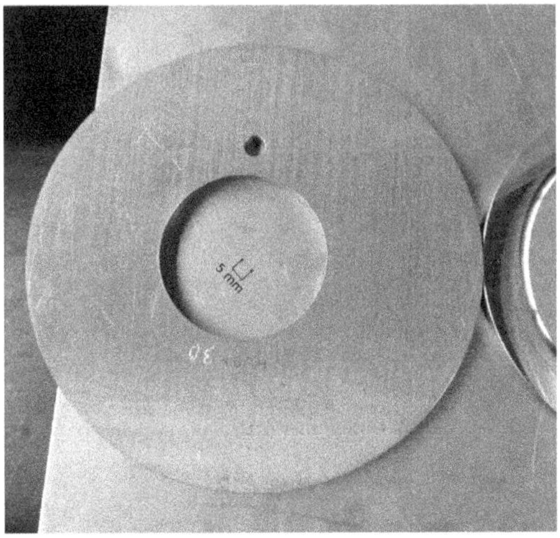

Figure 10. Metal blank with a hole of which the center deviates from the center of the blank (change in influencing factor 6).

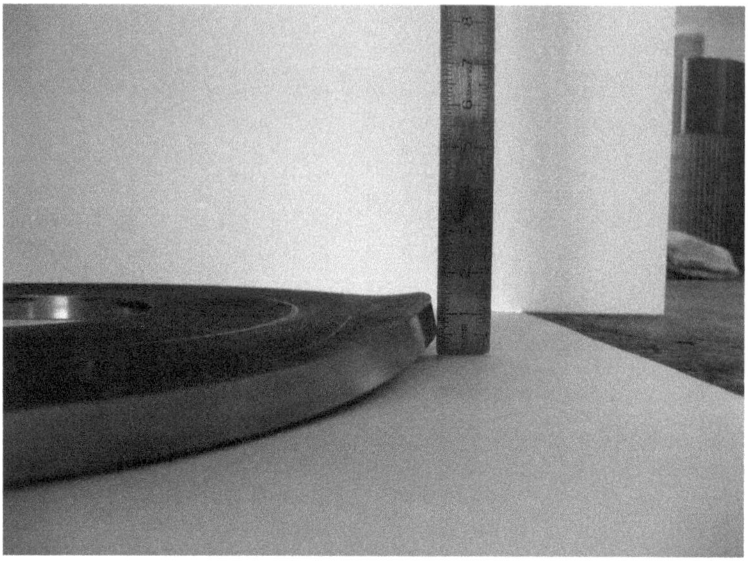

Figure 11. Metal blank with a defined deformation where the blank is not completely flat (change in influencing factor 7).

Figure 12. Metal blank with straight piece milled from it (change in influencing factor 8).

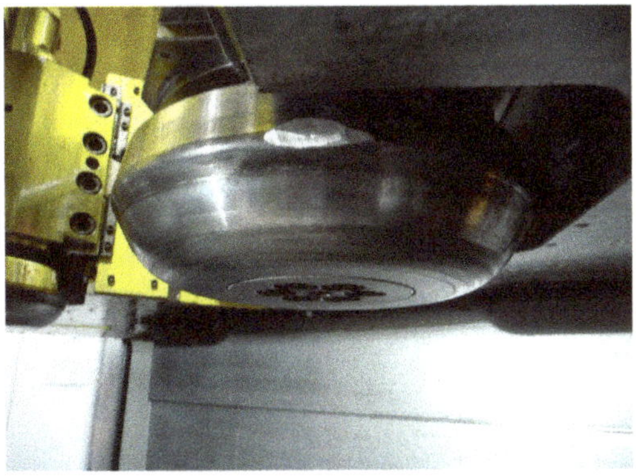

Figure 13. Roller that was damaged by grinding (change in influencing factor 10).

3.3. Experimental Design I

The plan for the first series of experiments is based on a two-level experimental design, i.e., each factor is tested on two settings or levels, with 16 experiments. The plan is constructed according to the method of Plackett and Burman and extended to 32 experiments by a foldover. This foldover, which is a plain copy of the original experimental design with simultaneous inversion, increases the resolution of the experimental design from level III to level IV [35]. This means that the main effects are much less aliased and the design is suitable for determining main effects reliably [35]. The primary purpose of this experimental design is to screen the factors. It provides a basis for the pre-selection of factors for the second series of experiments, in which these factors are examined with a higher number of experiments. The two levels examined per factor correspond to the respective minimum and maximum values for the numerical factors. This results in large distances between the tested levels, which is particularly advantageous in early investigation phases [35]. The factor of the alloy of the blank is set to two alloys common for the produced component. The remaining factors are binary and each has a defective and a non-defective setting.

Additionally, the experimental design includes six so-called reference points, resulting in a total of 39 experiments for this experimental design. Reference points are experiments that have identical factor settings corresponding to the nominal values and thus represent the reference process. Due to the identical factor settings, these experiments allow the experimental system to be monitored for changes over time [35] and to estimate experimental dispersion [35].

The order of the experiments is generally randomized to avoid systematic errors (see [35]). Deviating from this, the sequence contains later adjustments that minimize the setting changes of the two factors which are concerned with the condition of the machine, i.e., factors no. 9 and 10. These adjustments are necessary due to the high effort of the setting changes of these factors.

3.4. Experimental Design II

The second experimental design contains six of the original ten factors, no. 1, 2, 6, 7, 9 and 10. The main criteria for this selection decision, made in consultation with the process experts, are the expected practical relevance of the individual factors and their influence on the forming process observed in the first series of experiments.

Experimental design II is composed of a single-factor foldover (copy of the experimental design with inversion of one of the factors [36]) of experimental design I (32 experiments), an adapted face-centered central-composite design (22 experiments), "single-defect experiments" (13 experiments), additional random experiments (16 experiments), and reference experiments (10 experiments). A total of 93 experiments are performed in the second series of experiments.

With the help of the single-factor foldover, the alias chains between two-factor interactions remaining after the first foldover in experimental design I are eliminated. The Central-Composite-Design contains eleven experiments and includes the factors 1, 2 and 6, while the remaining factors remain in the nominal state. This decision is based on the screening of the factors in the first series of experiments. The blank for the central point experiment is prepared to a diameter of 200 mm (for practical reasons this slightly different value is used instead of the correct mean value for the diameter, 199 mm), a thickness of 7.25 mm and a displacement of the hole of 2.5 mm. Since the two-levelled part of the experimental design already exhausts the minimum and maximum limits of the factors, a face-centered central composite design is used. In addition, for the two factors diameter and thickness, further points are added between the respective star points and the central point, so that the factor space is sampled even more tightly in these two dimensions using five levels each. All experiments of the face-centered central composite design are repeated twice each. Ten additional experiments serve as reference points. Furthermore, there are 13 single-defect experiments, i.e., experiments that deviate from the nominal values in only a single factor, thus posing a special challenge for models for target variable-independent detection of deviations from the nominal process. The remaining 16 experiments are randomized experiments in which the factor settings are generated entirely by a pseudorandom number generator. Experimental design II is also randomized as far as possible.

3.5. Training Pipeline and Test Pipeline

The individual steps of preprocessing and model training are embedded in two pipelines, i.e., defined processing sequences—one for training the models and one for evaluating new data or test data. Each data point represents a single experiment. To avoid data leaks, training and test data are separated as early as possible and all processing steps are performed separately on the partial data sets. The evaluation method used in the present work is repeated k-fold cross-validation, so that the training pipeline is used for the k training data sets and the test pipeline is used for the k evaluation data sets created in each repetition of the cross-validation. However, the pipelines are designed to be agnostic to the sampling method, so other methods such as hold-out sampling or cross-validation could also be used with these pipelines.

The training pipeline consists of the steps initial data preparation, generation of additional "hand crafted" time series variables, a feature extraction for multivariate time series, feature selection, and model generation and model selection. In the first step of the training, the data are cleaned and transformations of time series variables are performed (see Section 3.5.1). This is followed by the construction of additional time series variables that are assumed to be able to support the predictions of the models (see Section 3.5.2). After this, interval features are generated on all time series variables using the feature extraction algorithm developed later in Section 3.5.3 for multivariate time series, STSFG. The set of interval features is then reduced using Recursive Feature Elimination (see Section 3.5.4). Lastly, a grid search is performed on the resulting feature set, in which the hyperparameters of the random forest algorithm are optimized. The left side of Figure 14 illustrates the flow of this pipeline.

The test pipeline flows in a generally similar way as the training pipeline. First, the data are cleaned and transformed. Then, the defined time series variables are generated and added to the data set. In contrast to the training pipeline, where STSFG feature extraction followed by feature selection is performed next, the next step here is generating interval features, but only considering those features selected during the training pipeline's feature selection. The final step is to evaluate the final random forest model on the test data. The final model is trained on the complete training data set and its hyperparameter settings correspond to the result of the grid search of the training pipeline. The right side of Figure 14 visualizes the flow of the test pipeline.

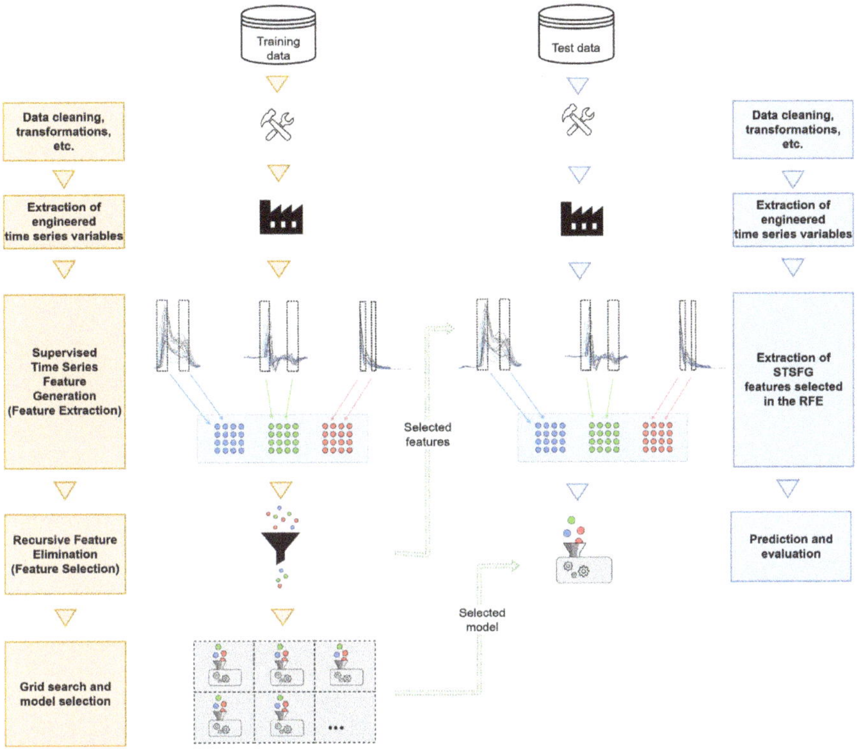

Figure 14. Training pipeline (**left**) and test pipeline (**right**). The training pipeline is used for the k training data sets and the test pipeline is used for the k evaluation data sets created in each repetition of the cross-validation. The different coloured dots represent the different features passing through the pipeline.

3.5.1. Data Cleaning and Transformations of Time Series Variables

In order to improve the data quality and to prepare the data for the machine learning algorithms, several data cleaning and transformation operations are performed on the time series variables:

- Extraction of in-process data: The raw data set of each experimental series consists of the continuous time series of all the time series variables over the complete course of the experimental series, including the time between experiments. Using the position data of the rollers, the data recorded during the actual forming processes are separated from the data recorded while the machine is idle. The latter are discarded.
- Removal of empty runs: If the CNC program is executed without a blank being inserted (which sometimes needs to be carried out for different reasons), the corresponding data are discarded.
- Removal of not useful time series variables: Some of the time series variables supplied by the PLC are redundant. These are discarded.
- One hot encoding: Some of the time series variables supplied by the PLC are nominal scaled. These are transformed using one-hot encoding.
- Imputation of missing data points: Missing data points are imputed using LASSO (Least Absolute Shrinkage and Selection Operator, see [37]) regression.
- Merging the data from the PLC with the DoE data: The time series data supplied by the PLC are merged with the corresponding influencing variables from the experimental design. For this purpose, a consecutive process counter from the PLC data and the experiment number from the experimental design are used. The assignment between

the process counter and the associated experiment number is documented while the experiments are carried out.

3.5.2. Engineering of Additional Time Series Variables

Several quotient features, i.e., features that are constructed as the quotient of other features, are constructed manually and added to the data set. This is particularly useful because the machine learning algorithm used in this work, random forest, is known to show bad performance at modeling these types of features [38].

$$QuotientPressureAxisX1 = \frac{PressureAxisX1SideA}{PressureAxisX1SideB} \tag{1}$$

$$QuotientPressureAxisX2 = \frac{PressureAxisX2SideA}{PressureAxisX2SideB} \tag{2}$$

$$QuotientPressureAxisX3 = \frac{PressureAxisX3SideA}{PressureAxisX3SideB} \tag{3}$$

$$QuotientPressureAxisZ1 = \frac{PressureAxisZ1SideA}{PressureAxisZ1SideB} \tag{4}$$

3.5.3. Feature Extraction Using the Methodology of the Randomized-Supervised Time Series Forest

Due to the multisensor approach and the various variables added by the machine's PLC as well as the engineered time series variables, each forming process is represented by a multivariate time series. The direct use of such a data set for machine learning is not reasonable due to the high dimensionality (in the present application, there are over 30,000 dimensions when using each measurement point of each time series variable as a single dimension) and the strong correlations between dimensions [17]. Moreover, such a view of the data ignores its time series nature, i.e., the fact that the data points have a defined temporal sequence. To reduce this high number of dimensions and to capture the temporal characteristics of the data, a feature extraction algorithm based on the randomized-Supervised Time Series Forest (r-STSF) is implemented.

The r-STSF is a machine learning algorithm for time series data described in [8] as an advanced successor of the Time Series Forest [39] and the Supervised Time Series Forest [40]. In the area of time series classification, the r-STSF achieves the state-of-the-art performance of complex and computationally expensive algorithms such as HIVE-COTE (Hierarchical Vote Collective of Transformation-Based Ensembles, see [41]) while being several orders of magnitude faster [8]. A feature extraction algorithm, which generates a pool of interval features from a set of time series data, is a key aspect of r-STSF.

The features generated by this feature extraction algorithm are interval features, i.e., features which are calculated by applying an aggregation function, such as the mean or the standard deviation on a consecutive subset of the values of a time series. Interval features are robust to shifts of the time series on the time axis and are able to capture the temporal characteristics of the data. Also, they are quick to calculate and intuitively interpretable. Intervall features can be calculated on the raw time series data as it is, but just as well on other time series representations, i.e., transformed versions of the time series. The feature extraction algorithm of r-STSF uses four time series representations, the raw time series, the frequency domain representation (calculated with discrete Fourier transformation), a derivative representation (first order difference of the time series) and an autoregressive representation (autoregression coefficients).

Algorithmically, the feature extraction is based on a recursive search for suitable interval features on successively restricted intervals within the time axis of the data set. The time series of the training data set are first divided into a left subset and a right subset at a randomly chosen point. Both subsets are then recursively subdivided randomly again and again until no further subdivision is possible. In each iteration of this recursion, a selected aggregation function, such as the arithmetic mean or the standard deviation, is applied to both subsets created during the subdivision, and the resulting interval feature is rated. As an estimation of the usefulness of each given interval feature for the classification task, the Fisher score (see [42]) is used as the metric for this rating. Of two interval features created at one devision, only the feature with the better score is included in the feature pool, and the recursion is continued only for the interval whose feature has the better score. This procedure is repeated for the chosen sets of aggregation functions and time series representations.

Due to the promising quality of the features for the reasons mentioned above as well as the speed of computation, this method is used as a basis for the feature extraction algorithm implemented in the present work, hereafter called Supervised Time Series Feature Generation or STSFG. Speed is an important criterion especially because of the large number of time series variables. By implementing the feature extraction functionality autonomously (instead of using the complete r-STSF algorithm), the following advantages are achieved:

- A feature selection can be incorporated between the feature extraction and the training of models on the generated features (see Figure 14).
- The Extra Tree algorithm is an integral part of the r-STSF as defined by [8]. By detaching the feature extraction functionality from the rest of r-STSF, it becomes possible to train arbitrary other machine learning algorithms on the generated features.
- In addition to the choice of learning algorithm, the choice of libraries and implementations used for it is also open (Cabello et al. [8] use the Python library sklearn). In the present work, the machine learning platform H2O is used to train the models.

Furthermore, the implemented STSFG algorithm integrates modifications and extensions to handle multivariate time series and metric target variables. Multivariate time series are not considered in [39,40], or [8], although such extensions are held out in prospect for future work in the latter two cases. The algorithm implemented here handles multivariate time series by using an additional loop over all time series variables and accumulating all the resulting features in the feature pool.

Due to the fact that some of the target variables to be predicted in the given application are metrically scaled, there is a need to be able to perform feature extraction for such target variables as well. Since the Fisher score proposed in [8,40] is only defined for classifications, Kendall's tau is used instead. As a rank correlation coefficient, Kendall's Tau is nonparametric, robust to outliers [43], and can also detect nonlinear correlations as long as they are monotonic [35]. Compared to Spearman's Rho, it is more robust and efficient [43].

The STSFG algorithm is implemented entirely in R. The pseudocode in Algorithms 1 and 2 describes the implementation, omitting certain implementation details for clarity, but without changing the logic of the algorithm.

In addition to the raw time series data, the derivative representation and the frequency domain representation are used. The autoregressive representation is not included yet but could be added without further modifications to the algorithm. Figure 15 illustrates different time series representations of the variable *VibrationS1*, which represents the strength of vibration at the spindle of the machine during the forming process. The aggregation functions used are the maximum and sum for binary features and the arithmetic mean, standard deviation, maximum, and root mean square (RMS) for metric features. The standard deviation is extended according to [39] so that it is also defined for intervals with only one element, in which case it is set to zero.

The STSFG algorithm has only one parameter, $stsfg_r$, which is the number of repetitions performed. Cabello et al. [8] investigate this parameter's influence in the r-STSF (called d there) and ultimately recommend setting it to 50 by default [8]. However, since in the STSFG the whole algorithm is additionally repeated for all time series variables, a parameter setting of $stsfg_r = 50$ leads to extremely long run times and feature sets that are no longer manageable due to their size. Therefore, the value $stsfg_r = 5$ is used here instead.

Algorithm 1: Supervised Time Series Feature Generation: Repeated iteration over all time series variables and time series representations.

```
Input: X: Set of |I| time series with |P| measuring points und |V| variables; y:
       Vector of labels of the |I| time series; A: Set of aggregation functions; f_r:
       Rating function for features; n_repeats: Number of repetitions (d in [8])
F ← ∅ ;
foreach time_series_representation do
    Ψ ← ConvertRepresentation(X; time_series_representation) ;
    foreach i = 1, ..., n_repeats do
        foreach a ∈ A do
            foreach v ∈ V do
                u ← CreateSeparationPoint(|P|) ;
                Ψ_L ← Ψ(I; v; 1, ..., u − 1) ;
                Ψ_R ← Ψ(I; v; u, ..., |P|) ;
                IntervalSearch(Ψ_L; y; A; f_r; F) ;
                IntervalSearch(Ψ_R; y; A; f_r; F) ;
            end
        end
    end
end
return F ;
```

Algorithm 2: IntervalSearch : Generation of interval features on a time series variable in a time series representation.

```
Input: Ψ_v: Set of |I| time series of a variable v in an arbitrary time series
       representation with |P'| values; y: Vector of labels of the |I| time series; a:
       Aggregation function; f_r: Rating function for features; F: List of already
       added features
s ← 1 ;
e ← |P'| ;
while s < e do
    u ← CreateSeparationPoint(|P|) ;
    Ψ_L ← Ψ(I; s, ..., u − 1) ;
    Ψ_R ← Ψ(I; u, ..., e) ;
    f_L ← a(Ψ_L) ;
    f_R ← a(Ψ_R) ;
    score_L ← f_r(f_L; y; F) ;
    score_R ← f_r(f_R; y; F) ;
    if score_L ≥ score_R then
        F ← F ∪ f_L ;
        e ← u − 1 ;
    else
        F ← F ∪ f_R ;
        s ← u ;
    end
end
return F ;
```

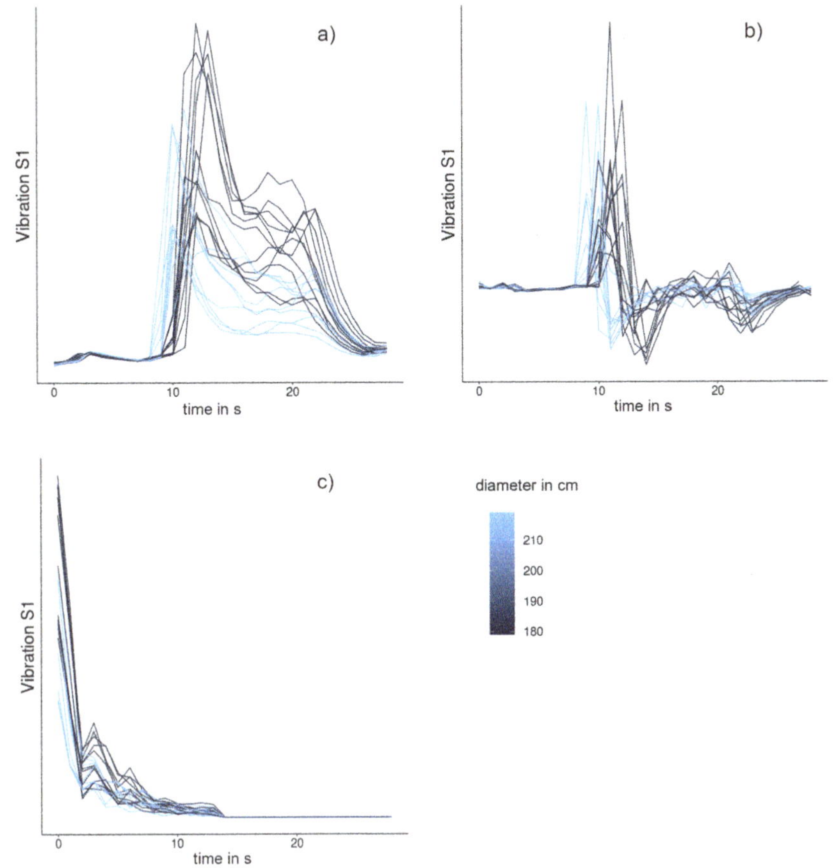

Figure 15. Vibrations of the spindle on three different time series representations. For a subset of the experiments, the figures show the strength of the vibration at the spindle (without unit) over time as raw time series data (**a**), in the time series representation of the derivatives (**b**) and at the frequency level (**c**). Each line represents a single experiment and the color of a time series depicts the diameter of the blank used. The sensor outputs values on a sensor-specific unitless scale, hence no labels are given on the y-axis.

3.5.4. Feature Selection Using Recursive Feature Elimination

The feature sets obtained from feature extraction are very large, usually several thousand features, and also suboptimal in terms of redundancy and relevance. This is primarily due to the fact that the time series variables on which the feature extraction is based already contain redundancy and many are only slightly relevant or not relevant at all for the individual target variables. Therefore, the next step in the preprocessing procedure is feature selection. In this step, the previously extracted set of features is reduced to a subset with as little redundancy and as high relevance as possible. This is intended to achieve:

- Improvement of the predictive ability of machine learning models trained on the resulting feature set. The experiments of [7] suggest that this is also possible for random forests, which are generally considered to be robust against irrelevant features.
- Minimization of the required sensors.
- Simplification of the models and thus improvement of their interpretability.
- Reduction in the time required to train the machine learning models.
- Reduction in the time required for the evaluation of new observations. This is particularly relevant for a possible live monitoring of the forming processes, where the time available for an evaluation of new data is limited.

For feature selection, a Recursive Feature Elimination (RFE) algorithm is implemented, which recursively trains random forest models and computes the permutation importance of the features to drop the least relevant ones. The comparison of the feature sets and the final selection of the best feature set is carried out using the out-of-bag errors of the random forests. The programming language used is R. For the model training and the calculation of permutation importances, functions of the software H2O are called.

RFE algorithms can be used effectively against both redundant and irrelevant features within a data set. Moreover, as, for example [44,45] show, they are also suitable for cases where the number of features is in the four-digit range or above and far exceeds the number of observations. However, the results of [45,46] also suggest that not too many features should be removed at once. In this way, the computational effort for the RFE increases, but this is less significant due to the rather small number of observations in the present use case—especially since the subsequent generation of the machine learning models (including the hyperparameter optimizations) is, in turn, considerably accelerated by the reduced number of features.

The parameter $rfe_{r_{perc}}$ determines how many features are to be removed per iteration of the RFE. For this parameter, the settings used in the literature differ widely. While in [47], only a single feature is removed per iteration (the simultaneous removal of multiple features is, however, mentioned as a possible option, see [47]), in other cases, up to 50 % of features are removed from the data set per RFE iteration (see, e.g., [44,46]). To keep the aforementioned risks of simultaneously eliminating too many features, a rather low value of $rfe_{r_{perc}} = 5\%$ is used here.

The parameters of the random forests only have small influence on the RFE [44,46] and care should only be taken that $rfe_{n_{tree}}$ is large enough for the metrics considered to stabilize sufficiently [46]. Therefore, these parameters are assigned default values from the literature.

3.5.5. Model Generation and Selection

In order to monitor the forming processes and the condition of the machine, machine learning models are trained that take all data gathered during a forming process as input to predict the influencing factors of this process. In principle, arbitrary machine learning models can be trained on the feature sets generated with STSFG and RFE. However, as, e.g., shown by the experiments in [47], the selection of the machine learning algorithm has far less impact on the predictive ability of the trained models than the quality of the features. Various works in the field of data-driven condition monitoring also suggest that the selection of the learning algorithm has a comparatively low relevance in the present use case (see Section 2.1). Therefore, only random forests are used here for the prediction of the influence variables. The main reasons for choosing the random forest algorithm are as follows:

- Random forests usually have a good predictive ability, which depends only little on the settings of the hyperparameters.
- Due to the small size of the training data set and the high number of dimensions, there is a high risk of generating overfitted models. Random forests are largely immune to this.
- The RFE algorithm used also employs random forests. Thus, the selected feature set is to some extent tuned for random forests.

According to the recommendation of [8], the trees of the random forests are trained with the Extra Tree algorithm. To optimize the parameter m_{try}, a grid search is employed in each run of the training pipeline (following the recommendation of [7], five values between 2 and the number of features are tested at even intervals). The number of trees n_{tree} is set to 2000 for all of the random forests. To generate the random forest models, functions of H2O are called.

4. Results

4.1. Evaluation Method

The methodology presented on Figure 14 is evaluated for all six of the factors present in experimental design II. In order to obtain estimators for the respective generalization errors that are as unbiased as possible, repeated k-fold cross-validation is used as the validation method. k is set to $k = 10$, following the recommendation of [48]. The number of repetitions is set to 5, which is comparatively small, to keep the total running time within acceptable limits. This is necessary because both feature extraction and feature selection are extremely computationally expensive and the entire procedure is to be evaluated for the six target variables. To avoid bias in the error estimates obtained with cross-validation, the whole process from initial preprocessing of time series variables to model generation and selection is repeated in its entirety for each partitioning within the cross-validation procedure. Without this, the supervised feature extraction, STSFG, as well as the feature selection with RFE would introduce a selection bias to the error estimation [46].

The parameters of RFE and STSFG are not optimized here, but are given default values (see Sections 3.5.3 and 3.5.4). Table 1 summarizes the parameter settings used for the RFE, STSFG and the not optimized parameters of the random forests.

Table 1. Parameters of the STSFG, the RFE and the random forests.

Parameter	Description	Value
$stsfg_r$	Repetitions of the STSFG	5
rfe_{perc}	Proportion of features removed in each RFE iteration	10%
$rfe_{n_{tree}}$	Number of trees of the random forests trained in the RFE	1000
$rfe_{m_{try}}$	Number of randomly selected features at each node of the random forests built in the RFE	$\sqrt{m}$, $\frac{m}{3}$
n_{tree}	Number of trees of the random forests	2000

4.2. Evaluation of Regression Models

For the regression models, the metrics RMSE, MAE, and q^2 are given. According to [49], q^2 is calculated from the MSE of the prediction and "SS, the sum of squares of the differences between observed values [...] and their mean [...]" in the validation data with

$$q^2 = 1 - \frac{MSE}{SS} = 1 - \frac{\sum_{i=1}^{N}(\hat{y}_i - y_i)^2}{\sum_{i=1}^{N}(y_i - \bar{y})^2} \quad (5)$$

Here, y is the vector of target variable values from the validation data, $\bar{y}$ is their mean, and $\hat{y}$ is the vector of predictions for y. q^2, unlike RMSE and MAE, does not have a unit, which is why it is useful as a metric to compare the results for different target variables. According to Equation (5), $q^2 = 0$ corresponds to a model that always predicts the target variable's mean from the validation data, and $q^2 = 1$ corresponds to a model that makes perfect predictions and therefore has an MSE of 0.

For each of the metric target variables, Table 2 contains the estimate of the generalization error determined by cross-validation. In addition to the metrics RMSE, MAE and q^2, the standard deviation of each metric is given as a comparison value.

The cross-validation RMSE is substantially below the respective standard deviation for all metric variables. Especially for factor 1 (diameter of the blank), a very low RMSE is achieved in relation to the standard deviation, which also is reflected in the high q^2-value of 0.96. The models for factors 2 (thickness of the blank) and 6 (displaced hole in the blank) have average q^2 values of 0.85 and 0.61, respectively.

Table 2. For each target variable, the average RMSE, MAE, and q^2 on the validation sets of the 10-fold cross-validation with 3 replicates are given. Also given are the standard deviations of the variables, which serve as reference values.

Target Variable	Ø RMSE	Ø MAE	Ø q^2	Standard Deviation
influencing factor 1 (diameter of the blank)	2.74 mm	1.57 mm	0.96	14.88 mm
influencing factor 2 (thickness of the blank)	0.21 mm	0.13 mm	0.85	0.66 mm
influencing factor 6 (displaced hole in the blank)	1.20 mm	0.86 mm	0.61	2.20 mm

4.3. Validation of Classification Models

For each of the binary target variables, Table 3 contains estimates of the generalization error obtained by cross-validation. In particular, the metrics MCC (Matthews Correlation Coefficient), F1, precision, recall and accuracy are shown in this table. Advantages of the MCC are that it provides reliable results even in the presence of unbalanced class distributions and that it is symmetric with respect to a swap of class definitions [50].

Among the classification models, a substantially higher accuracy is achieved for factors 7 (defined deformation) and 10 (damaged roller) than for factor 9 (flow rate of cooling liquid) in all metrics considered. The average MCC is 0.81 for factor 10, 0.76 for factor 7 and 0.45 for factor 9.

Table 3. Evaluation of the models for the binary target variables. The average MCC, F1, precision, recall and accuracy values on the validation sets of the 10-fold cross-validation with 3 repetitions are given.

Target Variable	Ø MCC	Ø F1	Ø Precision	Ø Recall	Ø Accuracy
influencing factor 7 (defined deformation)	0.76	0.89	0.81	0.92	0.91
influencing factor 10 (damaged roller)	0.81	0.87	0.82	0.94	0.93
influencing factor 9 (flow rate of cooling liquid)	0.45	0.72	0.66	0.82	0.72

4.4. Dimensionality Reduction

Over 16,000 of features are generated by the STSFG per target variable. Using the RFE, these large feature sets are reduced to less than 40 features on average. Table 4 shows, for each of the target variables, the average number of features before and after the RFE, as well as the percentages by which the feature sets are reduced on average.

Table 4. Evaluation of the dimensionality reduction by the RFE.

Target Variable	Ø Feature Count after STSFG	Ø Feature Count after RFE	Reduction
Influencing factor 1 (diameter of the blank)	16,719.7	21.1	99.87%
Influencing factor 2 (thickness of the blank)	17,113.3	33.0	99.81%
Influencing factor 6 (displaced hole in the blank)	16,787.8	18.6	99.89%
Influencingfactor 7 (defined deformation)	16,541.4	37.2	99.78%
Influencing factor 9 (flow rate of cooling liquid)	9294.3	30.0	99.68%
Influencing factor 10 (damaged roller)	16,329.6	12.7	99.92%

5. Discussion

The analysis of the state of research has shown that the data-driven monitoring of flow-forming processes has so far been a research gap. While similar manufacturing processes have been already well researched in this respect, it has not been clear whether such monitoring can be implemented for flow-forming processes, which are considered to be very complex. Therefore, one of the key findings of this work is that it is possible to detect unacceptable conditions in flow-forming processes using data-driven condition monitoring based on sensor data. Another finding is that the methodology used in the present work, including the data acquisition with DoE and the data preprocessing, is a promising way to approach such problems.

For companies using flow forming machines, this opens up the potential to optimise production through the implementation of automated process monitoring systems. By providing timely warnings, recommendations, or even halting the machine when necessary, these systems can significantly reduce the risk of financial losses associated with defective output and wasted material. Consequently, companies that manufacture these machines can gain a competitive advantage by equipping their machines with data-driven automatic monitoring capabilities, thereby providing their customers additional value.

The feature extraction method STSFG, which was derived from the r-STSF algorithm, has proved to be a promising means for generating features from multidimensional time series data in the present use case. Furthermore, it has been shown that a subsequent RFE can massively reduce the number of features needed for a prediction, which, among other things, reduces the training time and enables faster predictions. The latter is particularly relevant in the context of live monitoring of the forming processes. Also, the combination of STSFG and RFE may not only be useful for monitoring flow-forming processes. It stands to reason that this approach for generating and selecting features from time series data may also be an option to consider for monitoring other machines and processes.

Even apart from applications in the context of any kind of condition monitoring, to the author's knowledge, no work has yet been carried out that uses the still-very-young r-STSF algorithm in any practical application. The present work contributes to filling this gap and demonstrates the potential of this algorithm. The adaptation to multivariate time series and regression problems also expands the possible application areas. The implementation as a stand-alone feature extraction simplifies the integration into existing structures.

Nonetheless, the research conducted in the present work has several limitations that warrant acknowledgment. Firstly, the methodology used is rather complex, which may hinder reproducibility. Simplifying the methods could enhance accessibility and facilitate broader application. Secondly, only one example flow-forming process was examined, potentially limiting the generalizability of the findings to other instances of this manufacturing technique. Additionally, the accuracy of the models may not be sufficient depending on the specific application, highlighting the need for further validation in different contexts. Furthermore, as only specific defects were studied, it remains unclear how other defects may influence model predictions or whether general fault detection is feasible. Moreover, no satisfactory results were obtained for the influencing factors that were dropped after the first series of experiments, and the results for influencing factor 9, the flow rate of the cooling liquid, were less conclusive and require additional confirmation.

6. Future Work

While the present work fills research gaps, new research needs arise at the same time. Flow-forming processes have opened up a new area of application for data-driven process and machine monitoring. This offers various opportunities for further research. These include, in particular, the investigation of other, possibly more complex, flow-forming processes in order to examine to what extent the knowledge gained from the example flow-forming process can be transferred to other processes. Furthermore, ways to improve the prediction accuracy should be explored, e.g., by using a larger training data set or by implementing and comparing it with other methodologies and algorithms. For example, Cabello et al. [8] have already announced their own extension of r-STSF for multivariate time series (see [8,40]), the accuracy and computational efficiency of which could be compared to the approach used in the present work. Another approach to improve the prediction accuracy could be the utilization of additional sensors. For example, in [4], an ultrasonic monitoring system was demonstrated to be able to "to record changes in the tool-workpiece contact area, detect internal fractures in the part, and measure the thickness of parts spun in free air", although it had some limitations that may limit its practical applicability.

In the present work, the parameters of the used feature extraction and feature selection like $stsfg_r$ and rfe_{perc} were assigned default settings and not investigated in more detail. Optimization of these parameters as well as a sensitivity analysis may provide further insight into the influence of the parameters here and further improve the predictive ability of the models. A possible improvement of the prediction accuracy with the help of learning algorithms other than Extra Trees should also be investigated. The same applies to the use of the autoregressive time series representation and other aggregation functions not yet adopted here from the r-STSF.

Beyond the mere monitoring of the current state, research into the prediction of defects of the machine is also conceivable, e.g., to determine the time until a machine component fails ("Remaining Useful Life" [21]). Such systems, which belong to the research area "Predictive Maintenance", require a more comprehensive data acquisition, since the algorithms must be enabled to detect the development of a defect before it manifests itself [51].

Lastly, further research is also needed with respect to the r-STSF and the STSFG method derived from it in the present work. Future work can test these algorithms in other application areas and further investigate their potential for different production processes and for time series classification in general.

Author Contributions: Conceptualization, L.A. and C.E.; methodology, L.A.; software, L.A.; validation, L.A.; investigation, L.A.; resources, B.F.; writing—original draft preparation, L.A.; writing—review and editing, B.F. and C.M.F. and C.E.; visualization, L.A.; supervision, B.F. and C.M.F. and C.E.; project administration, C.E.; funding acquisition, B.F. and C.E. All authors have read and agreed to the published version of the manuscript.

Funding: This research was funded by the Federal Ministry for economic affairs and energy of Germany grant number ZIM ZF4038207HB8.

Institutional Review Board Statement: Not applicable.

Informed Consent Statement: Not applicable.

Data Availability Statement: Data is contained within the article.

Acknowledgments: Thanks go to Bijan Riesenberg for the help in all phases of the project and to WF Maschinenbau for providing the facilities and the machine to carry out the experiments as well as for contributing their knowledge about flow-forming processes.

Conflicts of Interest: The funders had no role in the design of the study; in the collection, analyses, or interpretation of data; in the writing of the manuscript; or in the decision to publish the results. Author Bodo Fink was the CEO of WF Maschinenbau at the time of the project.

Abbreviations

The following abbreviations are used in this manuscript:

CBM	Condition-based Maintenance
CM	Condition Monitoring
CNC	Computerized Numerical Control
DoE	Design of Experiments
FEA	Finite Element Analysis
HIVE-COTE	Hierarchical Vote Collective of Transformation-Based Ensembles
LASSO	Least Absolute Shrinkage and Selection Operator
MAE	Mean Absolute Error
MCC	Matthews Correlation Coefficient
PLC	Programmable Logic Controller
r-STSF	Randomized-Supervised Time Series Forest
RFE	Recursive Feature Elimination
RMS	Root Mean Square
RMSE	Root Mean Square Error
RUL	Remaining Useful Life
STSF	Supervised Time Series Forest
STSFG	Supervised Time Series Feature Generation
SVM	Support Vector Machine
TCM	Tool Condition Monitoring
TSF	Time Series Forest

Appendix A

Table A1. List of relevant measured variables.

Component	Measured Variable
spindle	revolutions per minute active current active torque active motor temperature vibration
feed of spindle	position position error pressure side A pressure side B
feed of roller 1 position	position error pressure side A pressure side b force side A force side B bearing temperature

Table A1. *Cont.*

Component	Measured Variable
feed of roller 2	position position error pressure side A pressure side b force side A force side B bearing temperature
feed of roller 3	position position error pressure side A pressure side b force side A force side B bearing temperature
tailstock traverse	pressure temperature traverse bearing force vibration
ejector	pressure force
hydraulic power unit	fill level temperature pressure
bearing lubrication spindle	fill level temperature
bearing lubrication tailstock traverse	fill level temperature

References

1. Jahazi, M.; Ebrahimi, G. The influence of flow forming parameters and microstructure on the quality of a D6ac steel. *J. Mater. Process. Technol.* **2000**, *103*, 362–366. [CrossRef]
2. Appleby, A.; Conway, A.; Ion, W. A novel methodology for in-process monitoring of flow forming. In Proceedings of the International Conference of Global Network for Innovative Technology and AWAM International Conference in Civil Engineering (IGNITE-AICCE'17): Sustainable Technology And Practice For Infrastructure and Community Resilience, 2017, AIP Conference Proceedings, Dublin, Ireland, 26–28 April 2017. [CrossRef]
3. Srinivasulu, M.; Komaraiah, M.; Rao, C.K.P. Application of Response Surface Methodology to Predict Ovality of AA6082 Flow Formed Tubes. *Int. J. Manuf. Mater. Mech. Eng.* **2013**, *3*, 52–65. [CrossRef]
4. Appleby, A. Monitoring of Incremental Rotary Forming. Ph.D. Thesis, University of Strathclyde (Department of Design, Manufacturing and Engineering Management), Glasgow, Scotland, 2019. [CrossRef]
5. Kononenko, I.; Kukar, M. *Machine Learning and Data Mining*, [repr. of the ed.] 2007, ed.; Woodhead Publishing: Chichester, England, 2013.
6. Bishop, C.M. *Pattern Recognition and Machine Learning*, 9 (corrected at 8th printing), ed.; Information science and statistics; Springer: Berlin/Heidelberg, Germany, 2009.
7. Kuhn, M.; Johnson, K. *Applied Predictive Modeling*, 1st ed.; Springer: New York, NY, USA, 2013. [CrossRef]
8. Cabello, N.; Naghizade, E.; Qi, J.; Kulik, L. Fast, accurate and explainable time series classification through randomization. *Data Min. Knowl. Discov.* **2023**, 38, 748–811. [CrossRef]
9. Goyal, D.; Pabla, B.S.; Dhami, S.S.; Lachhwani, K. Optimization of condition-based maintenance using soft computing. *Neural Comput. Appl.* **2017**, *28*, 829–844. [CrossRef]
10. Sobie, C.; Freitas, C.; Nicolai, M. Simulation-driven machine learning: Bearing fault classification. *Mech. Syst. Signal Process.* **2018**, *99*, 403–419. [CrossRef]
11. Kurukuru, V.S.B.; Haque, A.; Khan, M.A.; Tripathy, A.K. Fault classification for Photovoltaic Modules Using Thermography and Machine Learning Techniques. In Proceedings of the 2019 International Conference on Computer and Information Sciences (ICCIS), Sakaka, Saudi Arabia, 3–4 April 2019; pp. 1–6. [CrossRef]

12. Gong, C.S.A.; Su, C.H.S.; Tseng, K.H. Implementation of Machine Learning for Fault Classification on Vehicle Power Transmission System. *IEEE Sens. J.* **2020**, *20*, 15163–15176. [CrossRef]
13. Nazir, Q.; Shao, C. Online tool condition monitoring for ultrasonic metal welding via sensor fusion and machine learning. *J. Manuf. Process.* **2021**, *62*, 806–816. [CrossRef]
14. Kanawaday, A.; Sane, A. Machine Learning for Predictive Maintenance of Industrial Machines using IoT Sensor Data. In Proceedings of the 2017 8th IEEE International Conference on Software Engineering and Service Science (ICSESS), Beijing, China, 24–26 November 2017; pp. 87–90.
15. Costa, M.; Wullt, B.; Norrlöf, M.; Gunnarsson, S. Failure detection in robotic arms using statistical modeling, machine learning and hybrid gradient boosting. *Measurement* **2019**, *146*, 425–436. [CrossRef]
16. Tian, Z.; Wong, L.; Safaei, N. A neural network approach for remaining useful life prediction utilizing both failure and suspension histories. *Mech. Syst. Signal Process.* **2010**, *24*, 1542–1555. [CrossRef]
17. Abellan-Nebot, J.V.; Romero Subirón, F. A review of machining monitoring systems based on artificial intelligence process models. *Int. J. Adv. Manuf. Technol.* **2010**, *47*, 237–257. [CrossRef]
18. Breckweg, A. Automatisiertes und prozessüberwachtes Radialclinchen höherfester Blechwerkstoffe. Ph.D. Thesis, Universität Stuttgart (Fakultät für Maschinenbau), Stuttgart, Germany, 2007.
19. Choi, Y.J.; Park, M.S.; Chu, C.N. Prediction of drill failure using features extraction in time and frequency domains of feed motor current. *Int. J. Mach. Tools Manuf.* **2008**, *48*, 29–39. [CrossRef]
20. Cheng, F.; Dong, J. Monitoring tip-based nanomachining process by time series analysis using support vector machine. *J. Manuf. Process.* **2019**, *38*, 158–166. [CrossRef]
21. Lei, Y.; Li, N.; Gontarz, S.; Lin, J.; Radkowski, S.; Dybala, J. A Model-Based Method for Remaining Useful Life Prediction of Machinery. *IEEE Trans. Reliab.* **2016**, *65*, 1314–1326. [CrossRef]
22. Ren, L.; Cui, J.; Sun, Y.; Cheng, X. Multi-bearing remaining useful life collaborative prediction: A deep learning approach. *J. Manuf. Syst.* **2017**, *43*, 248–256. [CrossRef]
23. Drouillet, C.; Karandikar, J.; Nath, C.; Journeaux, A.C.; El Mansori, M.; Kurfess, T. Tool life predictions in milling using spindle power with the neural network technique. *J. Manuf. Process.* **2016**, *22*, 161–168. [CrossRef]
24. Bhuiyan, M.S.; Choudhury, I. Review of Sensor Applications in Tool Condition Monitoring in Machining. *Compr. Mater. Process.* **2014**, *13*, 539–569. [CrossRef]
25. Banerjee, P.; Hui, N.B.; Dikshit, M.K.; Laha, R.; Das, S. Modeling and optimization of mean thickness of backward flow formed tubes using regression analysis, particle swarm optimization and neural network. *SN Appl. Sci.* **2020**, *2*, 1353. [CrossRef]
26. Podder, B.; Banerjee, P.; Kumar, K.R.; Hui, N.B. Forward and reverse modeling of flow forming of solution annealed H30 aluminium tubes. *Neural Comput. Appl.* **2020**, *32*, 2081–2093. [CrossRef]
27. Riepold, M.; Arian, B.; Rozo Vasquez, J.; Homberg, W.; Walther, F.; Trächtler, A. Model approaches for closed-loop property control for flow forming. *Adv. Ind. Manuf. Eng.* **2021**, *3*, 100057. [CrossRef]
28. Bhatt, R.J.; Raval, H.K. An Investigation of forces during Flow-Forming Process. *Int. J. Adv. Manuf. Technol.* **2016**, *3*, 12–17.
29. Bhatt, R.J.; Raval, H.K. Optimization of process parameters during flow-forming process and its verification. *Mechanika* **2017**, *23*, 581. [CrossRef]
30. Bhatt, R.J.; Raval, H.K. In situ investigations on forces and power consumption during flow forming process. *J. Mech. Sci. Technol.* **2018**, *32*, 1307–1315. PII: 235, [CrossRef]
31. Abouelatta, O.B.; Madl, J. Surface roughness prediction based on cutting parameters and tool vibrations in turning operations. *J. Mater. Process. Technol.* **2001**, *118*, 269–277. [CrossRef]
32. Hoppe, F.; Hohmann, J.; Knoll, M.; Kubik, C.; Groche, P. Feature-based Supervision of Shear Cutting Processes on the Basis of Force Measurements: Evaluation of Feature Engineering and Feature Extraction. *Procedia Manuf.* **2019**, *34*, 847–856. [CrossRef]
33. Huang, P.T.; Chen, J. Neural network-based tool breakage monitoring system for end milling operations. *J. Ind. Technol.* **2000**, *16*, 2–7.
34. Akima, H. A New Method of Interpolation and Smooth Curve Fitting Based on Local Procedures. *J. ACM* **1970**, *17*, 589–602. [CrossRef]
35. Siebertz, K.; van Bebber, D.; Hochkirchen, T. *Statistische Versuchsplanung*, 2nd ed.; Springer: Berlin/Heidelberg, Germany, 2017. [CrossRef]
36. Montgomery, D.C.; Runger, G.C. Foldovers of 2 k-p Resolution IV Experimental Designs. *J. Qual. Technol.* **1996**, *28*, 446–450. [CrossRef]
37. Tibshirani, R. Regression Shrinkage and Selection via the Lasso. *J. R. Stat. Soc. Ser. B (Methodol.)* **1996**, *58*, 267–288.
38. Heaton, J. An empirical analysis of feature engineering for predictive modeling. In Proceedings of the SoutheastCon 2016, Norfolk, VA, USA, 30 March–3 April 2016; pp. 1–6. [CrossRef]
39. Deng, H.; Runger, G.; Tuv, E.; Vladimir, M. A time series forest for classification and feature extraction. *Inf. Sci.* **2013**, *239*, 142–153. [CrossRef]
40. Cabello, N.; Naghizade, E.; Qi, J.; Kulik, L. Fast and Accurate Time Series Classification Through Supervised Interval Search. In Proceedings of the 2020 IEEE International Conference on Data Mining (ICDM), Sorrento, Italy, 17–20 November 2020; pp. 948–953. [CrossRef]

41. Lines, J.; Taylor, S.; Bagnall, A. Time Series Classification with HIVE-COTE. *ACM Trans. Knowl. Discov. Data* **2018**, *12*, 1–35. [CrossRef]
42. Duda, R.O.; Hart, P.E.; Stork, D.G. *Pattern Classification*, 2nd ed.; John Wiley & Sons: Hoboken, NJ, USA, 2000.
43. Croux, C.; Dehon, C. Influence Functions of the Spearman and Kendall Correlation Measures. *SSRN Electron. J.* **2010**, *19*, 497–515. [CrossRef]
44. Díaz-Uriarte, R.; Alvarez de Andrés, S. Gene selection and classification of microarray data using random forest. *BMC Bioinform.* **2006**, *7*, 3. [CrossRef] [PubMed]
45. Darst, B.F.; Malecki, K.C.; Engelman, C.D. Using recursive feature elimination in random forest to account for correlated variables in high dimensional data. *BMC Genet.* **2018**, *19*, 65. [CrossRef] [PubMed]
46. Svetnik, V.; Liaw, A.; Tong, C.; Wang, T. Application of Breiman's Random Forest to Modeling Structure-Activity Relationships of Pharmaceutical Molecules. In *Multiple Classifier Systems*; Roli, F., Kittler, J., Windeatt, T., Eds.; Springer: Berlin/Heidelberg, Germany, 2004; pp. 334–343.
47. Guyon, I.; Weston, J.; Barnhill, S.; Vapnik, V. Gene Selection for Cancer Classification using Support Vector Machines. *Mach. Learn.* **2002**, *46*, 389–422. [CrossRef]
48. Kohavi, R. A Study of Cross-Validation and Bootstrap for Accuracy Estimation and Model Selection. In Proceedings of the 14th International Joint Conference on Artificial Intelligence—Volume 2, Montréal, QC, Canada, 20–25 August 1995; Morgan Kaufmann Publishers Inc.: Stanford, CA, USA, 1995; IJCAI'95, pp. 1137–1143.
49. Schüürmann, G.; Ebert, R.U.; Chen, J.; Wang, B.; Kühne, R. External Validation and Prediction Employing the Predictive Squared Correlation Coefficient Test Set Activity Mean vs. Training Set Activity Mean. *J. Chem. Inf. Model.* **2008**, *48 11*, 2140–2145. [CrossRef]
50. Chicco, D.; Jurman, G. The advantages of the Matthews correlation coefficient (MCC) over F1 score and accuracy in binary classification evaluation. *BMC Genom.* **2020**, *21*, 6. [CrossRef]
51. Ribeiro, R.P.; Pereira, P.; Gama, J. Sequential anomalies: A study in the Railway Industry. *Mach. Learn.* **2016**, *105*, 127–153. [CrossRef]

sensors

MDPI

Article

Impact of Measurement Uncertainty on Fault Diagnosis Systems: A Case Study on Electrical Faults in Induction Motors

Simone Mari *, Giovanni Bucci, Fabrizio Ciancetta, Edoardo Fiorucci and Andrea Fioravanti

Dipartimento di Ingegneria Industriale e dell'Informazione e di Economia, Università dell'Aquila, 67100 L'Aquila, Italy; giovanni.bucci@univaq.it (G.B.); fabrizio.ciancetta@univaq.it (F.C.); edoardo.fiorucci@univaq.it (E.F.); andrea.fioravanti@univaq.it (A.F.)

* Correspondence: simone.mari@univaq.it

Abstract: Classification systems based on machine learning (ML) models, critical in predictive maintenance and fault diagnosis, are subject to an error rate that can pose significant risks, such as unnecessary downtime due to false alarms. Propagating the uncertainty of input data through the model can define confidence bands to determine whether an input is classifiable, preferring to indicate a result of unclassifiability rather than misclassification. This study presents an electrical fault diagnosis system on asynchronous motors using an artificial neural network (ANN) model trained with vibration measurements. It is shown how vibration analysis can be effectively employed to detect and locate motor malfunctions, helping reduce downtime, improve process control and lower maintenance costs. In addition, measurement uncertainty information is introduced to increase the reliability of the diagnosis system, ensuring more accurate and preventive decisions.

Keywords: anomaly detection; artificial neural networks (ANNs); fault diagnosis; induction motors; measurement uncertainty

1. Introduction

With the development of Industry 4.0, there has been a significant increase in the complexity and efficiency of systems in the industrial sector. Electric motors provide the main energy source for machinery and must be in excellent condition over time to maintain high productivity levels. The various components in the motor may be subject to electrical and mechanical failures, causing slowdowns in the company's production process and significant economic losses [1–13]. Various maintenance strategies are implemented, which differ in machine downtime and cost, to verify the correct functioning of the motor and the condition of the components. By using the proper maintenance strategy and intervening only when necessary, machine productivity time can be optimized, costs can be reduced for the company, and operational safety can also be improved. A widely used maintenance approach is based on "reactive maintenance", which involves intervention only after the actual presence of a fault. Another approach is "preventive" or scheduled maintenance, in which machinery undergoes periodic inspections to replace worn components. The development of new continuous, non-intrusive monitoring systems, coupled with signal processing systems extracted from transducers, has enabled the development of "predictive maintenance", which ensures effective maintenance by locating and preventing failures [14]. Thanks to predictive maintenance and the reduction in the cost of transducers, there has been a significant improvement in machine downtime process control and a considerable decrease in maintenance costs previously caused by incorrect or ineffective maintenance. Among the parameters that indicate the health status of motors are motor temperature [15], absorbed current [16–18], mechanical vibrations [19–21], emitted acoustic noise [22–24], and frequency response analysis [25–28]. Vibrations are a fundamental parameter for analyzing the operation of rotating machines, allowing the diagnosis of any problems such as bearing wear, electrical issues, misalignment, or imbalance.

In recent proposals for fault diagnosis systems for induction motors, ref. [29] introduces a method based on functional dimensionality reduction techniques, specifically FPCA (Functional Principal Component Analysis) and FDM (Functional Data Modeling). The analysis is performed on both stator current and instantaneous active power to detect only two anomalous conditions: broken bars and low-frequency load oscillations. In [30], fault diagnosis of induction motors was carried out using models such as SVM, MNN, CNN, GBM, and XGBoost. An induction motor simulator was created to operate under normal conditions as well as with rotor and bearing faults. In [31], two learning models, the Deep Convolutional Generative Adversarial Network (DCGAN) and Convolutional Neural Network (CNN), are proposed for fault diagnosis in induction motors using vibration signals, addressing the issue of unbalanced training data. In [32], the focus was on diagnosing rotor bar breakage faults using five signal processing techniques: classic FFT, Park's vector approach, Hilbert transform, MUSIC method, and discrete wavelet technique. Although these works achieve good performance, they all rely on simulated data, which may not always reflect the real operating conditions of the motor. None of these studies address the impact that measurement uncertainty of the input can have on the performance of the diagnosis models. In practice, simulated data may not capture all the variability and imperfections of real-world data, such as noise and measurement errors, which can significantly affect the accuracy of the results. Additionally, uncertainty in input measurements can lead to inaccurate diagnoses, thereby limiting the reliability of fault diagnosis techniques in actual operating scenarios. Therefore, it is crucial to consider how measurement uncertainty can impact performance and improve methodologies to handle these effects in practice. In this article, we aim to analyze how, through monitoring motor vibrations transmitted via an accelerometer and analyzing the data collected in the frequency domain, it is possible to identify vibrations that are functional in detecting defects or faults in the motor. Specifically, we will analyze induction motors with a cage rotor, which have been artificially damaged, to identify them based on the analysis of vibrations induced by the damage. The onset of mechanical and electrical faults or signs of wear in the motor can give rise to acoustic noise and vibrations with different amplitude and frequency than those in the standard motor operation. For example, an electrical fault, such as a short circuit, can cause an increase in vibrations in the rotor area, while a power supply problem will cause the same in the stator area. Furthermore, it is essential to note that the human eye often does not interpret these vibration measurements. Therefore, an artificial neural network (ANN) [33,34] model is proposed to analyze raw vibration data and identify potential fault conditions. ANNs have proven effective in learning complex patterns from large datasets, making them suitable for detecting subtle changes in vibration patterns indicative of motor faults. Moreover, the information regarding the uncertainty of vibration measurements can be propagated within the ANN to establish a confidence interval for the output. This approach allows for a more robust assessment of the motor's condition, accounting for variability and potential errors in the vibration measurements.

2. The Diagnostics and Maintenance of Induction Motors

Diagnostic procedures allow information derived from the measurement of parameters and the collection of data related to a machine or a generic system to be converted into information concerning the actual faults of the machine under examination. The diagnostic procedure includes analysis and synthesis activities that, using measurements of specific physical quantities characteristic of the machine under analysis, allow the drawing of significant information about its conditions for evaluations and predictions on its reliability in the short and long term. The main objectives of diagnostics are fault detection, localization, and identification. The general problem of diagnostics is to define whether a specific fault is present based on available information, preferably without carrying out intrusive inspections of the machine [35]. A diagnostic program is developed following these steps: data acquisition, processing, and diagnosis determination. The methods for data acquisition and processing and for choosing the threshold that differentiates the "fault" situation from the "healthy" situation for a component can heavily influence the probability of making an error in the decision-making phase [36].

Maintenance refers to the combination of all technical and administrative actions, including supervision actions, aimed at maintaining or restoring an entity to a state where it can perform the required function. Four maintenance strategies can be distinguished:

(1) Reactive (or corrective) maintenance: Reactive maintenance occurs in response to the notification of a malfunction or a more severe event, such as a breakdown, resulting in the machine being shut down. Reactive maintenance often involves very high costs due to the loss of production and the repair of the machine itself. For example, a bearing failure causing the shaft to seize up can damage the shaft and supports, turning a minor repair into a costly and time-consuming intervention. Reactive maintenance can only be effective if there is significant redundancy and the shutdown of a single machine does not cause the failure of the entire system.
(2) Preventive (or periodic) maintenance: This type involves identifying critical components and replacing parts or components according to a schedule, regardless of their condition. Due to time and cost constraints, a significant limitation of this approach is that it is not possible to replace all mechanical components at risk cyclically. Additionally, interventions are often made on components that are still far from the end of their operational life. This approach has increased maintenance costs over time and has not always prevented significant failures. Also included in preventive maintenance are regular checks that cannot be very frequent (visual inspection of the machine, partial and general overhauls, and diagnostic activities), which result in extended downtime during maintenance phases. Therefore, this type of maintenance is recommended when its cost is lower than the consequences of a failure. The critical aspect of this type of maintenance is that revisions after long intervals still lead to undesirable failures while conducting them at too short intervals results in replacing still-efficient parts [37].
(3) Predictive (or condition-based) maintenance: It is a type of preventive maintenance that is carried out following the identification of one or more parameters that are measured and extrapolated, using appropriate mathematical models that allow the residual time before the failure to be identified [38]. Identifying anomalies in advance would allow for better operation and maintenance management, avoiding significant economic and environmental repercussions. For this reason, it is necessary to associate traditional maintenance practices with new strategies based on predictive actions. The analysis of the functioning of the machinery makes it possible to anticipate the occurrence of a failure and effectively direct maintenance resources.
(4) Proactive (or improvement-based) maintenance: A maintenance policy involves a revision intervention aimed at improving the value or performance of a system or part of it. Maintenance action is not dependent on malfunctions but rather stems from improvement needs expressed by both the user and the maintainer. The term improvement maintenance is opposed to corrective maintenance; conceptually, it is the opposite. Corrective maintenance refers to the maintenance actions that do not contribute to increasing the system's value or improving its performance but simply restoring the original state. In improvement maintenance, the maintenance action instead contributes to expanding the system's value and/or improving its performance. For this reason, it is considered the most profitable maintenance.

Generally, the maintenance solution adopted is a compromise between predictive and preventive maintenance [39], also calibrated thanks to the measurements taken for predictive maintenance, as shown in Figure 1. Induction motor failures can be mechanical or electrical and are caused by various factors, including overload, short circuits, insulation problems, wear of mechanical parts, overheating, etc. An important part of troubleshooting is understanding the typical causes of each type of failure. For example, among electrical failures, short circuits can be caused by damaged or worn insulation, while a damaged rotor or stator can cause open circuits. A short circuit can cause an increase in vibrations in the rotor area, while a power problem can cause an increase in the stator area. A lack of ventilation or dust or debris buildup can cause overheating. Machine overheating can lead to degradation of winding insulation, resulting in a decrease in their resistance. Low resistance results in excessive current flow inside the motor, compromising efficiency.

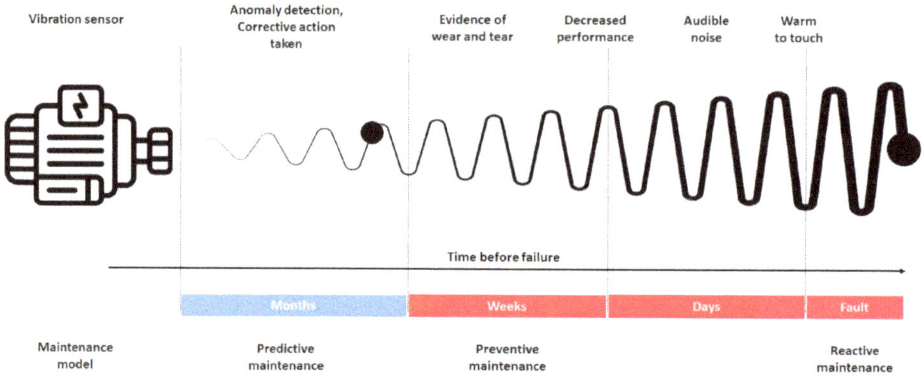

Figure 1. Maintenance models and intervention times.

Additionally, if the temperature of the windings exceeds the permissible temperature, the useful life of the insulation can be halved. Every electrical anomaly produces an asymmetry in the distribution of the magnetic flux at the air gap, which in turn creates a different distribution of the electromagnetic forces inside the electric machine, affecting the vibration behavior of the machine itself. Since the anomaly also causes a variation in the dispersed magnetic flux inside and outside the machine, vibration detection can also occur on the motor casing or the stator winding heads. On the other hand, bearing failure can be caused by a lack of lubrication, incorrect design, overheating, or misalignment among mechanical failures. It is also important to check the quality of the power supply and the protection devices connected to the motor. Overall, the key to solving electrical failures in asynchronous motors is to perform regular inspections and maintenance and use specialized sensors to diagnose and prevent problems quickly and effectively. Figure 2 shows the rates of common failures in induction motors [40].

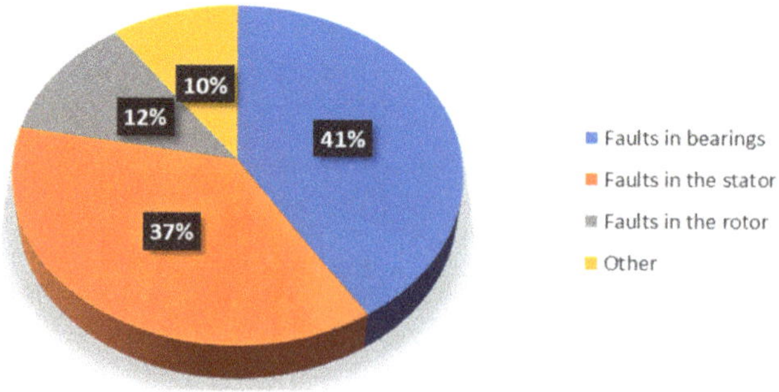

Figure 2. Common faults in induction motors.

3. Methodology

A rigorous methodological approach was followed to ensure detailed and accurate motor failure analysis through vibration measurements, including a selection of motors, configuration of failure conditions, and use of a high-performance accelerometer for vibration detection to deepen understanding of the phenomena involved. Next, a MEMS sensor was selected for the integrated system, and the data was used to train the ANN model.

3.1. Batch of Induction Motors and Analyzed Fault Conditions

Experimental tests were conducted on four new three-phase asynchronous motors, model BE 90 LA4 manufactured by Bonfiglioli [41]. These motors operate at 50 Hz and have the following characteristics:

(1) rated voltage of 230/400 V Δ/Y,
(2) current of 6.1/3.5 A Δ/Y,
(3) power of 1.5 kW,
(4) speed of 1430 rpm,
(5) power factor of 0.74,
(6) efficiency of 82.5% at full load, complying with IE2 standards according to IEC EN 60034 [42],
(7) insulation class F and IP 55 protection.

Three specific failures were reproduced on three of these motors, while one motor remained in its original condition to serve as a healthy reference:

(1) *Short Circuit Between Winding Turns*: This fault created a short circuit between two winding turns by scraping the insulation enamel and soldering the two copper conductors (Figure 3a). This fault can occur due to errors in the winding enameling or assembly process, potentially leading to critical failures. During the installation of the motor, the winding enameling process could be performed unevenly, or an error could occur during the winding process itself. It could lead to a short circuit between two winding turns, where the lack of insulation between conductors causes unwanted and potentially damaging current flow. This type of failure can seriously affect motor operation and require immediate maintenance.
(2) *Short-circuit Ring Damage*: This failure was induced by partially removing a section of the die-cast aluminum short-circuit ring of the rotor using a cutter (Figure 3b). In an industrial environment, a motor operating under high load conditions can suffer damage to the short-circuit ring due to vibration and mechanical stress. It can be caused by a defective material used for the ring or by a manufacturing accident that introduced a structural defect. Partial cutting of the ring can impair its strength and cause irregularities in motor vibration.
(3) *Rotor Bar Damage*: In this case, a 1 mm diameter hole was drilled in one of the rotor housings to damage an aluminum bar (Figure 3c). This type of damage is similar to those caused by casting defects such as voids or cracks. In an environment where motors are subject to intense vibration or high mechanical stress, such as in machinery operating in heavy industrial environments, the rotor bar could be damaged. It can be caused by accidental impacts, resonant vibrations, or manufacturing defects such as voids or micro-cracks in the rotor aluminum. The hole drilled in the damaged bar alters the balance and vibration of the rotor, adversely affecting the performance of the motor.

(a) (b) (c)

Figure 3. The different damage conditions analyzed: (**a**) short circuit between winding turns, (**b**) short-circuit ring damage, (**c**) rotor bar damage.

3.2. Experimental Setup and Phenomena Characterization

The test bench used for the experimental tests was designed to reproduce the actual operating conditions of the induction motors. The motors were powered through a Guasch MTL-B2B0060F12IXHF inverter (Guasch, Barcelona, Spain). This compact and ready-to-use inverter is designed for motor control and various inverter applications. The power stack includes IGBTs (2× CBI modules) with heat sink, opto-coupled drivers, output phase current sensors, DC-Link voltage sensors, and internal NTC temperature sensors. It is important to note that tests were conducted with various inverters at the nominal operating point of the motors, and no significant differences were observed in the harmonic content of the vibration measurements. As a result, the Guasch MTL-B2B0060F12IXHF inverter was selected for this study due to its flexibility in control, which provided precise regulation and greater versatility in managing the motor's operating conditions. The load was applied using a Magtrol HD 815 hysteresis dynamometer (Magtrol, Buffalo, NY, USA) equipped with a Magtrol TM 108 torque and speed transducer (Magtrol, Buffalo, NY, USA). In addition, the brake was controlled using a Magtrol DSP 6001 unit (Magtrol, Buffalo, NY, USA), which also acquired torque and rotational speed data.

Vibrations produced by the different motors under various operating conditions were measured using a triaxial ICP® accelerometer (PCB Piezotronics, Depew, NY, USA), with a sensitivity of 10 mV/g, a measurement range of ±500 g, and a frequency range of 2 to 7 kHz. The signals from the accelerometer were conditioned using a 4-channel, line-powered ICP® sensor signal conditioner with internally jumper-selectable gain settings of x1, x10, and x100. The Fluke 971 Temperature Humidity Meter measured (Fluke Corporation, Everett, WA, USA) ambient temperature and air humidity during each test. Motor temperature, on the other hand, was measured using the Fluke 62 Mini Thermometer (Fluke Corporation, Everett, WA, USA). The supply voltage and current drawn by the motor were measured using a Yokogawa WT 3000 Wattmeter (Yokogawa Electric Corporation, Tokyo, Japan). The test was repeated five times for each condition to correct any random errors in the measurement.

Despite the specifications of ISO 13373-9:2017 [43], which recommend placing accelerometers at strategic locations such as the bearings and motor shaft for accurate vibration measurement, it was chosen to measure vibration by placing the accelerometer on the motor casing using beeswax. The standard indicates that this method results in a reduction of the accelerometer bandwidth to about 30 kHz. However, this was found to be irrelevant since the harmonic content of the vibrations is exhausted around 5 kHz, as shown in Figure 4. Following this phenomena characterization, it was decided to implement a fault diagnosis system using the STEVAL-STWINKT1B board (STMicroelectronics, Geneva, Switzerland).

This solution incorporates a power-efficient ARM Cortex-M4 processor (ARM Ltd., Cambridge, UK) with a 120 MHz FPU and a 2048 kB flash memory (STM32L4R9), which is ideal for the processing needs required in fault diagnosis. The board is also equipped with the IIS3DWB MEMS sensor designed to measure vibration, with adequate sensitivity in the critical frequency range for mechanical fault diagnosis, ranging from DC to 6 kHz. The sensor also supports a user-selectable full-scale option, including ±2 g, ±4 g, ±8 g, and ±16 g.

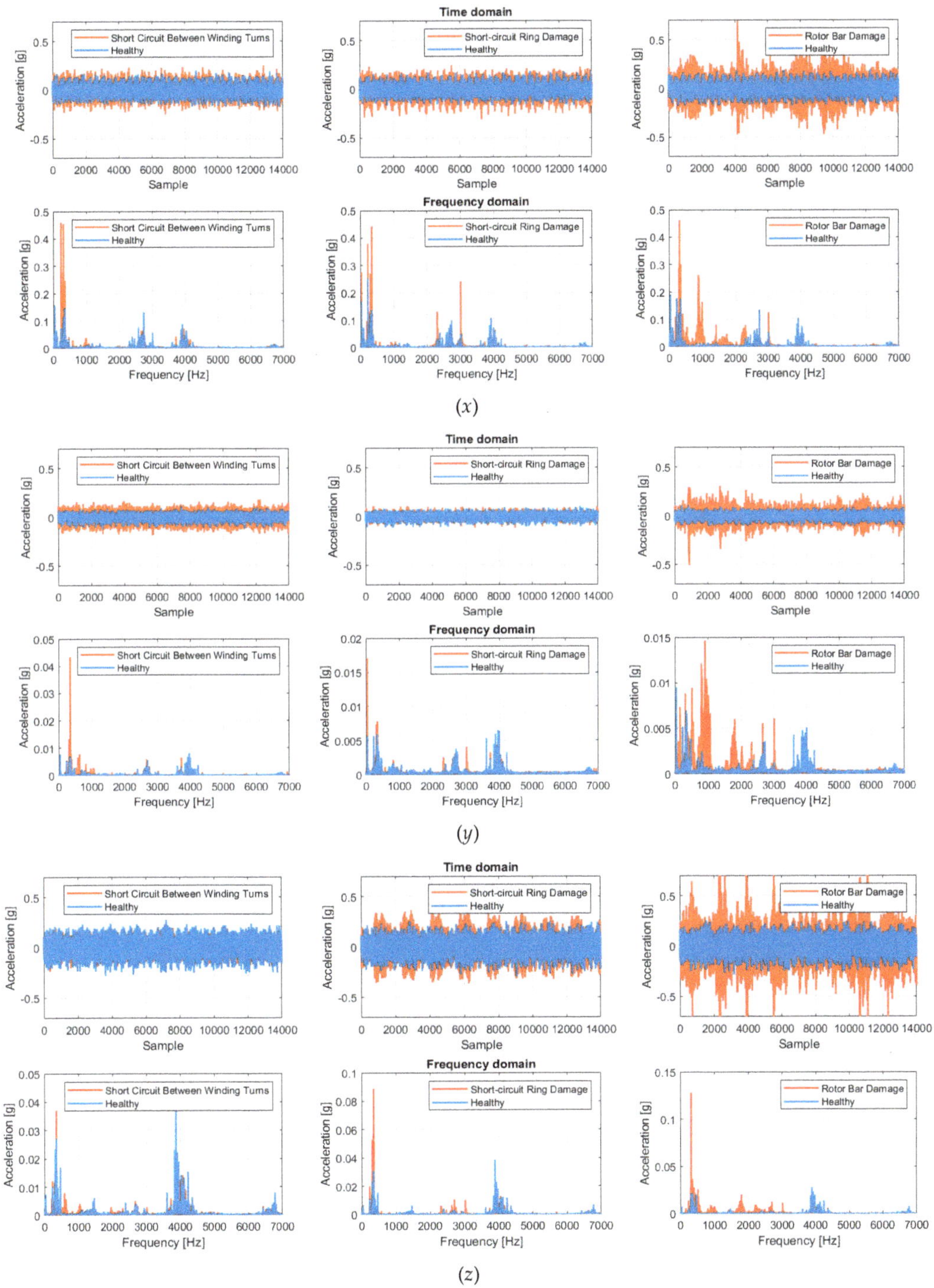

Figure 4. Vibration measurements acquired along the x, y, and z axes with their respective harmonic content.

3.3. *Artificial Neural Network Model*

Based on the phenomena characterization, the vibration signal was sampled at 10 kHz, indicating that relevant frequencies are within 5 kHz. According to the Nyquist theorem, the sampling frequency should be at least twice the maximum frequency of interest. Data were segmented into 1-s windows, equivalent to 10,000 samples, to facilitate processing. The algorithm was designed to classify these windows into multiple fault conditions or routine operations.

The ANN architecture employed, shown in Figure 5, consists of two hidden layers with 64 and 32 neurons, each using the Rectified Linear Unit (ReLU) activation Function (1).

$$ReLU(z) = \max(0, z) \tag{1}$$

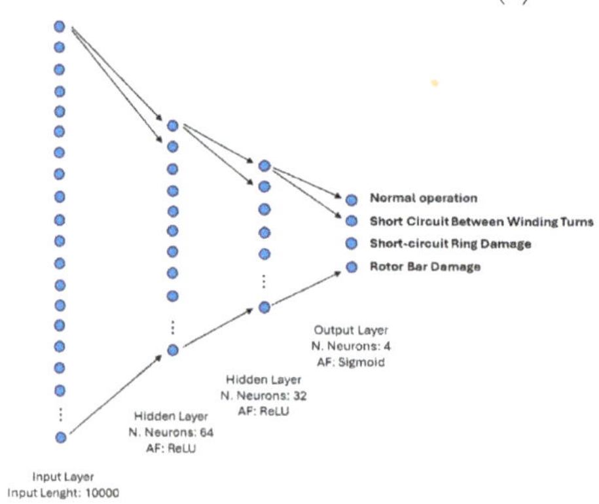

Figure 5. Implemented artificial neural network architecture.

In ANNs, a neuron computes its output z as in (2).

$$z = \mathbf{w} \cdot \mathbf{x} + b = \sum_{i=1}^{N} w_i x_i + b \tag{2}$$

where $\mathbf{x}$ is the input vector, $\mathbf{w}$ represents the weights, b is the bias term, and N is the number of input connections to the neuron. The ReLU function applies the non-linear activation by taking the maximum between zero and z, enabling the network to learn complex patterns from input data. The hidden layers extract and learn intricate patterns from vibration data, aiding in distinguishing fault signatures within the signals. Additionally, the network includes an output layer with four neurons and a sigmoid activation Function (3). The sigmoid function computes the probability $\sigma(z)$ that an input belongs to a particular class.

$$\sigma(z) = \frac{1}{1 + e^{-z}} \tag{3}$$

This architecture was implemented to leverage the ANN's capability to learn intricate relationships within data, thereby enhancing the accuracy and reliability of fault diagnosis in asynchronous motors based on vibration analysis.

The data collection phase using the STEVAL-STWINKT1B board required approximately 20 h, during which data were acquired for the four motors across 14 different speed conditions ranging from 0 up to the nominal value of 1430 rpm. Each operating condition was maintained for an acquisition period of approximately 20 min. This data collection effort resulted in a balanced dataset comprising 66,640 examples.

3.4. *Uncertainty Propagation in Artificial Neural Network*

Indirectly obtained measurements suffer from error propagation, affecting the reliability of the results [44]. This study explores how incorporating measurement uncertainty into the ANN can enhance the reliability of classification results by establishing confidence intervals. The propagation of uncertainty through the ANN is approached using the uncertainty propagation law outlined in the ISO Guide to the Expression of Uncertainty in Measurement (ISO GUM) [45–47]. The ISO GUM provides a framework for evaluating and expressing uncertainty in measurements, which is crucial for determining the confidence levels of the classification outcomes in the ANN model. Following it, the uncertainty associated with the input measurements is propagated through each layer of the ANN to quantify the overall uncertainty in the final outputs.

The first step involves quantifying the uncertainty in the input measurements. This study's uncertainty is based on the type B uncertainty provided by the sensor manufacturer, which is specified as ±2% of the full scale.

As the input vector **x** passes through the neural network, the uncertainty propagation follows the chain rule for derivatives to calculate the resulting output uncertainty. Considering the propagation of uncertainty, and given that the weights **w** are considered deterministic (i.e., they do not have associated uncertainty), the uncertainty in z, denoted as u_z, is influenced only by the uncertainties in the inputs **x**. Note (2), the combined uncertainty can be obtained using (4).

$$u_z = \sqrt{\sum_{i=1}^{n} \left(\frac{\partial z_i}{\partial x_i}\right)^2 u_x^2} = \sqrt{\sum_{i=1}^{n} w_i^2 u_x^2} \tag{4}$$

The ReLU function does not alter the uncertainty propagation significantly, as its derivative is a piecewise constant. However, it is essential to note that the ReLU activation will only propagate uncertainty if z is greater than zero. Thus, the uncertainty in the neuron's output will be zero when z is less than or equal to zero.

In the output layer, the sigmoid activation function $\sigma(z)$ is used to compute the probability that an input belongs to a particular class. Denoting by u_σ the uncertainty associated with the value from the output layer, where the sigmoid activation function is applied, this can be obtained with (5).

$$u_\sigma = \sqrt{\left(\frac{\partial \sigma(z)}{\partial z}\right)^2 u_z^2} = \sqrt{(\sigma(z)(1-\sigma(z)))^2 u_z^2} \tag{5}$$

By systematically applying these uncertainty propagation rules across each layer of the ANN, it is possible to associate uncertainty with the estimated probability of belonging to each class provided by the model.

4. Experimental Results

4.1. Training Setting and Metrics

The ANN was implemented using the TensorFlow and Keras libraries. Each data recording corresponding to different fault conditions and motor speeds was divided such that the first 70% of the data was used for training, and the remaining 30% was reserved for testing. The model was compiled with the Adam optimizer and binary cross-entropy loss function, with accuracy as the performance metric. The training was performed using an NVIDIA GeForce RTX 3060 GPU (Santa Clara, CA, USA). The training process included ten epochs and a batch size of 64; 20% of the training data was used for validation. The training data were mixed at the beginning of each epoch to ensure robust learning. For evaluating the model's performance, several key metrics were considered: precision, recall, F1-score, and accuracy. These metrics comprehensively understand how well the model performs across different classes.

Precision is the ratio of true positive (TP) predictions to the total number of positive predictions made by the model. It measures the accuracy of the positive predictions and is calculated as follows:

$$Precision = \frac{TP}{TP + FP} \quad (6)$$

where TP is the number of true positives and FP is the number of false positives. High precision indicates a low false positive rate.

Recall is the ratio of true positive predictions to the total number of actual positives. It measures the model's ability to identify all relevant instances and is calculated as follows:

$$Recall = \frac{TP}{TP + FN} \quad (7)$$

where FN is the number of false negatives. A high recall indicates a low false negative rate.

F1-Score is the harmonic mean of precision and recall, providing a single metric that balances both:

$$F1 - Score = 2 \times \frac{Precision \times Recall}{Precision + Recall} \quad (8)$$

Accuracy is the ratio of correctly predicted instances to the total number of instances and is calculated as:

$$Accuracy = \frac{TP + TN}{TP + TN + FP + FN} \quad (9)$$

where TN is the number of true negatives. It provides an overall measure of the model's performance across all classes.

Since the test dataset is balanced across different classes, it is appropriate to consider the arithmetic mean of the performance metrics (precision, recall, F1-score, and accuracy) obtained for each class as a general metric for the model [48]. This averaging approach ensures that the evaluation reflects the model's ability to classify each class accurately without being biased toward any particular class.

4.2. Achieved Results

Based on the distribution detailed in Section 4.1, the dataset of 66,640 examples was partitioned into 37,318 training examples, 9330 validation examples, and 19,992 test examples. The results obtained from the classification task along the x, y, and z-axes using the ANN are presented. The performance metrics and confusion matrices are detailed for each axis.

Figure 6 shows the error matrices obtained for training the ANN with measurements acquired along the x-axis (left) and y-axis (right).

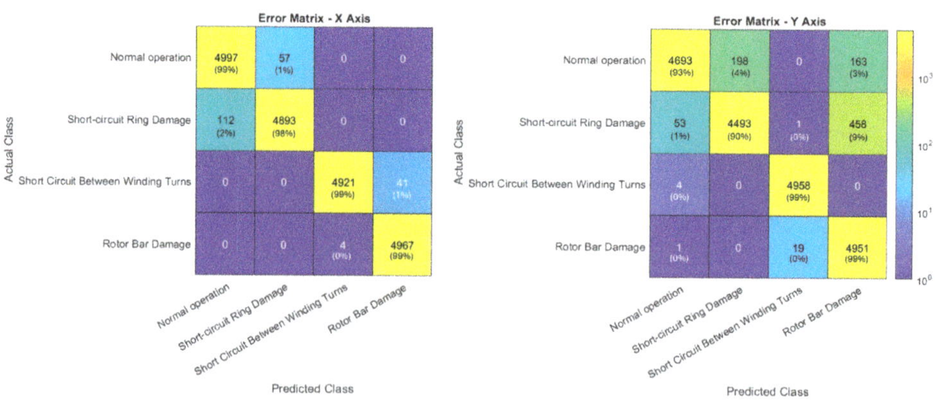

Figure 6. Error matrices obtained for training the ANN with measurements acquired, along the x-axis (**left**) and y-axis (**right**).

Following what was described in Section 4.1, Tables 1 and 2 show the metrics obtained by training and testing the ANN with measurements acquired along the x-axis and y-axis, respectively. Overall metrics were calculated using macro-average since the dataset has a balance of classes. The z-axis did not allow effective discrimination of faults, as evidenced by the low accuracy achieved (25%), and therefore, no further analysis was performed. Thus, no error matrices or performance metrics are presented for the z-axis.

Table 1. Performance obtained by training and testing the ANN with measurements acquired along the x-axis.

Classes	Precision	Recall	F1-Score	Accuracy
Normal Operation	0.98	0.99	0.98	0.99
Short-circuit Ring Damage	0.99	0.98	0.98	0.98
Short Circuit Between Winding Turns	1.00	0.99	1.00	0.99
Rotor Bar Damage	0.99	1.00	1.00	1.00
Overall	**0.99**	**0.99**	**0.99**	**0.99**

Table 2. Performance obtained by training and testing the ANN with measurements acquired along the y-axis.

Classes	Precision	Recall	F1-Score	Accuracy
Normal Operation	0.99	0.93	0.96	0.93
Short-circuit Ring Damage	0.96	0.90	0.93	0.90
Short Circuit Between Winding Turns	1.00	1.00	1.00	1.00
Rotor Bar Damage	0.89	1.00	0.94	1.00
Overall	**0.96**	**0.96**	**0.96**	**0.96**

The ANN model was designed as a feed-forward network with one, two, and three hidden layers, each with a variable number of neurons in the first hidden layer (16, 32, 64, and 128). In the cases of networks with two and three hidden layers, the number of neurons in the subsequent hidden layers is always half of the previous one. This decision was made to explore how network complexity affects the accuracy and reliability of fault classification. During the development phase, preliminary experiments were conducted with different configurations to balance model accuracy and computational complexity. The experiments were performed using vibration measurements along the x-axis as input data. As shown in Figure 7, while increasing the number of layers and neurons improved accuracy up to a certain point, the additional benefits became marginal beyond a certain threshold. For example, the model with two hidden layers and 64 neurons in the first hidden layer showed 0.99 accuracy, while adding more layers or neurons did not lead to significant improvements.

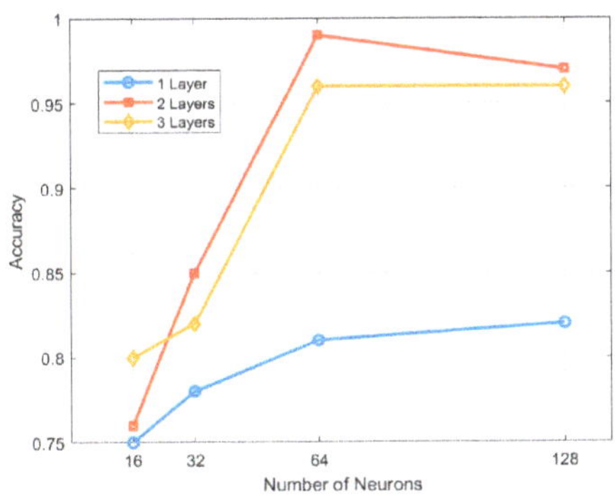

Figure 7. Model accuracy as a function of number of neurons for different configurations of hidden layers.

4.3. Considerations on Propagating Uncertainty through the Artificial Neural Network

Considerations of uncertainty propagation through the artificial neural network are crucial to improve the reliability of classifications. After propagating the uncertainty through the model, the confidence interval is determined by considering the uncertainty three times. This interval, which corresponds to about 99.7% of the area under the curve of a Gaussian distribution, represents a range within which the model output is expected to fall with high probability. In this analysis, the propagation of uncertainty through the ANN has been leveraged to evaluate the overlap of uncertainty bands and consequently improve the classification performance. This approach allows for identifying and excluding instances where uncertainty overlaps, thus potentially avoiding a significant number of incorrect classifications. In this sense, the classifications obtained on the test set were divided into four categories: unconfirmed correct classifications, incorrect classifications, confirmed correct classifications, and confirmed incorrect classifications. These values are shown in Table 3.

Table 3. Classification results based on uncertainty propagation.

Axis	Unconfirmed Correct Classifications	Unconfirmed Incorrect Classifications	Confirmed Correct Classifications	Confirmed Incorrect Classifications
x	175 (1% of correct outputs)	90 (42% of incorrect outputs)	19,603 (99% of correct outputs)	124 (58% of incorrect outputs)
y	838 (4% of correct outputs)	602 (67% of incorrect outputs)	18,257 (96% of correct outputs)	295 (33% of incorrect outputs)

Table 3 introduces four new categories of network outputs:

(1) **Unconfirmed correct classifications:** Occurs when the classification result is correct, but the highest probability of belonging to one of the two classes, considering the associated uncertainty band, overlaps with the uncertainty bands of the other probability values. In this case, the classification is considered invalid and the analysis must be repeated.

(2) **Unconfirmed incorrect classifications:** Occurs when the classification result is incorrect and the highest probability of belonging to one of the two classes, taking into account the uncertainty band, overlaps with the uncertainty bands of the other probability values. Again, the classification is considered invalid and the analysis must be repeated.

(3) **Confirmed correct classifications:** This occurs when the classification result is correct and the highest probability of belonging to one of the two classes, considering the uncertainty band, does not overlap with the uncertainty bands of the other probability values. In this scenario, the classification is confirmed.

(4) **Confirmed incorrect classifications:** Occurs when the classification result is incorrect and the highest probability of belonging to one of the two classes, taking into account the uncertainty band, does not overlap with the uncertainty bands of the other probability values. In this case, the classification is confirmed.

The results demonstrate that propagating uncertainty through the ANN and evaluating the overlap of uncertainty bands can significantly improve classification performance by avoiding many incorrect classifications. Specifically:

- On the x-axis, avoiding classifications with overlapping uncertainty bands could potentially reduce incorrect classifications by 42%.
- On the y-axis, this approach could potentially reduce incorrect classifications by 67%.

If the margin provided by the information on the propagation of measurement uncertainty is not taken into account, the errors of the models trained with the vibration measurements along the x- and y-axis amount to 214 and 897, respectively. However, by including the uncertainty information, the actual errors are reduced to 124 and 295, respectively, corresponding to the class of Confirmed incorrect classifications. This represents the number of errors that, despite the inclusion of uncertainty, could not be avoided. In cases where the uncertainty does not allow clear classification, the analysis must be repeated. As shown by the presence of cases in the category of Unconfirmed correct classifications, uncertainty considerations also lead to the exclusion of test examples that would otherwise have been classified correctly. However, these cases represent only a small percentage of the total correct classifications, as shown in Table 3. Therefore, the approach taken represents a good compromise between accuracy and reliability.

The importance of this analysis is highlighted in Figure 8, where two separate cases are compared. In the first case on the left, the uncertainty bands overlap, resulting in an otherwise incorrect classification that is not possible. In the second case on the right, the uncertainty bands do not overlap, thus confirming the correctness of the classification. The figure shows that the output uncertainty varies depending on the input sample, influenced by which neurons are activated during the classification process.

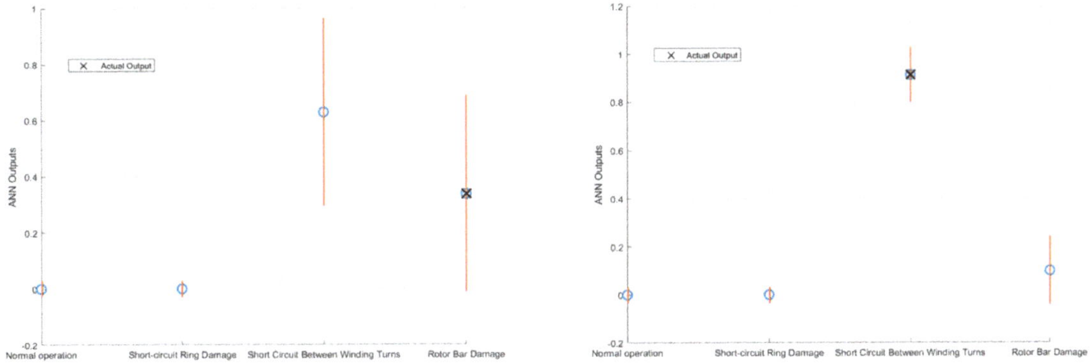

Figure 8. Impact of uncertainty on classification accuracy. On the left, the uncertainty bands overlap, resulting in an otherwise incorrect classification that is not possible. On the right, the uncertainty bands do not overlap, thus confirming the correctness of the classification.

By exploiting this method, we improve the robustness and reliability of the classification system, ensuring that uncertain predictions are handled carefully to minimize errors.

5. Conclusions

The study presented an electrical fault diagnosis system for induction motors based on an ANN model trained with vibration measurements. This approach uses vibration analysis to detect and localize motor malfunctions, helping to reduce downtime, improve process control, and lower maintenance costs. In addition, the introduction of measurement uncertainty information increases the reliability of the diagnosis system, ensuring more accurate and preventive decisions.

ANN demonstrated a high ability to detect and classify faults in induction motors based on vibration data. Specifically, the results obtained from measurements along the x-axis showed an overall F1-Score and accuracy of 99%, while for the y-axis, overall F1-Score and accuracy were 96%. However, the z-axis did not allow effective discrimination of faults, highlighting the importance of the choice of sensor orientation for data collection. Uncertainty propagation through the neural network proved crucial for improving classification reliability. Applying uncertainty propagation rules according to ISO GUM guidelines made it possible to establish confidence bands for model outputs. This approach made it possible to identify and exclude instances with overlapping uncertainties, thus reducing the risk of misclassifications by 42% and 67% for the model trained on x- and y-axis data, respectively.

The proposed diagnostic system has important implications for the predictive maintenance of electric motors, offering a non-intrusive and continuous method for monitoring motor conditions. Integrating measurement uncertainty information enables safer and more preventive decisions, reducing misclassification risks. Indeed, recent proposals for induction motor fault diagnosis systems have demonstrated high performance [29–32]. However, studies in the literature are primarily based on simulated data and do not address how the uncertainty of input measurements affects diagnostic accuracy. Simulated data often fail to capture real-world variabilities and imperfections, such as noise and measurement errors, which can significantly affect diagnostic performance. This paper is notable for being the first to consider the impact of measurement uncertainty on fault diagnosis. This approach can be extended to other types of industrial machinery, improving the overall effectiveness of predictive maintenance. The importance of this work goes far beyond the specific field of induction motors. In all areas where ML-based classification systems are applied, such as industrial equipment status monitoring or medical diagnosis, adopting methods that can handle input data uncertainty can significantly improve the reliability of operational decisions. Reducing false alarms avoids the waste of time and money associated with unnecessary or inappropriate interventions.

However, it is important to recognize that model performance could be affected by variations in the faults themselves, such as different sizes or geometries than those used for training. For example, a short-circuit ring defect with dimensions different from those in the training data could generate different vibration frequencies, impairing the model's ability to recognize and classify it correctly. Similarly, damage to the rotor bar with different characteristics could affect the accuracy of the model. This study mainly dealt with the propagation of uncertainty, but the problem of generalization with respect to defect variations represents an additional type of uncertainty that could affect the accuracy of diagnosis. It is essential to consider this challenge in future studies, exploring how such variables can be evaluated and incorporated in the context of predictive diagnosis.

Author Contributions: Conceptualization, S.M. and A.F.; methodology, S.M.; software, S.M.; validation, S.M., F.C. and E.F.; formal analysis, S.M.; investigation, S.M.; resources, S.M.; data curation, S.M.; writing—original draft preparation, S.M.; writing—review and editing, E.F. and A.F.; visualization, E.F.; supervision, G.B.; project administration, G.B. All authors have read and agreed to the published version of the manuscript.

Funding: This research received no external funding.

Institutional Review Board Statement: Not applicable.

Informed Consent Statement: Not applicable.

Data Availability Statement: Data are contained within the article.

Conflicts of Interest: The authors declare no conflicts of interest.

References

1. Henao, H.; Capolino, G.-A.; Fernandez-Cabanas, M.; Filippetti, F.; Bruzzese, C.; Strangas, E.; Pusca, R.; Estima, J.; Riera-Guasp, M.; Hedayati-Kia, S. Trends in fault diagnosis for electrical machines: A review of diagnostic techniques. *IEEE Ind. Electron. Mag.* **2014**, *8*, 31–42. [CrossRef]
2. Thomson, W.T. Online motor current signature analysis prevents premature failure of large induction motor drives operating in the North Sea oil and gas industry. In Proceedings of the 9th European Fluid Machinery Congress Applying Latest Technology to New and Existing Process Equipment, Institution of Mechanical Engineers, Fluid Machinery Group, The Hague, The Netherlands, 23–26 April 2006; pp. 263–272.
3. Thomson, W.T.; Fenger, M. Industrial application of current signature analysis to diagnose faults in 3-phase squirrel cage induction motors. In Proceedings of the IEEE Conference Record of Annual Pulp and Paper Industry Technical Conference, Atlanta, GA, USA, 19–23 June 2000; pp. 205–211.
4. Samsi, R.; Ray, A.; Mayer, J. Early detection of stator voltage imbalance in three-phase induction motors. *Electr. Power Syst. Res.* **2009**, *79*, 239–245. [CrossRef]
5. Wadhwani, S.; Gupta, S.P.; Kumar, V. Fault classification for rolling element bearing in electric machines. *IETE J. Res.* **2008**, *54*, 264–275. [CrossRef]
6. Jin, X.; Chow, T.W.S. Anomaly detection of cooling fan and fault classification of induction motor using Mahalanobis-Taguchi system. *Expert Syst. Appl.* **2013**, *40*, 5787–5795. [CrossRef]

7. Pietrowski, W. Detection of time-varying inter-turn short-circuit in a squirrel cage induction machine by means of generalized regression neural network. *COMPEL Int. J. Comput. Math. Electr. Electron. Eng.* **2017**, *36*, 289–297. [CrossRef]
8. Arkan, M.D.; Kostic-Perovic, P.J. Unsworth Modelling and simulation of induction motors with inter-turn faults for diagnostics. *Elect. Power Syst. Res.* **2005**, *75*, 57–66. [CrossRef]
9. Ciancetta, F.; Fiorucci, E.; Ometto, A.; Fioravanti, A.; Mari, S.; Segreto, M.-A. A Low-Cost IoT Sensors Network for Monitoring Three-Phase Induction Motor Mechanical Power Adopting an Indirect Measuring Method. *Sensors* **2021**, *21*, 754. [CrossRef] [PubMed]
10. Siyambalapitiya, D.; McLaren, P. Reliability improvement and economic benefits of online monitoring system for large induction machines. *IEEE Trans. Ind. Applicat.* **1990**, *26*, 1018–1025. [CrossRef]
11. Trutt, F.C.; Sottile, J.; Kohler, J.L. Online condition monitoring of induction motors. *IEEE Trans. Ind. Appl.* **2002**, *38*, 1627–1632. [CrossRef]
12. Morales-Perez, C.; Rangel-Magdaleno, J.; Peregrina-Barreto, H.; Amezquita-Sanchez, J.; Valtierra-Rodriguez, M. Incipient broken rotor bar detection in induction motors using vibration signals and the orthogonal matching pursuit algorithm. *IEEE Trans. Instrum. Meas.* **2018**, *67*, 2058–2068. [CrossRef]
13. Thomson, W.T. A review of online condition monitoring techniques for three-phase squirrel-cage induction motors-past present and future. In Proceedings of the IEEE SDEMPED, Seattle, WA, USA, 9–12 September 1999; pp. 1–3.
14. Çınar, Z.M.; Abdussalam Nuhu, A.; Zeeshan, Q.; Korhan, O.; Asmael, M.; Safaei, B. Machine Learning in Predictive Maintenance towards Sustainable Smart Manufacturing in Industry 4.0. *Sustainability* **2020**, *12*, 8211. [CrossRef]
15. Paramasivam, A.; Abimanyu, J.P.; Pittu, P.S.K.R.; Vijayalakshmi, S.; Kaushal Mayur, M.K. The Design and Development of an Internet of Things-Based Condition Monitoring System for Industrial Rotating Machines. *Eng. Proc.* **2023**, *58*, 71. [CrossRef]
16. Mari, S.; Credo, A.; Bucci, G.; Ciancetta, F.; Fiorucci, E.; Fioravanti, A. Current Measurements for Fault Diagnosis in Induction Motors. In Proceedings of the 26th IMEKO TC4 International Symposium and 24th International Workshop on ADC and DAC Modeling and Testing (IWADC), Pordenone, Italy, 20–21 September 2023.
17. Hoang, D.T.; Kang, H.J. A Motor Current Signal-Based Bearing Fault Diagnosis Using Deep Learning and Information Fusion. *IEEE Trans. Instrum. Meas.* **2020**, *69*, 3325–3333. [CrossRef]
18. Thomson, W.T.; Fenger, M. Current signature analysis to detect induction motor faults. *IEEE Ind. Appl. Mag.* **2001**, *7*, 26–34. [CrossRef]
19. Skowron, M.; Frankiewicz, O.; Jarosz, J.J.; Wolkiewicz, M.; Dybkowski, M.; Weisse, S.; Valire, J.; Wyłomańska, A.; Zimroz, R.; Szabat, K. Detection and Classification of Rolling Bearing Defects Using Direct Signal Processing with Deep Convolutional Neural Network. *Electronics* **2024**, *13*, 1722. [CrossRef]
20. Mari, S.; Bucci, G.; Ciancetta, F.; Fiorucci, E.; Fioravanti, A. Machine Learning for Anomaly Detection in Induction Motors. In Proceedings of the 2023 IEEE International Conference on Metrology for eXtended Reality, Artificial Intelligence and Neural Engineering (MetroXRAINE), Milano, Italy, 25–27 October 2023; pp. 1057–1062.
21. Apeiranthitis, S.; Zacharia, P.; Chatzopoulos, A.; Papoutsidakis, M. Predictive Maintenance of Machinery with Rotating Parts Using Convolutional Neural Networks. *Electronics* **2024**, *13*, 460. [CrossRef]
22. Lin, S.-L. Application of Machine Learning to a Medium Gaussian Support Vector Machine in the Diagnosis of Motor Bearing Faults. *Electronics* **2021**, *10*, 2266. [CrossRef]
23. Burriel-Valencia, J.; Puche-Panadero, R.; Martinez-Roman, J.; Sapena-Bano, A.; Pineda-Sanchez, M.; Perez-Cruz, J.; Riera-Guasp, M. Automatic Fault Diagnostic System for Induction Motors under Transient Regime Optimized with Expert Systems. *Electronics* **2019**, *8*, 6. [CrossRef]
24. Wang, L.; Bao, X.; Di, C.; Zhou, Y. Influence on Vibration and Noise of Squirrel-Cage Induction Machine with Double Skewed Rotor for Different Slot Combinations. *IEEE Trans. Magn.* **2016**, *52*, 8104404. [CrossRef]
25. Bucci, G.; Ciancetta, F.; Fioravanti, A.; Fiorucci, E.; Mari, S.; Silvestri, A. Online SFRA for Reliability of Power Systems: Characterization of a Batch of Healthy and Damaged Induction Motors for Predictive Maintenance. *Sensors* **2023**, *23*, 2583. [CrossRef]
26. Aranda Carmona, A.J. Diagnostic Method of Electrical Rotors by Applying the Sweep Frequency Response Analyzer (SFRA). In Proceedings of the 44th International Conference on Large High Voltage Electrical Systems, Paris, France, 1–9 August 2012.
27. Bucci, G.; Ciancetta, F.; Fioravanti, A.; Fiorucci, E.; Mari, S.; Silvestri, A. Online SFRA Characterization of a Batch of Induction Motors for Predictive Maintenance. In Proceedings of the 25th IMEKO TC4 International Symposium, Brescia, Italy, 12–14 September 2022.
28. Brandt, M.; Gutten, M.; Kaščák, S. Diagnostic of Induction Motor Using SFRA Method. In Proceedings of the Conference on Diagnostics in Electrical Engineering, Pilsen, Czech Republic, 1–4 September 2016.
29. Barroso, M.; Carazo, A.; Gómez, C.; Fernández, M.A.; González, J.A.; Silva, J.; García, J.L. Fault Detection in Induction Motors Using Functional Dimensionality Reduction Methods. *arXiv* **2023**, arXiv:2306.09365.
30. Kim, M.-C.; Lee, J.-H.; Wang, D.-H.; Lee, I.-S. Induction Motor Fault Diagnosis Using Support Vector Machine, Neural Networks, and Boosting Methods. *Sensors* **2023**, *23*, 2585. [CrossRef] [PubMed]
31. Chang, H.-C.; Wang, Y.-C.; Shih, Y.-Y.; Kuo, C.-C. Fault Diagnosis of Induction Motors with Imbalanced Data Using Deep Convolutional Generative Adversarial Network. *Appl. Sci.* **2022**, *12*, 4080. [CrossRef]

32. Abdelhak, G.; Sid Ahmed, B.; Djekidel, R. Fault Diagnosis of Induction Motors Rotor Using Current Signature with Different Signal Processing Techniques. *Diagnostyka* **2022**, *23*, 2022201. [CrossRef]
33. Zhang, Y.; Huang, Y.; Li, W.; Lin, M. Application of artificial neural network and DS algorithm to calibration transfer of rice protein powder. In Proceedings of the 6th International Conference on Instrumentation & Measurement, Computer, Communication and Control (IMCCC), Harbin, China, 21–23 July 2016; pp. 822–825.
34. Jung, D.; Nguyen, M.D.; Park, M.; Kim, J.; Mun, K. Multiple classification of gait using time-frequency representations and deep convolutional neural networks. *IEEE Trans. Neural Syst. Rehabil. Eng.* **2020**, *28*, 997–1005. [CrossRef] [PubMed]
35. Jin, X.; Weiss, B.A.; Siegel, D.; Lee, J. Present status and future growth of advanced maintenance technology and strategy in US manufacturing. *Int. J. Prognost. Health Manag.* **2016**, *7*, 12. [CrossRef]
36. Zonta, T.; da Costa, C.A.; da Rosa Righi, R.; de Lima, M.J.; da Trindade, E.S.; Li, G.P. Predictive maintenance in the Industry 4.0: A systematic literature review. *Comput. Ind. Eng.* **2020**, *150*, 106889. [CrossRef]
37. Zhou, N.; Xu, Y.; Cho, S.; Wee, C.T. A Systematic Review for Switchgear Asset Management in Power Grids: Condition Monitoring, Health Assessment, and Maintenance Strategy. *IEEE Trans. Power Del.* **2023**, *38*, 3296–3311. [CrossRef]
38. Bousdekis, A.; Apostolou, D.; Mentzas, G. Predictive Maintenance in the 4th Industrial Revolution: Benefits, Business Opportunities, and Managerial Implications. *IEEE Eng. Manag. Rev.* **2020**, *48*, 57–62. [CrossRef]
39. Rahman, H.U.; Raza, M.; Afsar, P.; Khan, H.U.; Nazir, S. Analyzing Factors That Influence Offshore Outsourcing Decision of Application Maintenance. *IEEE Access* **2020**, *8*, 183913–183926. [CrossRef]
40. Nandi, S.; Toliyat, H.A.; Li, X. Condition Monitoring and Fault Diagnosis of Electrical Motors—A Review. *IEEE Trans. Energy Convers.* **2005**, *20*, 719–729. [CrossRef]
41. Bonfiglioli. BE 90LA 4 230/400-50 IP55 CLF B5. Datasheet. 8U09030001. Available online: https://shop.bonfiglioli.com/store/italy/it/datasheet/download/product-datasheet/8U09030001 (accessed on 13 February 2023).
42. *IEC 60034-1*; Rotating Electrical Machines—Part 1: Rating and Performance. International Electrotechnical Commission: London, UK, 2017.
43. *ISO 13373-9:2017*; Condition Monitoring and Diagnostics of Machines—Vibration Condition Monitoring—Part 9: Diagnostic Techniques for Rotor Unbalance, Misalignment, and Faults in Journal Bearings and Gears from Vibration Measurements. ISO: Geneva, Switzerland, 2017.
44. Mari, S.; Credo, A.; Bucci, G.; Ciancetta, F.; Fiorucci, E.; Fioravanti, A.; Petrov, I.; Pyrhonen, J. Uncertainty Evaluation in Inductance Measurement of Synchronous Reluctance Motors. In Proceedings of the Conference Record—IEEE Instrumentation and Measurement Technology Conference (I2MTC 2024), Glasgow, UK, 20–23 May 2024.
45. *JCGM 100:2008*; Evaluation of Measurement Data—Guide to the Expression of Uncertainty in Measurement GUM 1995 with Minor Corrections. Joint Committee for Guides in Metrology (JCGM): Paris, France, 2008.
46. *JCGM 101:2008*; Evaluation of Measurement Data—Supplement 1 to the 'Guide to the Expression of Uncertainty in Measurement'—Propagation of Distributions Using a Monte Carlo Method. Joint Committee for Guides in Metrology (JCGM): Paris, France, 2008.
47. *JCGM 102:2008*; Evaluation of Measurement Data—Supplement 2 to the 'Guide to the Expression of Uncertainty in Measurement'—Extension to any Number of Output Quantities. Joint Committee for Guides in Metrology (JCGM): Paris, France, 2011.
48. Koyejo, O.; Natarajan, N.; Ravikumar, P.K.; Dhillon, I.S. Consistent multilabel classification. *Proc. NIPS* **2015**, *28*, 3321–3329.

sensors MDPI

Article

A Calculation Method of Bearing Balls Rotational Vectors Based on Binocular Vision Three-Dimensional Coordinates Measurement

Wenbo Lu [1], Junpeng Xue [1,2,*], Wei Pu [1,2], Hongyang Chen [3], Kelei Wang [1] and Ran Jia [1]

1 School of Aeronautics and Astronautics, Sichuan University, Chengdu 610065, China; 2022226210008@stu.scu.edu.cn (W.L.); pwei@scu.edu.cn (W.P.); wangkelei_@stu.scu.edu.cn (K.W.); 2023226210008@stu.scu.edu.cn (R.J.)
2 Robotic Satellite Key Laboratory of Sichuan Province, Chengdu 610065, China
3 Xichang Satellite Launch Center, Xichang 615000, China; chy89728@163.com
* Correspondence: jpxue@scu.edu.cn

Abstract: The rotational speed vectors of the bearing balls affect their service life and running performance. Observing the actual rotational speed of the ball is a prerequisite for revealing its true motion law and conducting sliding behavior simulation analysis. To address the need for accuracy and real-time measurement of spin angular velocity, which is also under high-frequency and high-speed ball motion conditions, a new measurement method of ball rotation vectors based on a binocular vision system is proposed. Firstly, marker points are laid on the balls, and their three-dimensional (3D) coordinates in the camera coordinate system are calculated in real time using the triangulation principle. Secondly, based on the 3D coordinates before and after the movement of the marker point and the trajectory of the ball, the mathematical model of the spin motion of the ball was established. Finally, based on the ball spin motion model, the three-dimensional vision measurement technology was first applied to the measurement of the bearing ball rotation vector through formula derivation, achieving the analysis of bearing ball rolling and sliding characteristics. Experimental results demonstrate that the visual measurement system with the frame rate of 100 FPS (frames per second) yields a measurement error within $\pm 0.2\%$ over a speed range from 5 to 50 RPM (revolutions per minute), and the maximum measurement errors of spin angular velocity and linear velocity are 0.25 °/s and 0.028 mm/s, respectively. The experimental results show that this method has good accuracy and stability in measuring the rotation vector of the ball, providing a reference for bearing balls' rotational speed monitoring and the analysis of the sliding behavior of bearing balls.

Keywords: vision measurement; motion model construction; rotational vectors measurement

1. Introduction

The bearing structure is the core support component of many rotating machines and is widely used in large, heavy-duty equipment such as wind turbines and aircraft engines. Bearing balls are susceptible to heat and slip during high-speed rotational motion, leading to early failure of the bearing well before it reaches its design life [1,2]. In addition, turbines and engines are becoming increasingly integrated and complex in industrial production. Accordingly, mechanical components such as bearings, gears, and couplings are more prone to failures attributed to frequent shocks and vibrations [3]. The rotational speed and direction of the balls have a significant impact on the service life and normal operation of the bearings. Hence, observing the actual rotational speed of the balls can provide accurate and real physical data for the simulation and analysis of the motion law and sliding behavior. Currently, there is a lack of measurement methods for the real data of rotational speed in the field of rotating machinery. Many measurement methods are only based on the observation of the rotational speed of rotating parts, and there is a lack of deep study on the motion parameters related to the revolution and spin of the ball. It is difficult to measure the spin angular velocity of the ball at high speeds. Therefore, it is urgent to carry out research on the measurement of spin angular velocity during the high-speed motion of balls.

At present, many scholars have studied the motion laws and modeling of the rolling ball. Oktaviana investigated the effect of ball bearing pressure at different speed ranges and the effect of ball rotational speed on bearing slip behavior according to the Hirano criterion [4,5]. Gupta developed a complete set of kinetic models to analyze the motion and sliding of the balls considered in the bearing [6]. Chen gave a quasi-static model of a ball bearing, establishing creep-slip and spin ratios to accurately define the relative motion between the balls and the inner and outer raceways [7]. Ostensen modeled the dynamics of ball motion [8]. Imado proposed a hall element method for detecting ball motion in ball bearings [9]. Selvaraj developed a test rig for cylindrical roller bearings to measure the rotational speed of bearing elements under various operating conditions [10]. Wang investigated the dynamic characteristics of ball bearings and gave a method for determining the angular velocity and pitch angle of the balls [11]. In these studies, the motion model of the ball was considered and the motion law of the ball was analyzed; however, the rotation vectors, such as the spin angular velocity of the ball, were not actually measured.

The binocular vision measurement technology has been widely used in the fields of object three-dimensional coordinate measurement and morphology estimation due to its non-contact, low impact from the external environment, and strong anti-interference ability. Ye combined the high-speed rotation image acquisition of low-speed cameras with the association of 3D digital images to search for uncoded targets between two rotating images. The 3D motion parameters obtained through 3D reconstruction and singular value decomposition enable full-field 3D motion and deformation measurement of high-speed rotating objects [12]. Bao introduced rotor speed as a Kalman filter parameter and combined it with prior information on rotating machinery to construct an extended Kalman filter model, achieving dynamic axial clearance measurement of high-speed rotating machinery [13]. Wang decomposed the deformed image and reference image into multi-resolution layers and converted the two-dimensional template into a one-dimensional grayscale value sequence. The correlation coefficient between the two comparison images was calculated in the low-resolution layer, achieving the estimation of the initial values of the translation and rotation speed of the rotating object [14]. Jia proposed a method for calibrating projector parameters in a DCP system with high accuracy, which uses the binocular camera to uniquely determine the coordinates of 3D feature points [15]. Yu proposed an adaptive binocular fringe dynamic projection method that can avoid image saturation by adaptively adjusting the projection intensity and achieving higher accuracy for high dynamic range three-dimensional vision measurement [16]. Xu proposed a contactless measurement method based on vision technology and realized the measurement of the rotation angle of power transformer winding coils [17]. In these studies, the three-dimensional coordinates of the feature points and the rotation angle of the components are measured using visual and image methods; however, the rotation vector of the components is not further measured.

In recent years, binocular vision measurement technology has been applied to the field of measuring the rotational speed of rotating components. Furuno proposed a rotation rate estimation algorithm based on a rotation matrix, which measured the rotation rate and translation speed of the ball [18]. Liao proposed a method of using cameras and marker tracking to measure the instantaneous speed of wind turbine blades, which achieved better measurement accuracy and speed compared to image correlation methods [19]. Wang proposed a rotation speed measurement system based on a low-cost imaging device, which uses structural similarity and two-dimensional correlation algorithms to process the image and complete the measurement of rotor rotation speed [20]. Wang proposed an imaging-based instantaneous speed measurement system and a periodic determination method based on the Chirp-Z transform and parabolic interpolation autocorrelation algorithm, which can effectively measure the rotor revolution speed [21–23]. Wang proposed a high-speed stereo vision system that can detect balls and identify marker points from highly underexposed low-resolution images under normal indoor lighting conditions without additional strobes, enabling the measurement of the rotational speed of a golf ball during high-speed motion [24]. Jung proposed a camera-based multi-exposure image acquisition sensor system and ball-tracking algorithm to capture multi-exposure images from two cameras simultaneously using a synchronization controller. In addition, the 2D center point of the ball is extracted from the separated and overlapped images, and the 3D (three-dimensional) coordinate reconstruction of the extracted feature points is performed using the direct linear transformation method to measure the speed of the golf ball [25]. Seo proposed a 3D ball motion estimation system based on an infrared scanning sensor, which can estimate the motion speed and motion angle of a golf ball [26]. Zhang proposed a dual-camera high-speed stereo vision table tennis tracking system in which a computer receives the image coordinates of the ball from the camera and calculates its 3D position in the working frame. The flight trajectory of the ball is estimated and predicted based on the measured position of the ball and the ball motion model. The direction of the velocity of the ball during the motion can be estimated [27]. Xie extracted the features from the left and right images obtained using the binocular camera and a three-dimensional measurement method based on an improved feature-matching algorithm is proposed to improve the measurement accuracy for objects with sparse or weak textures [28]. These studies used visual measurement methods to measure the instantaneous speed and rotational speed of spheres such as ping-pong balls and tennis balls. However, during the movement of these spheres, the surface of these spheres is not blocked by other devices, making it impossible to reproduce the actual application scenarios of bearing structures. In the practical application of bearing structure, the ball inside the bearing will be blocked by various devices, which makes it more difficult to obtain the actual motion image of the ball. Therefore, the rotational vectors of the bearing ball cannot be measured directly by the existing research methods.

Zhang utilized a binocular stereo vision system to track the position of the ball and a frame difference technique as well as a third panning camera to measure the spin angular velocity of the ball, which allowed them to identify the brand of moving table tennis under normal lighting conditions. Then, the 3D attitude of the ball is fitted in the ball-centered coordinate system, and the rotation state of the table tennis was estimated by using a planar fitting method based on the consensus of weighted random samples. The trajectory prediction of the moving table tennis is realized by extended Kalman filtering [29]. Li proposed a high-precision rotation angle measurement method using machine vision to measure the rotation angle, where the coordinates of the points on the calibration map are obtained from the visual image of the calibration map acquired by a CCD camera. The mathematical model of the aberration error of the CCD camera was established while the measurement of the rotation angle of the spot array was realized by using the coordinate rotation measurement equation [30]. Zhong proposed an infrared spectroscopic measurement system based on non-projective stripe vision, in which the designed LVD-FP was completely pasted on the surface of the rotating shaft along the direction of stripe density variation. The streak period density of each imaged LVD-FP intensity changes due to the rotational axis, which realizes the measurement of the rotation angle, the determination of the direction of the rotational axis, and the measurement of the instantaneous angular velocity between two adjacent frames [31]. Kim pasted RGB color patterns on the rotational axis to be tested, and a high-speed camera was used to image the color marker points. By detecting the speed of the RGB color change, the measurement of the rotational speed of the rotary axis is realized [32]. Guo proposed a Lucus–Kanade algorithm based on template matching to accomplish the motion tracking of rotating objects by aligning template images in a video sequence. For the special case of non-planar cylindrical objects, a nonlinear transformation model for rotational tracking is designed to measure the rotational speed of a cylindrical coupling by tracking a rotating object in a video sequence [33].

In summary, when the ball is engaged in high-speed reciprocating motion, it is also engaged in spin motion during revolution, making it impossible to directly measure the spin speed of the ball. The above study used visual measurement methods to measure the rotational speed of moving objects or rotating components but did not accurately measure the spin angular velocity and linear velocity of high-speed moving balls. Thus, aiming at the need for accurate measurement of ball spin vector in mechanical field, and the problem that it is difficult to measure the ball spin angular velocity directly with existing research methods, a mathematical model for calculating the ball spin angular velocity is constructed for the first time. Moreover, a method of measuring bearing ball rotation vector based on marked point vision three-dimensional coordinate measurement is proposed, and the accurate measurement of each bearing ball rotation vector is finally realized, which provides a reference for the monitoring of the bearing running state and analysis of ball sliding behavior.

2. Proposed Theories and Methods

2.1. Overview of the Principle of Ball Rotational Speed Measurement

In order to analyze the sliding behavior of bearing balls, we need to calculate the sliding velocity of the balls. However, the premise for calculating the sliding velocity of the ball is to know the revolution angular velocity and spin angular velocity of the ball. As shown in Figure 1, the measurement of ball rotational speed is divided into two parts: revolution and spin. For spin measurement, we measure the spin angular velocity, spin linear velocity, and sliding velocity of the ball. For revolution measurement, we measure the revolution angular velocity and linear velocity value of the ball.

In the spin measurement part of the ball, a series of marker points with positional configurations are preset on the surface of the ball to be measured first. Secondly, a binocular camera is used to capture the motion process of the ball and analyze the images of adjacent frames to locate and extract the 2D coordinates of the marker points. Then, the 3D coordinates of the marker points can be measured using the triangulation principle. Further, the spatial position of the ball is determined based on the 3D coordinates of the marker points, and the coordinates of the ball center are determined. A spatial circle formed by the movement trajectories of the different marker points is established in the course of the spin movement of the ball. We determined the three kinds of geometric constraints on the marker points made by the ball rotor axis and the ball surface so as to construct a mathematical model of the high-speed movement of the ball and establish a system of nonlinear equations corresponding to the geometric constraints. Finally, the 3D coordinates of the marker points, the coordinates of the ball center, and the radius of the ball are substituted into the constructed nonlinear model for optimization and solution in order to obtain the coordinates of the center of each spatial circle and the radius of the corresponding spatial circle. The spin angle and direction of each marker point per unit time are further calculated to obtain the direction and numerical results of the ball spin angular velocity.

In the revolution measurement part of the ball, the transverse coordinates of the ball in the pixel coordinate system are extracted first. Second, according to the specific correspondence between the transverse coordinates of the ball and the movement time, a linear equation is established to solve for the revolution period of the ball, so as to calculate the revolution speed.

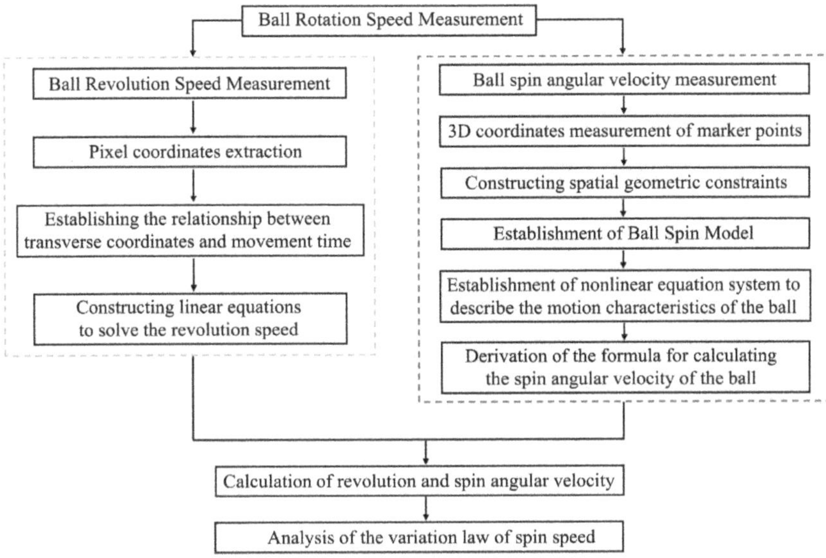

Figure 1. Steps for measuring the rotational vectors of ball bearings.

2.2. Proposed Ball Spin Measurement Principle

In the spin measurement of the ball, a series of circular marker points with relative positional configurations were laid on the ball. As shown in Figure 2, the three marker points P_1, P_2, and P_3, are marked on the ball surface by a specific graphene material. This specific graphene material can ensure that the marker points will not be easily worn out during repeated friction. In addition, there is a certain relative positional configuration between the three marker points, which ensures that the selected three marker points can be distinguished at any time. The three marker points before the ball movement are labeled as P_1, P_2, and P_3, and the corresponding three marker points after the ball movement are labeled as P_1', P_2', and P_3'. The 3D coordinates of the six points P_1, P_2, P_3, P_1', P_2', and P_3' are calculated using binocular vision principles. Firstly, extract the 2D coordinates of these six points in the image. Then, calculate the 3D coordinates of these six points based on the pre-calibration parameters of binocular vision.

Tests have shown that the spin angular velocity of the ball is much greater than the revolution angular velocity [34]. Thus, the distance of the ball revolution is much smaller than the distance of the marker point's movement in a short time interval, and the following reasonable assumptions can be made: during a short time of movement, the displacement changes of the marker point on the surface of the ball can be regarded as the displacement change caused by the spin motion of the ball. At this time, the mixed movement of the ball can be regarded as a spin movement only.

As shown in Figure 3, for the marker point P_1, there is only one plane perpendicular to the axis of rotation l and intersecting with l at point A over the line segment P_1P_1'. This plane intersects with the surface of the ball to form the space circle $\odot A$. Then, the points P_1 and P_1' are located on the space circle $\odot A$ centered at the point A and radiused at $AP_1 = AP_1'$. For the marker points P_2, P_2', and P_3, P_3', the space circle $\odot B$ and circle $\odot C$ are obtained in the same way.

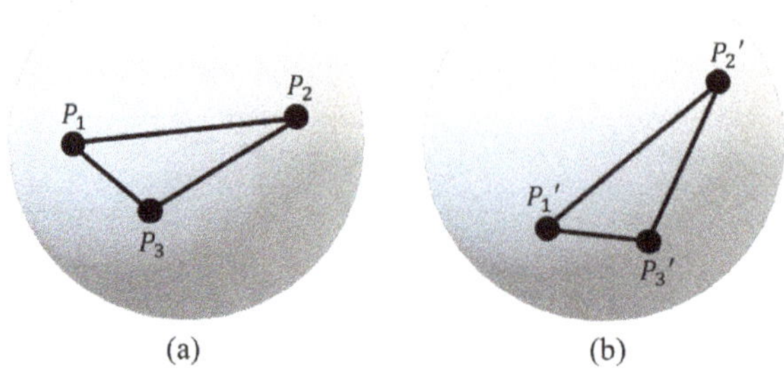

Figure 2. Schematic diagram before and after ball movement: (**a**) Position of the marker points before ball movement. (**b**) Position of the marker points after ball movement.

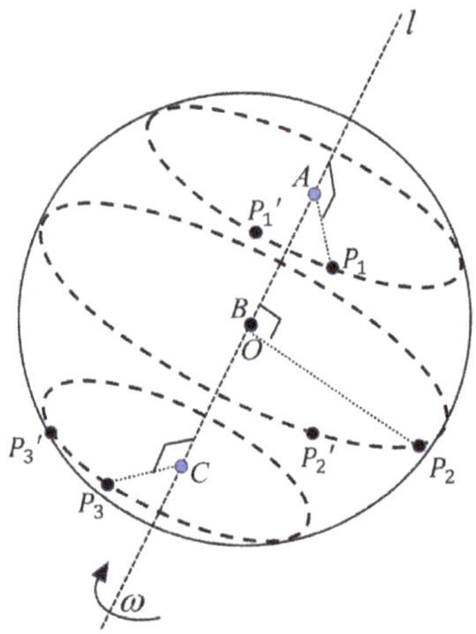

Figure 3. Ball spin motion model.

Let the center of the ball be $O(x_o, y_o, z_o)$, the coordinates of the centers of the space circles be $\odot A$, $\odot B$, $\odot C$ be $A(x_A, y_A, z_A)$, $B(x_B, y_B, z_B)$, and $C(x_C, y_C, z_C)$, respectively. The radius of the space circles $\odot A$, $\odot B$, and $\odot C$ are given as D, E, and F, respectively. The coordinates of two points on the space circle $\odot A$ are $P_1(x_{P_1}, y_{P_1}, z_{P_1})$, $P_1'(x_{P_1'}, y_{P_1'}, z_{P_1'})$, the coordinates of two points on the space circle $\odot B$ are $P_2(x_{P_2}, y_{P_2}, z_{P_2})$, $P_2'(x_{P_2'}, y_{P_2'}, z_{P_2'})$, and the coordinates of two points on the space circle $\odot C$ are $P_3(x_{P_3}, y_{P_3}, z_{P_3})$, $P_3'(x_{P_3'}, y_{P_3'}, z_{P_3'})$. For this ball spin angular velocity measurement model, the following geometric constraints exist:

(1) Points P_1, P_1' lie on the space circle $\odot A$, points P_2, P_2' lie on the space circle $\odot B$, and points P_3, P_3' lie on the space circle $\odot C$.

(2) The rotary axis l is perpendicular to the plane in which the space circles $\odot A$, $\odot B$, and $\odot C$ are located.

(3) The centers of the circle A, B, C, and the center of the ball O share the same line.

Based on the above geometric constraints, a nonlinear equation system can be constructed to describe the motion characteristics of the ball.

(1) Let the spatial circular equations of circles $\odot A$, $\odot B$, and $\odot C$ be:

$$(x - x_A)^2 + (y - y_A)^2 + (z - z_A)^2 = {R_A}^2 \quad (1)$$

$$(x - x_B)^2 + (y - y_B)^2 + (z - z_B)^2 = {R_B}^2 \quad (2)$$

$$(x - x_C)^2 + (y - y_C)^2 + (z - z_C)^2 = R_C{}^2 \tag{3}$$

Substituting the 3D coordinates of two points P_1, P_1' on the space circle A into Equation (1):

$$\begin{cases} (x_{P_1} - x_A)^2 + (y_{P_1} - y_A)^2 + (z_{P_1} - z_A)^2 = R_A{}^2 \\ (x_{P_1'} - x_A)^2 + (y_{P_1'} - y_A)^2 + (z_{P_1'} - z_A)^2 = R_A{}^2 \end{cases} \tag{4}$$

Similarly, for the space circles $\odot B$ and $\odot C$, substituting the 3D coordinates of P_2 and P_2', P_3, and P_3' into Equations (2) and (3), respectively:

$$\begin{cases} (x_{P_2} - x_B)^2 + (y_{P_2} - y_B)^2 + (z_{P_2} - z_B)^2 = R_B{}^2 \\ (x_{P_2'} - x_B)^2 + (y_{P_2'} - y_B)^2 + (z_{P_2'} - z_B)^2 = R_B{}^2 \\ (x_{P_3} - x_C)^2 + (y_{P_3} - y_C)^2 + (z_{P_3} - z_C)^2 = R_C{}^2 \\ (x_{P_3'} - x_C)^2 + (y_{P_3'} - y_C)^2 + (z_{P_3'} - z_C)^2 = R_C{}^2 \end{cases} \tag{5}$$

(2) The rotary axis l is perpendicular to the plane in which the space circles $\odot A$, $\odot B$, and $\odot C$ are located, respectively:

$$l \perp P_1 P_1' A \tag{6}$$

$$l \perp P_2 P_2' B \tag{7}$$

$$l \perp P_3 P_3' C \tag{8}$$

According to Equation (6), it is obtained from the line–plane perpendicularity theorem:

$$\begin{cases} \overrightarrow{OA} \perp \overrightarrow{P_1 A} \\ \overrightarrow{OA} \perp \overrightarrow{P_1' A} \end{cases} \tag{9}$$

Substituting the 3D coordinates of the points O, A, P_1, P_1' into Equation (9):

$$\begin{cases} (x_A - x_{P_1})(x_A - x_O) + (y_A - y_{P_1})(y_A - y_O) + (z_A - z_{P_1})(z_A - z_O) = 0 \\ (x_A - x_{P_1'})(x_A - x_O) + (y_A - y_{P_1'})(y_A - y_O) + (z_A - z_{P_1'})(z_A - z_O) = 0 \end{cases} \tag{10}$$

Similarly, according to Equations (7) and (8):

$$\begin{cases} (x_B - x_{P_2})(x_B - x_O) + (y_B - y_{P_2})(y_B - y_O) + (z_B - z_{P_2})(z_B - z_O) = 0 \\ \left(x_B - x_{P_2'}\right)(x_B - x_O) + \left(y_B - y_{P_2'}\right)(y_B - y_O) + \left(z_B - z_{P_2'}\right)(z_B - z_O) = 0 \\ (x_C - x_{P_3})(x_C - x_O) + (y_C - y_{P_3})(y_C - y_O) + (z_C - z_{P_3})(z_C - z_O) = 0 \\ \left(x_C - x_{P_3'}\right)(x_C - x_O) + \left(y_C - y_{P_3'}\right)(y_C - y_O) + \left(z_C - z_{P_3'}\right)(z_C - z_O) = 0 \end{cases} \tag{11}$$

(3) The collinear of the four points at the center of the circle, A, B, C, and the center of the ball O can be obtained from the properties of the straight lines in space:

$$\begin{cases} \overrightarrow{OA} \parallel \overrightarrow{OB} \\ \overrightarrow{OA} \parallel \overrightarrow{OC} \end{cases} \tag{12}$$

Substitute the 3D coordinates of points A, B, C, and O into Equation (12):

$$\begin{cases} \frac{x_A - x_O}{x_B - x_O} = \frac{y_A - y_O}{y_B - y_O} = \frac{z_A - z_O}{z_B - z_O} \\ \frac{x_A - x_O}{x_C - x_O} = \frac{y_A - y_O}{y_C - y_O} = \frac{z_A - z_O}{z_C - z_O} \end{cases} \tag{13}$$

Organize Equation (13):

$$\begin{cases} (x_A - x_O)(y_B - y_O) - (y_A - y_O)(x_B - x_O) = 0 \\ (x_A - x_O)(z_B - z_O) - (z_A - z_O)(x_B - x_O) = 0 \\ (y_A - y_O)(z_B - z_O) - (z_A - z_O)(y_B - y_O) = 0 \\ (x_A - x_O)(y_C - y_O) - (y_A - y_O)(x_C - x_O) = 0 \\ (x_A - x_O)(z_C - z_O) - (z_A - z_O)(x_C - x_O) = 0 \\ (y_A - y_O)(z_C - z_O) - (z_A - z_O)(y_C - y_O) = 0 \end{cases} \tag{14}$$

In the above equations, the unknowns are the center of the ball $O(x_o, y_o, z_o)$, the center of the circle $A(x_A, y_A, z_A)$, $B(x_B, y_B, z_B)$, $C(x_C, y_C, z_C)$, and the radius of the spatial circle R_A, R_B, R_C—15 unknown parameters in total. If the coordinates $P_1(x_{P_1}, y_{P_1}, z_{P_1})$ and $P_1'(x_{P_1'}, y_{P_1'}, z_{P_1'})$, $P_2(x_{P_2}, y_{P_2}, z_{P_2})$ and $P_2'(x_{P_2'}, y_{P_2'}, z_{P_2'})$, $P_3(x_{P_3}, y_{P_3}, z_{P_3})$ and $P_3'(x_{P_3'}, y_{P_3'}, z_{P_3'})$ of the moments before and after each marker point are known, the above system of nonlinear equations can be solved and optimized.

First of all, after using P_1, P_1', P_2, P_2', P_3, P_3'—six points of the 3D coordinates to determine both the center of the ball coordinates $O(x_o, y_o, z_o)$ and the radius of the ball uniquely—and substituting $P_1(x_{P_1}, y_{P_1}, z_{P_1})$, $P_1'(x_{P_1'}, y_{P_1'}, z_{P_1'})$, $P_2(x_{P_2}, y_{P_2}, z_{P_2})$, $P_2'(x_{P_2'}, y_{P_2'}, z_{P_2'})$, $P_3(x_{P_3}, y_{P_3}, z_{P_3})$, and $P_3'(x_{P_3'}, y_{P_3'}, z_{P_3'})$ into the system of nonlinear equations, then, the system of nonlinear equations can be solved and optimized to obtain the coordinates of the center of the circle A, B, C, and the radius of the spatial circle R_A, R_B, R_C, and then the spatial circle $\odot A$, $\odot B$, and $\odot C$ with known parameters can be constructed in the space.

The coordinates of the center and radius of each spatial circle are known; therefore, the ball spin direction and the magnitude of the spin angular velocity can be solved. First, it is known that the four points A, B, C, and O are collinear, and let (x, y, z) be any point on the rotation axis l; then, the linear equation of l in space is:

$$\frac{x - x_A}{x_B - x_A} = \frac{y - y_A}{y_B - y_A} = \frac{z - z_A}{z_B - z_A} \tag{15}$$

Then find the angular velocity of the spin of the ball. In Figure 4, to solve the spin angular velocity of the marker point P_3, for example, where ∂ denotes the spin angle of the marker point P_3, r denotes the radius of the spatial circle $\odot C$, and d denotes the distance between the marker point P_3 and P_3', the spin angle of the ball in unit time can be expressed as:

$$\partial = 2\arcsin\left(\frac{d}{2r}\right) \tag{16}$$

Let the time interval between the moments before and after the movement of the ball be t, according to Equation (16), the spin angular velocity of the marker point can be found as:

$$\omega = \frac{2\arcsin\left(\frac{d}{2r}\right)}{t} \tag{17}$$

The distance d can be solved according to the formula for the distance between any two points in space:

$$\begin{cases} d_{P_1P_1'} = \sqrt{\left(x_{P_1} - x_{P_1'}\right)^2 + \left(y_{P_1} - y_{P_1'}\right)^2 + \left(z_{P_1} - z_{P_1'}\right)^2} \\ d_{P_2P_2'} = \sqrt{\left(x_{P_2} - x_{P_2'}\right)^2 + \left(y_{P_2} - y_{P_2'}\right)^2 + \left(z_{P_2} - z_{P_2'}\right)^2} \\ d_{P_3P_3'} = \sqrt{\left(x_{P_3} - x_{P_3'}\right)^2 + \left(y_{P_3} - y_{P_3'}\right)^2 + \left(z_{P_3} - z_{P_3'}\right)^2} \end{cases} \tag{18}$$

Substituting Equation (18) into Equation (17), the spin angular velocity at each marker point is solved for:

$$\begin{cases} \omega_{P_1} = \dfrac{2\arcsin\left(\dfrac{\sqrt{\left(x_{P_1}-x_{P_1'}\right)^2+\left(y_{P_1}-y_{P_1'}\right)^2+\left(z_{P_1}-z_{P_1'}\right)^2}}{2R_A}\right)}{t} \\ \omega_{P_2} = \dfrac{2\arcsin\left(\dfrac{\sqrt{\left(x_{P_2}-x_{P_2'}\right)^2+\left(y_{P_2}-y_{P_2'}\right)^2+\left(z_{P_2}-z_{P_2'}\right)^2}}{2R_B}\right)}{t} \\ \omega_{P_3} = \dfrac{2\arcsin\left(\dfrac{\sqrt{\left(x_{P_3}-x_{P_3'}\right)^2+\left(y_{P_3}-y_{P_3'}\right)^2+\left(z_{P_3}-z_{P_3'}\right)^2}}{2R_C}\right)}{t} \end{cases} \tag{19}$$

The spin angular velocities of points P_1, P_2, and P_3 are ω_{P_1}, ω_{P_2}, and ω_{P_3}, respectively. For the remaining points P_4, P_5, P_6, ..., P_n, the corresponding spin angular velocities ω_{P_4}, ω_{P_5}, ω_{P_6}, ..., ω_{P_n} can be calculated for each point in the same way. Let the spin angular velocity of the ball be ω. In order to reduce the measurement error of the ball spin angular velocity, the average value of the spin angular velocity of each point is taken as the measurement result of the ball spin angular velocity:

$$\omega = \frac{\sum\limits_{n=1} \omega_{P_n}}{n} \tag{20}$$

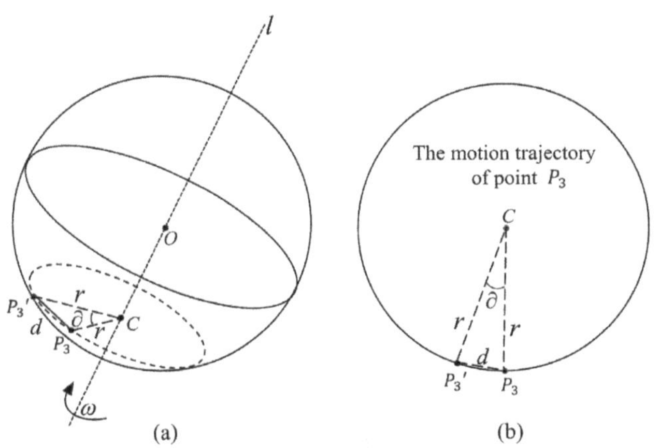

Figure 4. Ball spin angle calculation: (**a**) Schematic diagram of ball spin angle. (**b**) The vertical view of the motion trajectory of point P_3.

2.3. Ball Revolution Speed Measurement Principle

The time series method is used to measure the revolution speed of the ball as shown in Figure 5. In Figure 5b, r is the radius of the turntable, r_B is the radius of the ball, R is the radius of the ball's revolution, and ω_C is the angular velocity of the turntable. Firstly, record the position of the ball in any image as the initial position and extract the pixel coordinates of that initial position. The shooting time of each image can be clearly given by computer instructions; therefore, the time t when the ball is in its initial position can be recorded. If the time t_1 for the ball to reach its initial position again after one revolution can be calculated, then the time difference $(t_1 - t)$ is the period of the ball's revolution.

Since the camera uses discrete imaging, the position of the ball in the image after a cycle of rotation deviates from the initial position, making it impossible to record the time t_1 directly. To address this problem, select multiple images where the ball appears in the camera's field of view after one revolution and extract the pixel coordinates of the ball in these images to preliminarily determine the corresponding image when the ball returns to its initial position. Furthermore, select the images with the balls located on the left and right sides of the initial position, and record the corresponding camera shooting time. Finally, a linear equation is set up to solve the equation and to calculate the ball revolution period so that the ball revolution speed can be calculated.

As shown in Figure 6, the time at which the ball first reaches the initial position is noted as t, and the time at which it returns to the initial position again after a cycle of rotation is noted as t_1. The transverse coordinate of the ball in the pixel coordinate system is labeled as X_1. The times when the ball is located to the left and to the right of its initial position after a cycle of rotation are labeled as t_2 and t_3, respectively. And t_2 and t_3 are the corresponding shooting times of the camera, which can be directly displayed by computer commands. The transverse coordinates are labeled as X_2 and X_3, respectively.

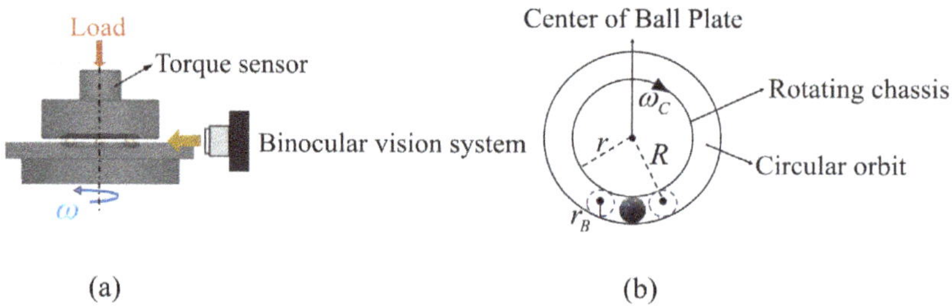

Figure 5. Measurement of ball bearing revolution speed: (**a**) Schematic diagram of ball bearing revolution. (**b**) The top view of the ball bearing revolution.

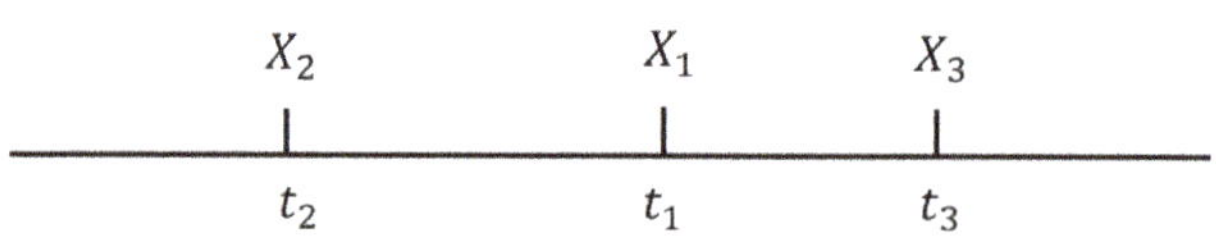

Figure 6. Ball revolution speed measurement principle.

Then, the lateral coordinates of the ball and the movement time satisfy the following relationship:

$$\frac{X_3 - X_1}{t_3 - t_1} = \frac{X_1 - X_2}{t_1 - t_2} \tag{21}$$

And t_1 can be obtained:

$$t_1 = \frac{t_2(X_3 - X_1) + t_3(X_1 - X_2)}{X_3 - X_2} \tag{22}$$

The time t at which the ball first reaches its initial position is known; therefore, the ball revolution period is:

$$T = t_1 - t = \frac{t_2(X_3 - X_1) + t_3(X_1 - X_2)}{X_3 - X_2} - t \tag{23}$$

Thus, the ball revolution speed ω_R can be obtained:

$$\omega_R = \frac{2\pi}{T} = \frac{2\pi(X_3 - X_2)}{t_2(X_3 - X_1) + t_3(X_1 - X_2) - t(X_3 - X_2)} \tag{24}$$

Then, the sliding velocity V_s of the ball can be calculated using the following formula:

$$V_s = \omega_C r - (\omega r_B + \omega_R r) \tag{25}$$

3. Experimental Measurement of the Proposed Method

The proposed on-line measurement method of ball rotation vectors was verified on the ball vision measurement platform. As shown in Figure 7, the device includes two black-and-white industrial cameras, a general-purpose mechanical tester, and a computer. The industrial camera model ids UI-3160CP Rev.2.1 has a resolution of 1920 × 1200 pixels and a frame rate of up to 169 FPS (frames per second). The main test stand for ball rotational speed measurement is the Bruker UMT TriboLab. The computer is a Windows 10 operating system with an Intel(R) Core (TM) i5-7400 (The manufacturer is Intel Corporation of Santa Clara, CA, USA) processor. In this experiment, two black-and-white industrial cameras are used to take real-time pictures of the moving balls, and the Bruker Universal Mechanical Tester controls the periodic rotation of the balls at a constant revolution speed. The diameter of the ball used in the experiment is 12.7 mm and the diameter of the rotating chassis is 100 mm.

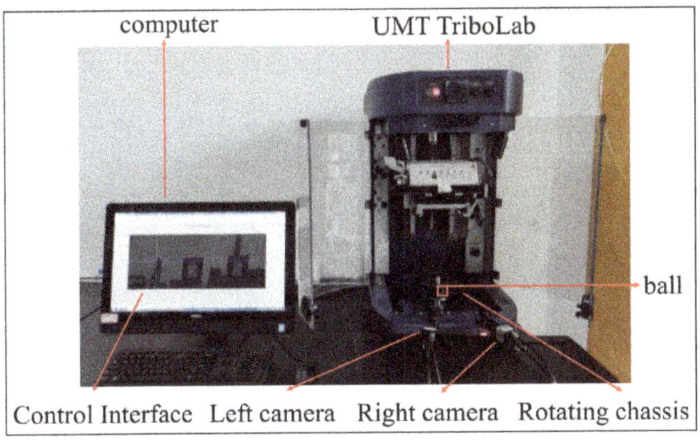

Figure 7. Ball rotational speed measurement platform.

3.1. Experimental Environment Setting

Before using the camera to take pictures of the movement of the ball, the following operations were performed:

(1) Blackening treatment was applied to the edges of the turntable, and interference light in the environment was appropriately blocked to solve the problem that the marker points could not be identified due to the serious reflection on the surface of the metal ball.

(2) A series of marker points are made on the surface of the ball with obvious color differences so that the marker points captured by the camera are clearer. In addition, the configuration of the marks is similar to the intersection of the longitude and latitude lines on the surface of the globe. This ensures that the left and right cameras see enough marker points in the image captured at any given moment.

3.2. Experimental Results and Analysis

In order to analyze the motion pattern of the ball over a period of time, several images were taken from the moment the ball entered the field of view to the moment it left the field of view. Use the left and right cameras to synchronously capture images of the ball moving in the raceway as shown in Figures 8 and 9. The moment when the ball first appears completely within the field of view was recorded as 0 and captured a frame rate of 100 FPS for a total of 12 images. Among them, 0 s, 0.02 s, 0.04 s, 0.06 s, 0.08 s, and 0.10 s are the previous moments of the ball movement, and 0.01 s, 0.03 s, 0.05 s, 0.07 s, 0.09 s, and 0.11 s are the subsequent moments of the ball movement. The time interval before and after the ball movement is 0.01 s.

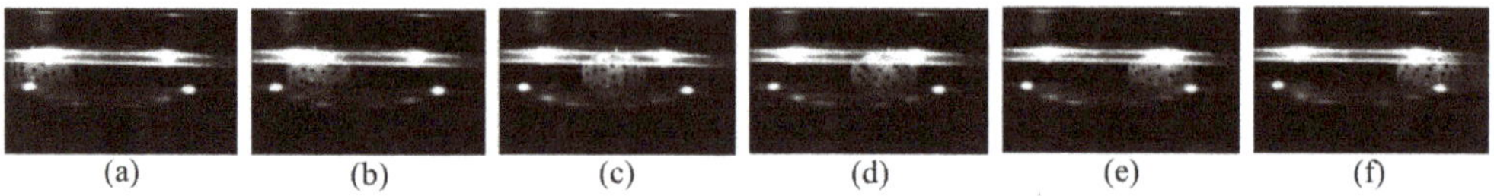

Figure 8. Images taken by the left camera at the moment before the ball movement: (**a**) t = 0 s. (**b**) t = 0.02 s. (**c**) t = 0.04 s. (**d**) t = 0.06 s. (**e**) t = 0.08 s. (**f**) t = 0.10 s.

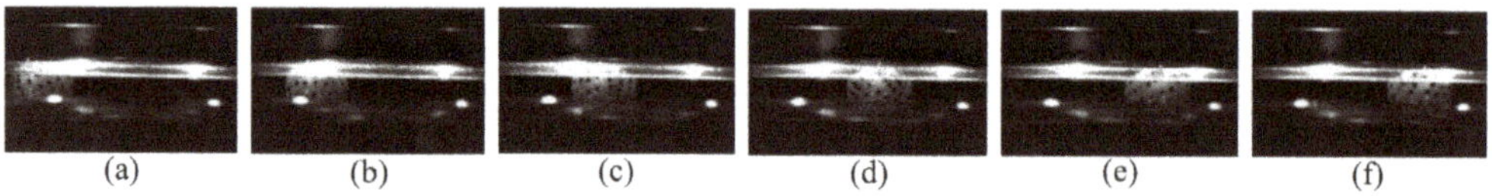

Figure 9. Images taken by the right camera at the moment before the ball movement: (**a**) t = 0 s. (**b**) t = 0.02 s. (**c**) t = 0.04 s. (**d**) t = 0.06 s. (**e**) t = 0.08 s. (**f**) t = 0.10 s.

At t = 0.04 s, the distribution of marker points P_1, P_2, and P_3 in the left and right images are shown in Figure 10. The 2D coordinates of these points in the left and right images were extracted separately, and the results are shown in Table 1.

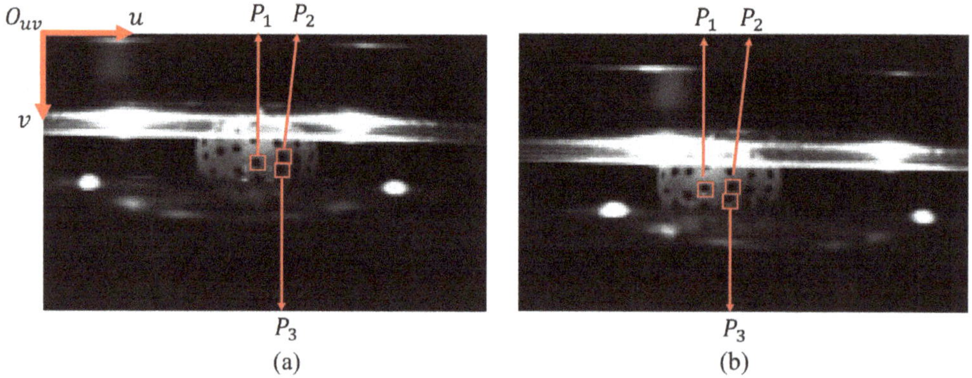

Figure 10. In the pixel coordinate system of the image, the distribution of marker points P_1, P_2, and P_3 in the left and right images: (**a**) Left image. (**b**) Right image.

Table 1. Two-dimensional coordinates of marker points in the left and right image.

Point	2D Coordinates in Left Image	2D Coordinates in Right Image
P_1	(929.75, 566)	(796.25, 674)
P_2	(1043.75, 540)	(911.75, 666.5)
P_3	(1036.25, 593)	(899.75, 719)

The rotational speed of the turntable is set to 5 rpm. Calculating the 3D coordinates of these six points based on the pre-calibration parameters of binocular vision [35]. The 3D coordinates of the marker points in the coordinate system of the left camera can be measured at the moment before and the moment after the ball movement. The calculation results are shown in Table 2.

Table 2. Three-dimensional coordinates of marker points.

Point	*X* Coordinate/mm	*Y* Coordinate/mm	*Z* Coordinate/mm
P_1	−5.46	−7.18	307.62
P_2	−2.14	−8.90	307.17
P_3	2.94	−9.58	307.91
P_1'	−1.86	−6.22	306.74
P_2'	0.10	−8.18	306.96
P_3'	4.12	−9.02	308.22

The 3D coordinates of the marker points P_1, P_1', P_2, P_2', P_3, P_3' are known. According to (8), (9), (14), (15) and (18), the Levenberg–Marquardt algorithm was used to optimize and solve the 3D coordinates $A(x_A, y_A, z_A)$, $B(x_B, y_B, z_B)$, $C(x_C, y_C, z_C)$, and $O(x_o, y_o, z_o)$ [36]. The calculation results are shown in Table 3.

Table 3. Center coordinates and radius of the ball.

Point	Three-Dimensional Coordinate	Radius
A	(−1.24, −5.90, 312.51)	6.30
B	(−2.56, −7.21, 311.64)	5.85
C	(−4.50, −9.16, 308.56)	4.48
O	(−0.79, −5.45, 312.81)	6.37

In order to verify whether the solution of the center coordinates satisfies the three spatial constraints determined in Section 2.2, the rotation axis l, the center of the spatial circle A, B, C, the ball center O, and the spatial position of the ball are displayed, as shown in Figure 11. Obviously, the centers A, B, and C of the three space circles are collinear with the ball center coordinates O at four points, and the actual measurement results conform to the three geometric constraints identified in Section 2.2.

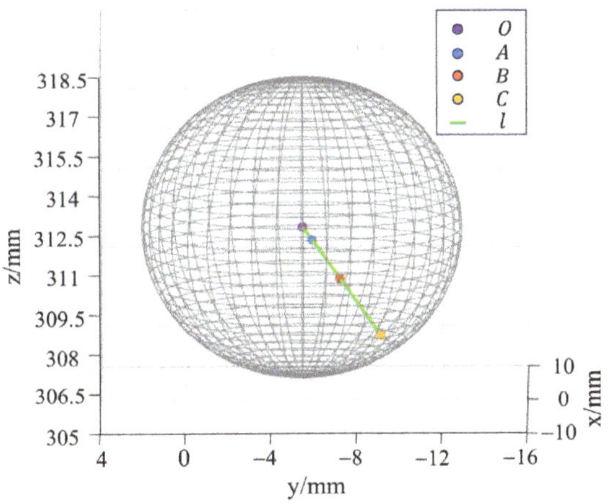

Figure 11. Rotation axis, space circle center, ball center, and the spatial position of the ball.

Knowing the 3D coordinates of the marker points P_1, P_1', P_2, P_2', P_3, P_3' and the radii of the spatial circle R_A, R_B, and R_C, according to (23) and (24), the magnitude of the spin angular velocities of the ball can be calculated at 0 s, 0.02 s, 0.04 s, 0.06 s, 0.08 s, and 0.10 s, and the spin angular velocity of the ball is decomposed along the three axis directions of the spherical coordinate system. The spherical coordinate system is based on the origin of the ball center O as shown in Figure 12.

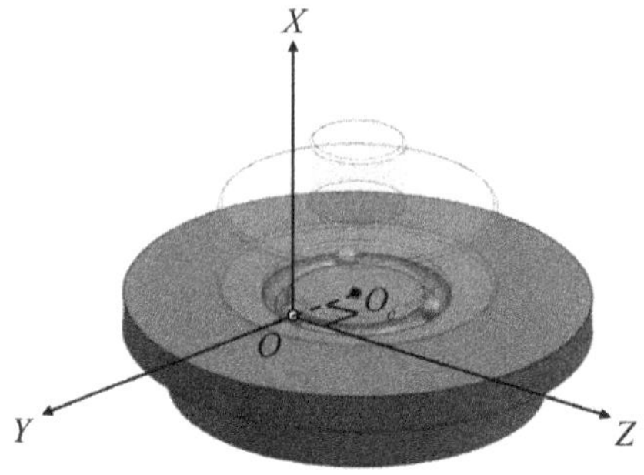

Figure 12. Spherical center coordinate system.

The X-axis is perpendicular to the plane where the turntable is located and faces upwards. The Z-axis is tangent to the contact point between the ball and the raceway and points towards the direction of the ball's revolution speed. The Y-axis is perpendicular to the plane where the X-axis and Z-axis are located and satisfies the right-hand rule. In this way, the spin angular velocities of the ball around the X-axis, Y-axis, and Z-axis can be obtained. When the chassis speed is 5 rpm, the measurement results of the spin angular velocity of the ball bearings are shown in Figure 13.

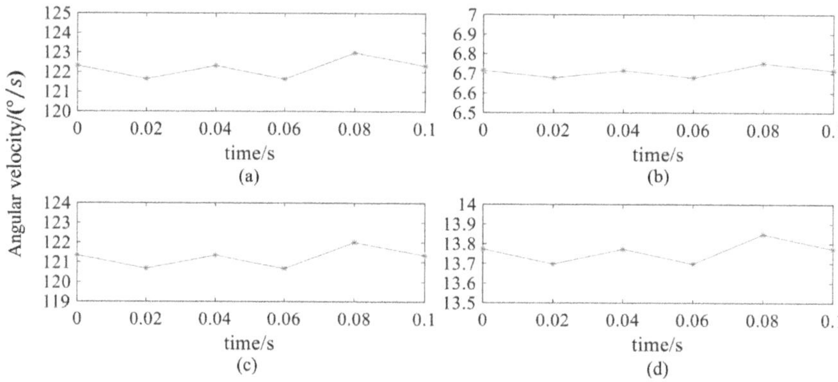

Figure 13. When the chassis speed is 5 rpm, the magnitude of the spin angular velocity of the ball and the decomposition of the spin angular velocity of the ball along the X, Y, and Z axis: (**a**) Total angular velocity. (**b**) X-axis. (**c**) Y-axis. (**d**) Z-axis.

Figure 13a shows that the spin angular velocity of the ball is almost the same during the motion in 0~0.10 s. The spin angular velocity of the ball shows a consistent increasing or decreasing trend, and the change is relatively stable, with small fluctuations around 122 °/s. Figure 13b–d shows that the spin angular velocities around the X, Y, and Z axis of the spherical coordinate system are 6.7 °/s, 121.0 °/s, and 13.7 °/s, respectively. Moreover, the spin angular velocity presents a relatively consistent variation rule with that shown in Figure 13a. Especially, the spin angular velocity has the largest component around the Y-axis in the spherical coordinate system, while the variation rule of the angular velocity is most similar to that in Figure 13a. The spin angular velocity has the smallest component around the X-axis in the spherical coordinate system, while the changing trend of the angular velocity is the most stable. The spin angular velocity around the X-axis and Z-axis is much smaller than the spin angular velocity around the Y-axis, which means that the ball moves at its maximum speed in the direction tangential to the chassis and at its minimum speed in the direction perpendicular to the chassis plane. This is consistent with the expected theoretical results.

The measurement results of the spin angular velocities are shown in Figure 14. When the chassis speed is 10 rpm, 20 rpm, 30 rpm, 40 rpm, and 50 rpm respectively, the spin angular velocity of the ball is 241, 482, 740, 1001, and 1262 °/s, respectively. It is obvious that with the increase in the chassis speed multiple, the spin angular velocity increases linearly. Moreover, for every 10 rpm increase in chassis speed, the spin angular velocity increases by about 260 °/s.

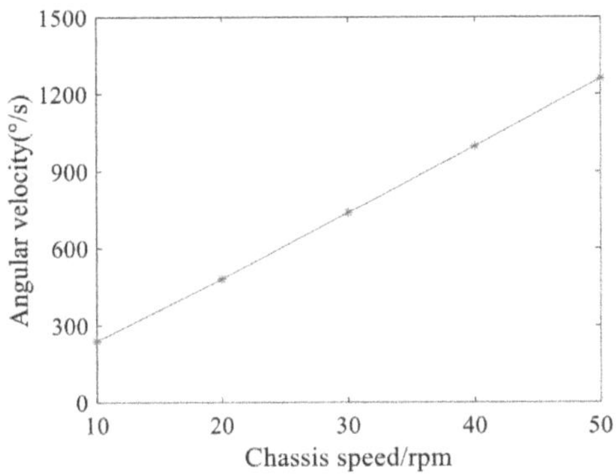

Figure 14. Spin angular velocities of the ball with different chassis speeds.

At 0.06 s, with the chassis speed of 10 rpm, 20 rpm, 30 rpm, 40 rpm, and 50 rpm respectively, the spin angular velocity of the ball is decomposed along the three axes directions of the spherical coordinate system to obtain the spin angular velocity of the ball around the *X*, *Y*, and *Z* axes, as shown in Figure 15. As the chassis speed multiplies, the spin angular velocity of the ball around the *X*, *Y*, and *Z* axes all show a linear upward trend. Figure 15a shows that when the chassis speed is 10 rpm, the spin angular velocity of the ball around the *X*-axis is 13.2 °/s; when the chassis speed is 50 rpm, the spin angular velocity of the ball around the *X*-axis is 69.3 °/s. For every 10 rpm increase in chassis speed, the spin angular velocity of the ball around the *X*-axis increases by 14.2 °/s. In addition, the spin angular velocity of the ball around the *X* and *Z* axes is much smaller than the spin angular velocity around the *Y* axis, which is consistent with the expected theoretical results. Figure 15b,c shows that the spin angular velocity of the ball around the *Y* and *Z* axes have the same variation rule as shown in Figure 15a. Furthermore, when the chassis speed increases by 10 rpm, the spin angular velocity around the *Y* and *Z* axes increases by 257.9 °/s and 29.3 °/s, respectively.

When the chassis speed is set to 50 rpm, calculate the spin angular velocity of the ball after stable motion for the 1–5th turn and t = 0.06 s. The measurement results are shown in Figure 16. Obviously, the spin angular velocity of the ball stabilizes around 1262 °/s at 0.06 s per revolution, indicating that the total spin angular velocity of the ball is almost constant. The spin angular velocity in the second circle is the highest at 1262.20 °/s, while, in the third circle, the spin angular velocity is the lowest, at 1261.95 °/s. The maximum difference in angular velocity and measurement error are 0.25 °/s and 0.18%, respectively.

As shown in Figure 16b–d, the spin angular velocities of the ball around the *X*, *Y*, and *Z* axes are different at different turns. In addition, the spin angular velocity variation of the ball around the *X*-axis is the smallest, with a maximum angular velocity difference of 0.38 °/s. The spin angular velocity variation around the *Z*-axis is the largest, with a maximum angular velocity difference of 60.65 °/s. The maximum angular velocity difference around the *Y*-axis is 6.44 °/s, and the variation pattern of the angular velocity component of the ball around the *Y*-axis is consistent with the overall angular velocity. It is worth noting that whenever the spin angular velocity component of the ball around the *Y*-axis increases, the angular velocity components around the *X* and *Z* axes both decrease and vice versa. Analyzing the reasons for the above phenomenon, the total spin angular velocity of the ball is almost constant; however, as the ball continues to move, the direction of its spin's angular velocity constantly changes. This results in a constant change in the angular velocity component of the ball around each coordinate axis, which further confirms the change in the direction of spin of the ball during its motion.

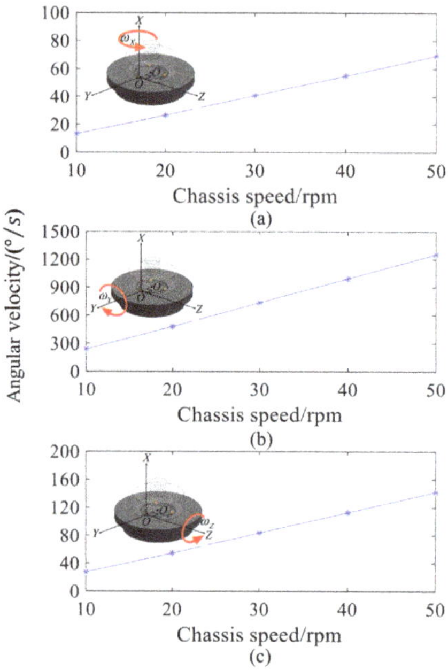

Figure 15. The ball spin angular velocity is decomposed along the axis of the spherical coordinate system: (**a**) X-axis. (**b**) Y-axis. (**c**) Z-axis.

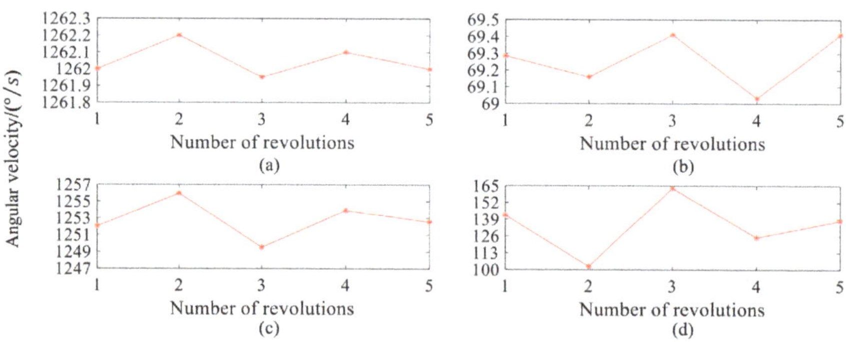

Figure 16. The spin angular velocity and the decomposition of spin angular velocity around the X, Y, and Z axes of the ball at different turns: (**a**) Total angular velocity. (**b**) X-axis. (**c**) Y-axis. (**d**) Z-axis.

When the chassis speed is 50 rpm, the measured revolution speed of the ball is 26.8 rpm, which is 160.8 °/s. Calculate the revolution linear velocity, total spin linear velocity and spin linear velocity along the Z-axis for the 1~5 turns after stable ball motion. The measurement results are shown in Figure 17. The red line represents the revolution linear velocity of the ball, the blue line represents the spin linear velocity of the ball, and the green line represents the spin linear velocity along the tangent direction of motion. It shows that the revolution linear velocity of the ball is a fixed value of 140.7 mm/s and the spin linear velocity of the ball is also stable at 140.2 mm/s. Moreover, the maximum difference and measurement error of the ball spin linear velocity is 0.028 mm/s and 0.19%, respectively, indicating that the measurement accuracy of the proposed method can reflect micrometer-level errors and achieve good measurement results.

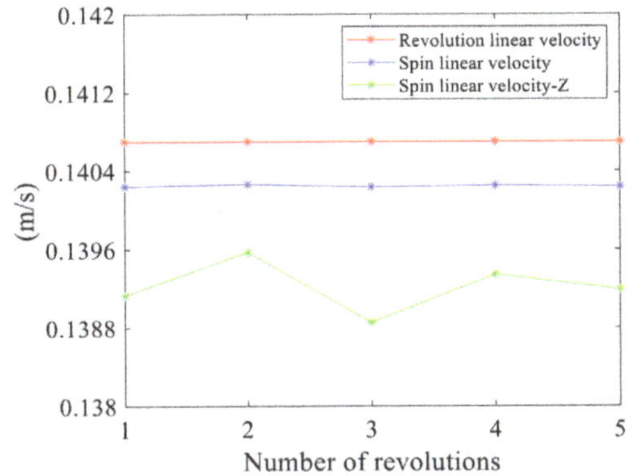

Figure 17. The revolution linear velocity, spin linear velocity, and spin linear velocity along the Z-axis in the spherical coordinate system at different turns.

On the other hand, the spin linear velocity of the ball along the tangent direction of motion fluctuates slightly around 139.2 mm/s. Considering the reason for this difference, due to the limited frame rate of the camera used, it is impossible to guarantee that the position of the ball in the image is completely consistent after each revolution. The above part has proven that when the ball moves to a different position, the spin direction of the ball changes, and the scalar of the spin linear velocity of the ball is constant. This causes the magnitude of the spin linear velocity, which is decomposed into the tangent direction, which is also changing. What is more, the spin linear velocity of the ball along the tangent direction of motion is slightly smaller than the revolution linear velocity, while the difference between them is 1.5 mm/s.

4. Discussion

Analyze the sliding phenomenon of the ball during movement. When the turntable speed is set to 50 rpm, the revolution speed of the ball is 26.8 rpm. At this point, the revolution linear velocity of the ball is a fixed value of 140.7 mm/s, and the spin linear velocity of the ball is also stable at 140.2 mm/s. According to Equation (25), the sliding velocity of the ball is calculated, and the result is shown in Figure 18.

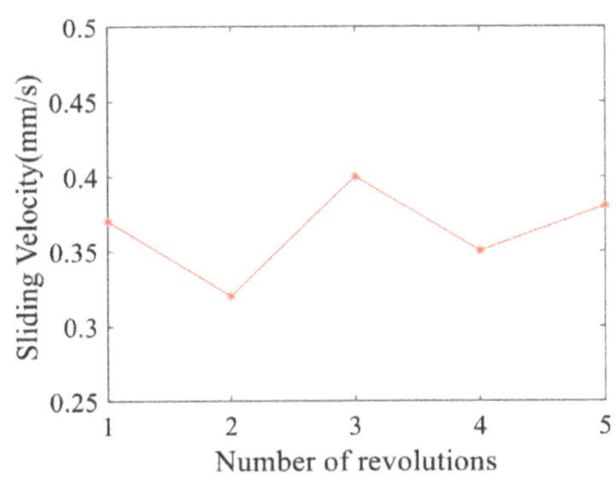

Figure 18. The sliding velocity of the ball at different turns.

According to Equation (25), as the spin angular velocity of the ball increases, the sliding velocity of the ball will decrease, and vice versa. It can be inferred from this that the variation law of ball sliding velocity is completely opposite to the variation law of ball spin angular velocity. Combining Figures 16a and 18 for analysis, the variation law of the two are indeed opposite. This further proves the effectiveness of this method for measuring the sliding velocity of bearing balls, which can achieve monitoring of the operating status of bearings. Given that the speed of the ball is 26.8 revolutions per minute, which means the rotation period of the ball is 2.24 s, the sliding distance between the ball and the chassis after a cycle of rotation is 0.67 mm.

At present, we have achieved the measurement of the rotational vectors of the bearing ball on the Bruker friction and wear equipment. In practical applications, we have carried out measurements of bearing balls in real operating conditions using control moment gyroscope (CMG) test system (CMG test system built according to the real operating environment). As long as the partial area of the ball is visible in the section plane, the high precision spin angular velocity and linear velocity can be measured according to the short rolling time of the ball. Then, the sliding behavior and running state of the ball are analyzed and monitored, respectively.

In practical applications, the selection of lubricating oil will also affect the sliding of the bearing balls. Different lubricating oil has different viscosity, and the viscosity of lubricating oil is one of several factors that affect the traction force of ball movement and thus the sliding behavior of bearing balls. If the viscosity of the lubricating oil is too small, the tractive force will decrease, which will make the contact interface oil film easy to damage; consequently, both the direct contact slide of the interface and the friction will increase. In general, viscosity increases traction and decreases sliding. However, when the viscosity is too large, the wear of the interface and lubricant will also be greatly increased. In addition, temperature is also one of the factors affecting the sliding behavior of the bearing ball. The ball constantly reciprocates in the bearing, and there is a certain friction between the ball and the raceway, and the contact interface temperature rises. This results in a decrease in the viscosity of the lubricating oil, which in turn increases the relative slide between the ball and the raceway. This paper mainly focused on the establishment of the ball spin motion model and the measurement of the ball rotational vectors under real rotation conditions, proving the correctness and applicability of the method. Therefore, the sliding characteristics under different lubricating oil, temperature, humidity, and pressure conditions were not further analyzed. However, these factors will become the main work of our next research.

5. Conclusions

We develop a novel method for measuring the spin angular velocity of bearing balls based on the three-dimensional coordinates of feature points, and the binocular vision technology is first applied to the measurement of the rotation vectors of the bearing balls. Through the binocular vision measurement system and the ball spin motion model constructed, the motion vectors of the ball were measured effectively and accurately, and the sliding behavior of the ball was analyzed. The experimental results show that the measurement error range of spin angular velocity is [0.21 °/s, 0.25 °/s], and the measurement error range of linear velocity is [0.023 mm/s, 0.028 mm/s]. Specifically, by measuring several groups of bearing balls at different chassis speeds, we control the measurement error within ±0.2%, which indicates that the measurement system has good measurement accuracy. The proposed method shows that the real-time measurement of rotational vectors such as spin angular velocity of bearing balls can be realized by the measurement of three-dimensional coordinates of binocular vision feature points. This is of great significance for the analysis of the sliding behavior of bearing balls in the mechanical field and the observation of the operating status of bearings.

Author Contributions: Methodology, W.L.; Software, W.L.; Supervision, J.X., W.P. and H.C.; Validation, W.L., R.J. and K.W.; Writing—original draft, W.L. All authors have read and agreed to the published version of the manuscript.

Funding: This research was funded by the Sichuan Science and Technology Program, grant number 2023YFG0181.

Institutional Review Board Statement: Not applicable.

Informed Consent Statement: Not applicable.

Data Availability Statement: Data are contained within the article.

Conflicts of Interest: The authors declare no conflicts of interest.

References

1. Han, Q.; Chu, F. Nonlinear dynamic model for skidding behavior of angular contact ball bearings. *J. Sound Vib.* **2015**, *354*, 219–235. [CrossRef]
2. Gao, S.; Han, Q.; Pennacchi, P.; Chatterton, S.; Chu, F. Dynamic, thermal, and vibrational analysis of ball bearings with over-skidding behavior. *Friction* **2023**, *11*, 580–601. [CrossRef]
3. Yu, Y.; He, Y.; Karimi, H.R.; Gelman, L.; Cetin, A.E. A two-stage importance-aware subgraph convolutional network based on multi-source sensors for cross-domain fault diagnosis. *Neural Netw.* **2024**, *179*, 106518. [CrossRef]
4. Oktaviana, L.; Tong, V.C.; Hong, S.W. Skidding analysis of angular contact ball bearing subjected to radial load and angular misalignment. *J. Mech. Sci. Technol.* **2019**, *33*, 837–845. [CrossRef]
5. Xu, T.; Xu, G.; Zhang, Q.; Hua, C.; Tan, H.; Zhang, S.; Luo, A. A preload analytical method for ball bearings utilising bearing skidding criterion. *Tribol. Int.* **2013**, *67*, 44–50. [CrossRef]
6. Gupta, P.K. Dynamics of rolling-element bearings part IV: Ball bearing results. *J. Tribol.* **1979**, *101*, 319–326. [CrossRef]
7. Chen, W.; Ma, Z.; Gao, L.; Li, X.; Pan, J. Quasi-static analysis of thrust-loaded angular contact ball bearings part I: Theoretical formulation. *Chin. J. Mech. Eng.* **2012**, *25*, 71–80. [CrossRef]
8. Östensen, J.; Åström, H.; Höglund, E. Analysis of a grease-lubricated roller bearing under arctic conditions. *Proc. Inst. Mech. Eng. Part J J. Eng. Tribol.* **1995**, *209*, 213–220. [CrossRef]
9. Imado, K. Fundamental study of hall element method to detect ball motion in ball bearings. *Tribol. Trans.* **2000**, *43*, 66–73.
10. Selvaraj, A.; Marappan, R. Experimental analysis of factors influencing the cage slip in cylindrical roller bearing. *Int. J. Adv. Manuf. Technol.* **2011**, *53*, 635–644. [CrossRef]
11. Wang, W.Z.; Hu, L.; Zhang, S.G.; Zhao, Z.Q.; Ai, S. Modeling angular contact ball bearing without raceway control hypothesis. *Mech. Mach. Theory* **2014**, *82*, 154–172. [CrossRef]
12. Ye, M.; Liang, J.; Li, L.; Qian, B.; Ren, M.; Zhang, M.; Lu, W.; Zong, Y. Full-field motion and deformation measurement of high speed rotation based on temporal phase-locking and 3D-DIC. *Opt. Lasers Eng.* **2021**, *146*, 106697. [CrossRef]
13. Bao, R.; Duan, F.; Fu, X.; Yu, Z.; Liu, W.; Guo, G. Frequency-scanning interferometry for axial clearance of rotating machinery based on speed synchronization and extended Kalman filter. *Opt. Lasers Eng.* **2023**, *164*, 107515. [CrossRef]
14. Wang, L.; Bi, S.; Li, H.; Gu, Y.; Zhai, C. Fast initial value estimation in digital image correlation for large rotation measurement. *Opt. Lasers Eng.* **2020**, *127*, 105838. [CrossRef]
15. Jia, R.; Xue, J.; Lu, W.; Song, Z.; Xu, Z.; Lu, S. A Coupled Calibration Method for Dual Cameras-Projector System with Sub-Pixel Accuracy Feature Extraction. *Sensors* **2024**, *24*, 1987. [CrossRef] [PubMed]
16. Yu, C.; Ji, F.; Xue, J.; Wang, Y. Adaptive binocular fringe dynamic projection method for high dynamic range measurement. *Sensors* **2019**, *19*, 4023. [CrossRef]
17. Xu, J.; Zheng, S.; Sun, K.; Song, P. Research and application of contactless measurement of transformer winding tilt angle based on machine vision. *Sensors* **2023**, *23*, 4755. [CrossRef]
18. Furuno, S.; Kobayashi, K.; Okubo, T.; Kurihara, Y. A study on spin-rate measurement using a uniquely marked moving ball. In Proceedings of the 2009 ICCAS-SICE, Fukuoka, Japan, 18–21 August 2009; IEEE: Piscataway, NJ, USA, 2009; pp. 3439–3442.
19. Liao, Y.H.; Wang, L.; Yan, Y. Instantaneous rotational speed measurement of wind turbine blades using a marker-tracking method. In Proceedings of the 2022 IEEE International Instrumentation and Measurement Technology Conference (I2MTC), Ottawa, ON, USA, 16–19 May 2022; IEEE: Piscataway, NJ, USA, 2022; pp. 1–5.
20. Wang, Y.; Wang, L.; Yan, Y. Rotational speed measurement through digital imaging and image processing. In Proceedings of the 2017 IEEE International Instrumentation and Measurement Technology Conference (I2MTC), Turin, Italy, 22–25 May 2017; IEEE: Piscataway, NJ, USA, 2017; pp. 1–6.
21. Wang, T.; Wang, L.; Yan, Y.; Zhang, S. Rotational speed measurement using a low-cost imaging device and image processing algorithms. In Proceedings of the 2018 IEEE International Instrumentation and Measurement Technology Conference (I2MTC), Houston, TX, USA, 14–17 May 2018; IEEE: Piscataway, NJ, USA, 2018; pp. 1–6.
22. Wang, T.; Yan, Y.; Wang, L.; Hu, Y. Rotational speed measurement through image similarity evaluation and spectral analysis. *IEEE Access* **2018**, *6*, 46718–46730. [CrossRef]
23. Wang, T.; Yan, Y.; Wang, L.; Hu, Y.; Zhang, S. Instantaneous rotational speed measurement using image correlation and periodicity determination algorithms. *IEEE Trans. Instrum. Meas.* **2019**, *69*, 2924–2937. [CrossRef]
24. Wang, S.; Xu, Y.; Zheng, Y.; Zhu, M.; Yao, H.; Xiao, Z. Tracking a golf ball with high-speed stereo vision system. *IEEE Trans. Instrum. Meas.* **2018**, *68*, 2742–2754. [CrossRef]

25. Jung, J.; Park, H.; Kang, S.; Lee, S.; Hahn, M. Measurement of initial motion of a flying golf ball with multi-exposure images for screen-golf. *IEEE Trans. Consum. Electron.* **2010**, *56*, 516–523. [CrossRef]
26. Seo, S.W.; Kim, M.; Kim, Y. Optical and acoustic sensor-based 3D ball motion estimation for ball sport simulators. *Sensors* **2018**, *18*, 1323. [CrossRef]
27. Zhang, Z.; Xu, D.; Tan, M. Visual measurement and prediction of ball trajectory for table tennis robot. *IEEE Trans. Instrum. Meas.* **2010**, *59*, 3195–3205. [CrossRef]
28. Xie, Z.; Yang, C. Binocular Visual Measurement Method Based on Feature Matching. *Sensors* **2024**, *24*, 1807. [CrossRef]
29. Zhang, Y.; Xiong, R.; Zhao, Y.; Wang, J. Real-time spin estimation of ping-pong ball using its natural brand. *IEEE Trans. Instrum. Meas.* **2015**, *64*, 2280–2290. [CrossRef]
30. Li, W.; Jin, J.; Li, X.; Li, B. Method of rotation angle measurement in machine vision based on calibration pattern with spot array. *Appl. Opt.* **2010**, *49*, 1001–1006. [CrossRef]
31. Zhong, J.; Zhong, S.; Zhang, Q.; Peng, Z.; Liu, S.; Yu, Y. Vision-based measurement system for instantaneous rotational speed monitoring using linearly varying-density fringe pattern. *IEEE Trans. Instrum. Meas.* **2018**, *67*, 1434–1445. [CrossRef]
32. Kim, H.; Yamakawa, Y.; Senoo, T.; Ishikawa, M. Visual encoder: Robust and precise measurement method of rotation angle via high-speed RGB vision. *Opt. Express* **2016**, *24*, 13375–13386. [CrossRef]
33. Guo, J.; Zhu, C.; Lu, S.; Zhang, D.; Zhang, C. Vision-based measurement for rotational speed by improving Lucas–Kanade template tracking algorithm. *Appl. Opt.* **2016**, *55*, 7186–7194. [CrossRef]
34. Ma, S.; Yin, Y.; Chao, B. A real-time coupling model of bearing-rotor system based on semi-flexible body element. *Int. J. Mech. Sci.* **2023**, *245*, 108098. [CrossRef]
35. Lu, X.; Xue, J.; Zhang, Q. High camera calibration method based on true coordinate computation of circle center. *Chin. J. Lasers* **2020**, *47*, 0304008.
36. Moré, J.J. The Levenberg-Marquardt algorithm: Implementation and theory. In *Proceedings of the Numerical Analysis: Proceedings of the Biennial Conference, Dundee, Scotland, 28 June–1 July 1977*; Springer: Berlin/Heidelberg, Germany, 2006; pp. 105–116.

sensors

MDPI

Review

Fault Types and Diagnostic Methods of Manipulator Robots: A Review

Yuepeng Zhang [1,2], Jun Wu [1,2,*], Bo Gao [1,2], Linzhong Xia [1,2], Chen Lu [2], Hui Wang [2] and Guangzhong Cao [3]

1 School of Sino-German Robotics, Shenzhen Institute of Information Technology, Shenzhen 518172, China; ypzhang@sziit.edu.cn (Y.Z.); gaob@sziit.edu.cn (B.G.); xialz@sziit.edu.cn (L.X.)
2 Inovance Industrial Robot Reliability Technology Research Institute, Shenzhen Institute of Information Technology, Shenzhen 518172, China; luchen@sziit.edu.cn (C.L.); wanghui@sziit.edu.cn (H.W.)
3 Guangdong Key Laboratory of Electromagnetic Control and Intelligent Robots, Shenzhen University, Shenzhen 518060, China; gzcao@szu.edu.cn
* Correspondence: wujun@sziit.edu.cn

Abstract: Manipulator robots hold significant importance for the development of intelligent manufacturing and industrial transformation. Manufacturers and users are increasingly focusing on fault diagnosis for manipulator robots. The voltage, current, speed, torque, and vibration signals of manipulator robots are often used to explore the fault characteristics from a frequency perspective, and temperature and sound are also used to represent the fault information of manipulator robots from different perspectives. Technically, manipulator robot fault diagnosis involving human intervention is gradually being replaced by new technologies, such as expert experience, artificial intelligence, and digital twin methods. Previous reviews have tended to focus on a single type of fault, such as analysis of reducers or joint bearings, which has led to a lack of comprehensive summary of various methods for manipulator robot fault diagnosis. Considering the needs of future research, a review of different fault types and diagnostic methods of manipulator robots provides readers with a clearer reading experience and reveals potential challenges and opportunities. Such a review helps new researchers entering the field avoid duplicating past work and provides a comprehensive overview, guiding and encouraging readers to commit to enhancing the effectiveness and practicality of manipulator robot fault diagnosis technologies.

Keywords: manipulator robot; type of fault; signal acquisition; fault diagnosis

1. Introduction

With the rapid development of intelligent manufacturing, manipulator robots are gradually replacing traditional manual operations, significantly improving production efficiency while reducing labor costs [1–4]. They have become one of the key technologies in the intelligent manufacturing industry due to their high efficiency, high precision, and high stability. However, in automated production, robots at each stage have a clear division of labor and close collaborative relationships [5–7]. When a robot malfunctions, it will inevitably affect the progress and safety of production [8]. The field of manipulator robotics is experiencing a notable increase in demand for research into fault diagnosis [9]. Firstly, there is a demand for increased production efficiency. Manipulator robots, as key production tools in modern manufacturing, directly impact production efficiency and product quality due to their operational status [8,10]. Through fault diagnosis in manipulator robots, potential issues can be detected promptly, thereby preventing sudden equipment failures that lead to production downtime and consequently enhancing production efficiency [11].

Secondly, there is a need to reduce maintenance costs. Fault diagnosis of manipulator robots enables the extension of maintenance and overhaul intervals, thereby reducing the operation and maintenance costs for enterprises [12,13]. Thirdly, there is a demand for the intelligent upgrading of production. Fault diagnosis technology for manipulator robots is a key element in this process. It can enhance the technological autonomy and controllability, as well as the level of intelligence of manipulator robots, thereby satisfying the ever-increasing production demands of enterprises [14]. Fourthly, there is a demand for safety assurance. The operational environments of manipulator robots are typically complex and challenging. Through fault diagnosis, monitoring the operational status and providing early warnings for abnormal conditions can prevent the occurrence of safety accidents [15].

The application of advanced fault diagnosis techniques to manipulator robot systems not only improves the operational efficiency and performance of the system [16,17], but also enhances its reliability by optimizing the fault diagnosis process [18,19]. Fault diagnosis methods for manipulator robots include data mining, machine learning, neural networks, and digital twins, among others, and have yielded some research outcomes [20–25]. For instance, Wei Bo et al. [24] conducted an oil sample analysis by extracting the mixed lubricating oil from the Rotary Vector (RV) reducer to investigate the relationship between rotational speed and load with the properties of the lubricating oil, thereby determining the operational status of the robot. Yan Bei et al. [25] employed a combination of horizontal-level functions and gas volume techniques to estimate multiple performance indicators of manipulator robot lubricating oil.

Some robot companies have conducted long-term research and practice in the field of fault diagnosis [26]. For example, well-known robot manufacturers such as FANUC, KUKA, ABB, etc. have leading technologies and solutions in fault diagnosis of manipulator robots [27,28]. In addition, some universities and research institutions have also explored the fault diagnosis of manipulator robots [29–34]. Ref. [32] focuses on the potential actuator faults in the joints of heavy-duty manipulator robots during operation. It investigates a data-driven intelligent fault diagnosis algorithm, which involves injecting various faults into the established dynamics model of the heavy-load manipulator robot to obtain unbalanced samples. Ultimately, a composite neural network is employed to achieve the diagnosis and identification of faults. Ref. [33] investigates fault detection methods for manipulator robot gearboxes based on vibration signals. This method employs the time-domain averaging of phase to remove estimated deterministic signals from the measured vibration signals, thereby calculating the residual signal and achieving fault diagnosis. Ref. [34] investigates a fault diagnosis method for manipulator robot joint bearings based on acoustic signal analysis. This method extracts defect frequency peaks and changes in bearing rotational speed from the original acoustic signals using the envelope spectrum and then employs a convolutional neural network (CNN) to extract features and classify the defects.

The fault diagnosis of manipulator robots is meaningful for improving the development of industrial automation. For fault diagnosis in manipulator robots, there are some questions worth considering: What are the faults of existing manipulator robots? What are the methods and techniques for fault diagnosis? How can existing technologies be combined to solve data quality, generalization ability, and interpretability issues, to improve the accuracy and practicality of fault diagnosis? What are the advantages and limitations of existing fault diagnosis methods? What are the future expectations of manipulator robot fault diagnosis? This paper provides a systematic review of the types of failures in manipulator robots, as well as the methods and techniques used for fault diagnosis. Compared to other reviews, this review has two differences:

(1) Previous reviews have not classified the types of faults and the development of diagnostic methods, but instead mostly analyzed the faults of a certain component of manipulator robots. For example, Huang et al. [35] discussed the diagnosis of compound faults in rotating machinery, and Nandi et al. [36] reviewed the state monitoring and fault diagnosis of motors, but did not analyze diagnostic methods and techniques. This review provides general guidance on this topic.
(2) Previous reviews have not discussed and analyzed the overall structure of manipulator robots and the types of failures that can occur in each component. This review supplements this content.

As a comprehensive review dedicated to summarizing and discussing the types of faults and fault diagnosis methods in manipulator robots, this paper has made the following contributions: (1) This paper analyzes the composition of manipulator robot systems and the types of faults in various components, including Rotary Vector (RV) reducers, balance cylinders, motors, and joint bearings. (2) This paper systematically summarizes the methods and techniques involved in fault diagnosis of manipulator robots, categorizing them into traditional and modern approaches based on their development history. Furthermore, it provides a detailed classification and discussion of these methods based on their characteristics. (3) This paper macroscopically explored the advantages, challenges, and development trends of fault diagnosis technology for manipulator robots.

2. Composition and Fault Types of Manipulator Robots

2.1. Faulty Components and Fault Types in Manipulator Robots

The precise structure of the core components of manipulator robots allows them to maintain stable performance under various working conditions and to handle industrial tasks that require high precision and high intensity. Once a critical component of a manipulator robot fails, it not only reduces the service life but also accelerates the degradation cycle of the robot. The basic components of a manipulator robot system involve reducers, counterbalance cylinders (only for heavy-duty manipulator robots), servo motors, robotic arms, encoders, control systems, cables and connectors, and lubrication systems, as shown in Figure 1. Among these components, servo motors, RV reducers, and joint bearings are the most common and susceptible to failure in manipulator robots. When the failure occurs, it directly impacts production efficiency and progress.

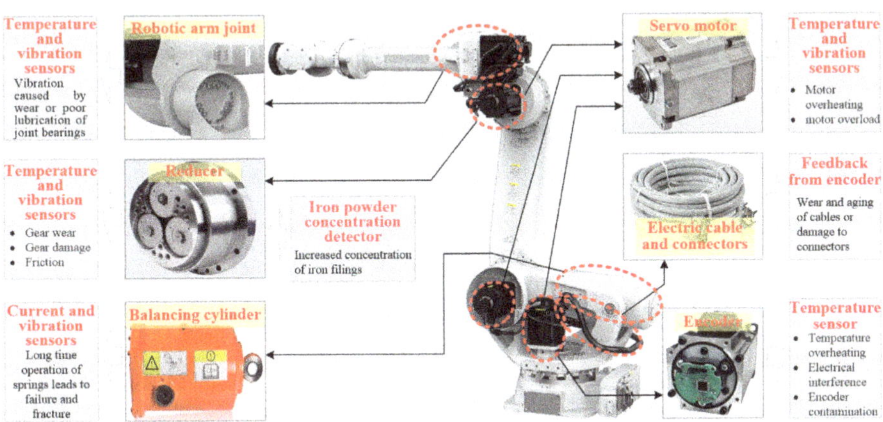

Figure 1. Components and fault characteristics of manipulator robots.

2.2. Fault Types of Manipulator Robots

The development process of fault diagnosis for manipulator robots has been of concern since the invention and manufacture of robots [37,38]. Initially, robot manufacturers and users relied on simple protections, such as inspections conducted by experienced engineers or regular production line shutdowns [39,40]. As the tasks performed by manipulator robots become increasingly complex, it has become very important to diagnose faults before they occur, as robot downtime outside of project execution plans can disrupt the overall project schedule and cause serious economic losses. In fault diagnosis of manipulator robots, there are indeed complex interactions between different fault types, which have a significant impact on the operating status and diagnostic strategies of the robot. The coupling fault between mechanical components and electrical systems leads to abnormal vibration of the entire system, which increases the complexity of fault diagnosis. The operating conditions of manipulator robots can affect the manifestation of faults, which puts higher demands on fault diagnosis models. Manipulator robot failures often do not occur in isolation, but gradually trigger chain failures of other components over time. For manipulator robot systems, the main causes of component failures are as follows, and summarized in Table 1.

Reducer: In the long-term operational state of the manipulator robot, the reducer in the joints may experience gear wear or damage due to frequent high/low-speed transitions, high torque loads, and prolonged transmission [41–46]. Rohan et al. [43] proposed a method for detecting and diagnosing mechanical faults in RV reducers using current signals from embedded control units. This method involves decomposing the original signal into wavelets and extracting, selecting, and injecting various types of features from them to enhance the accuracy of machine learning-based classifiers. Yang et al. [44] proposed a composite fault diagnosis method for harmonic reducers based on a deep capsule graph convolutional network (GCN). The method acquires fault signal spectra from multi-sensor data and constructs a relationship graph using an adjacency matrix. It then utilizes both a deep capsule network and a GCN to learn fault vector representations and the relationships between different single-label faults. Yang [45] proposed a method based on generative adversarial networks to address the issue of data imbalance. This method involves collecting vibration data and using a multi-scale CNN to achieve fault diagnosis of harmonic reducers. Xing et al. [46] proposed a one-dimensional CNN adaptive method based on matrix kernels, aiming at the characteristics of sampling signals in manipulator robots. This method adjusts the size of the convolutional kernels to process the data of multichannel fusion and to extract contextual features from continuous time-domain data.

Joint bearing: Poor lubrication of the joint bearings, rotor asymmetry, defects in the bearing seat rings, and defects of rolling elements can lead to unsmooth motion or vibration in manipulator robots [47–51]. Yazici et al. [47] reported an adaptive statistical time-frequency method for detecting bearing faults. This method is tested on defective bearings with scratches on the outer race, balls, and cages, and all defects are correctly classified. Xia et al. [48] proposed a deep perception adversarial domain adaptive method for fault diagnosis of manipulator robot bearings under variable operating conditions, which is superior to CNN and conditional domain adversarial networks. McInerny et al. [49] diagnosed bearing faults using envelope detection technology and demonstrated that this method is superior to traditional spectral analysis methods. Raouf et al. [50] used a feature aggregation network with a hierarchical information aggregation mechanism for fault detection of manipulator robot bearings and demonstrated its adaptability and robustness in different experimental scenarios. Zhang et al. [51] considered the rolling elements and bearing seats of bearings, established a dynamic model of rolling bearings compound fault based on time-varying displacement, and discussed the relationship between the vibration response of compound faults and three different operating conditions.

Servo motor: Servo motors are the power source for manipulator robots. Due to prolonged high-load operation, it may experience a fault due to overheating, overload, or damage to the motor encoder [52,53]. Junior et al. [52] measured acceleration data in two different directions, and used one-dimensional CNN to detect and diagnose six different types of motor faults. Suawa et al. [53] collected multi-source data such as motor vibration and sound, and analyzed and solved the problem of predictive maintenance of motors through artificial intelligence methods.

Electric cable: The abrasion and aging of cables will lead to electrical connection failure, affecting the normal work of manipulator robots [54,55]. Aiming at the damage of conductor or insulator in the cable due to different conditions, which leads to faults, Kim et al. [54] studied a new method for soft faults diagnosis of control cable based on detection, using the three-phase current to calculate the sum and ratio of current, and taking it as the input of diagnosis network to detect the cause of soft faults. Kim [55] calculated the three-phase current through the automatic encoder, calculated the difference fraction to detect the cable fault, and used the fault method combining long short-term memory (LSTM) and attention mechanism to estimate the fault severity through time-related information.

Screw: Due to the long time and many times of carrying objects, the wear of robot screw leads to the deterioration of health. Yang et al. [56] used the motor current signal to diagnose the health state of the ball screw of the manipulator robot, the Fisher score for feature selection, and logical regression and k-nearest neighbor for fault diagnosis.

Compound fault: The manipulator robot system contains many components, which are spatiotemporally related and need to cooperate to complete the work. Due to the poor working environment, improper operation, long-term work, and other reasons, the functional components of manipulator robots are prone to damage, resulting in compound faults in the robot system [35,57–59]. For example, He et al. [58] proposed an adaptive redundant multi-wavelet compound fault diagnosis method to address the complexity and diversity of compound faults and combined it with Hilbert transform demodulation analysis to effectively detect compound faults in gearboxes. Zhou et al. [59] proposed a compound fault diagnosis algorithm for manipulator robots based on multimodal feature extraction and fusion. This method uses a multi-head self-attention enhanced CNN module and an LSTM module to simultaneously learn fault-related features from different perspectives, fuse the extracted features, and perform compound fault classification.

In addition, sensors used in manipulator robots, such as encoders and force sensors, may also experience faults due to temperature changes, mechanical shocks, or electrical interference. For heavy-duty manipulator robots, the spring inside the balancing cylinder may also break due to prolonged operation. Table 1 summarizes the component positions, fault diagnosis data types, and diagnostic methods of common faults in manipulator robots.

However, in the actual working environment of manipulator robots, compound faults are the most common and strongly correlated with individual faults, rather than simply linear superposition. In addition, compound faults in manipulator robots are caused by the coupling of multiple subsystems and various types of noise, which makes it difficult to generalize them into common patterns, thus rendering their fault diagnosis more challenging [60,61].

Table 1. Fault locations, data types, and diagnostic methods for manipulator robots in common.

Position of Faults	Types of Data	Methods for Fault Diagnosis	References
Reducer	Current signal data	Based on wavelet features and statistical analysis	[43]
	Vibration signal data	Deep capsule graph convolutional network	[44]
	Vibration signal data	Generative adversarial networks and multi-scale convolutional neural networks	[45]
	Vibration signal data	Adaptive method of one-dimensional convolutional neural network	[46]
Joint bearing	Current data	Adaptive statistical time-frequency method	[47]
	Acceleration data	Deep perception adversarial domain adaptive method	[48]
	Vibration signal data	Envelope detection technology	[49]
	Current signal data	Feature aggregation network based on hierarchical information aggregation mechanism	[50]
	Vibration signal data	Four-degree-of-freedom dynamic model of compound local fault rolling bearings based on time-varying displacement	[51]
Servo motor	Vibration signal data (IMI uniaxial accelerometer)	One-dimensional convolutional neural network	[52]
	Vibration signal data, sound data	Deep convolutional neural networks, long-short term memory methods, and convolutional neural networks–long short-term memory	[53]
Electric cable	Current signal data	Calculate the total sum and ratio of three-phase currents	[54]
	Current signal data	Calculate the calculation difference fraction of the three-phase current	[55]
Screw	Current signal data	Fisher score, logistic regression, and k-nearest neighbor method	[56]
Compound fault	Acceleration data	Combining adaptive redundant multi-wavelet packet with Hilbert transform demodulation analysis	[58]
	Current signal data	Multi-head self-attention enhanced convolutional neural network module and long short-term memory network	[59]

2.3. *Types of Fault Signal Acquisition*

The common signals for collecting fault data of manipulator robots are mainly divided into two forms based on their collection paradigm and signal source. One type is the data generated by the robot itself during its operation, such as the current, voltage, output torque, angle, angular velocity of the robot joint motor, and the concentration of iron filings of the reducer wear [62]. Another type is to deploy sensors externally and collect available signals according to demand, such as robot vibration signals collected through high-precision accelerometers (vibration sensors), motor and reducer oil temperature collected through temperature sensors, and joint noise signals [63–67]. Among them, vibration signals are commonly used in the fault analysis of manipulator robots [68,69]. For instance, Islam et al. [70] collected data from manipulator robots using vibration sensors, preprocessed the vibration data, and then employed the gradient tree-based XGBoost machine learning classification algorithm to achieve fault diagnosis of manipulator robots. There is a distinct physical relationship between the faults of manipulator robots and vibration signals. When a robot faults, the vibrations generate additional torque components that affect the current spectrum, resulting in specific features in the frequency domain, which can be represented as [71]:

$$f_{bec} = |f_s \pm k f_r| \quad (1)$$

where f_r is the rotational frequency of the analyzed axis.

The faults can be modeled as additional torque components with amplitudes greater than the load at mechanical characteristic frequencies. This component is an additional component of the load torque T_L. The instantaneous torque applied to the motor is expressed as:

$$T_{load}(t) = T_L + \Delta T_{load}(t) \quad (2)$$

where T_{load} is the load torque.

It is worth considering that although vibration signals contain abundant information about the operational status of robots, the need for external sensors to measure vibrations and the requirement for higher-performance sensors due to frequency effects significantly increase costs. In addition, each system of manipulator robots contains multiple rigid and flexible components, and each joint requires multiple measurements, resulting in inaccurate diagnostic results and tedious work [72]. Some manipulator robots cannot use or install external sensors in actual industrial production environments, making it difficult to use vibration signals for manipulator robot fault analysis. Similarly, only collecting data from manipulator robots, such as voltage, current, speed, etc., cannot fully reflect fault information [73,74].

3. Fault Diagnosis Methods for Manipulator Robots

Fault diagnosis technology, as an important component of predictive maintenance, has become a crucial tool for ensuring the efficient, stable, and safe operation of robot systems in automated production. The fault diagnosis of manipulator robots is of great significance for the transformation and upgrading of the modern manufacturing industry.

Manipulator robot systems exhibit a wide variety of faults, and relying on conventional signal detection methods and fault diagnosis techniques has significant limitations. Faults in different components and locations cause varying changes, and their external performances are also diverse, representing a complex nonlinear mapping relationship. With the rapid development of data analysis technologies such as artificial intelligence and deep learning, obtaining high-precision robot fault signals through various sensors and measurement schemes is a basic prerequisite for fault diagnosis. The increase in methods for collecting various data information has led to a surge of heterogeneous information, there is a phenomenon of a large influx of heterogeneous information, characterized by overlap, contradiction, multi-source, and a wide variety of structures. There are significant differences in precision, detail, structure, and extraction methods, and these uncertain factors pose challenges to manipulator robot fault diagnosis.

There are various methods for diagnosing faults in manipulator robots, which can be divided into several stages based on the development process of diagnostic technology:

(1) The first method involves comparing the operational status parameters of manipulator robots, such as abnormal temperatures, vibrations, and noises, with their historical normal operating parameters to make maintenance decisions. This diagnostic method is simple and convenient, but accuracy is relatively low.

(2) The second method is to use the manipulator robot's own state recognition and fault diagnosis mode to reflect fault information based on incorrect codes, making it easier for engineers to directly locate the type and location of the fault.
(3) The third method is to construct digital twins of virtual, utilizing actual manipulator robots and mapped proportional digital models, and implementing remote monitoring, providing fault diagnosis services, and maintenance support through the network.

In addition, the research on fault diagnosis of manipulator robots can be classified according to their working principles and application technologies, and the application technologies can be further divided into traditional methods and modern methods, as shown in Figure 2. The traditional method usually utilizes expert experience, statistics, and other methods to use the knowledge and experience of early technical engineers to determine the cause of manipulator robot failures. Modern methods use new methods and technologies, such as machine learning, deep learning, and other artificial intelligence methods, as well as information fusion and digital twin to diagnose robot faults.

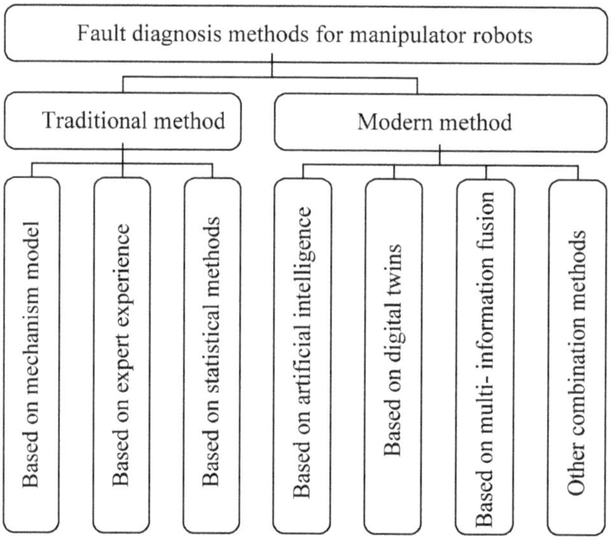

Figure 2. Classification of fault diagnosis methods for manipulator robots.

3.1. *Based on Traditional Methods*

3.1.1. Method Based on Expert Experience System

The expert experience method is an early used method for manipulator robot fault diagnosis [31,75–77]. Its core is to use the experience and knowledge of experts with rich experience and professional knowledge to achieve fault diagnosis of manipulator robots, as shown in Figure 3. Liu et al. [76] first estimated the motion state of the robot using a Kalman filter and calculated the probability of each motion state corresponding to different fault modes, providing preliminary diagnostic results. They then selected appropriate knowledge from the expert system's knowledge base to infer and judge the diagnostic results. Sun et al. [77] proposed a novel fault diagnosis method based on experimental data-driven stochastic fuzzy evidence acquisition and intuitionistic fuzzy set fusion. This method involves constructing a fuzzy expert system to match the samples under test. The intuitionistic fuzzy set fusion employs binary vector arrays of intuitionistic fuzzy data sets to represent the likelihood values of samples, capturing uncertainty information to enhance the reliability and accuracy of fault diagnosis.

Expert experience systems, while not requiring the establishment of precise mathematical models, heavily rely on expert empirical knowledge.

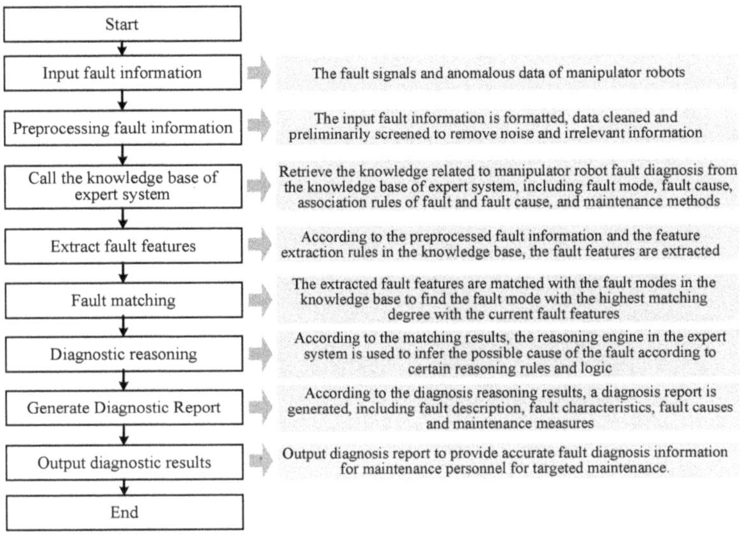

Figure 3. Fault diagnosis process of a manipulator robot based on an expert experience system.

3.1.2. Method Based on Statistical Models

By employing operations such as dimensionality reduction during data processing, redundant data can be effectively eliminated, significantly reducing the computational load. For instance, using multivariate statistical analysis methods, fault diagnosis can be performed based on the correlations between variables. Wu et al. [78] collected vibration signals from manipulator robots under three operating conditions as raw data. They employed principal component analysis, locally preserving projection, and isometric feature mapping to extract three-dimensional features from the vibration signals and used these features as the input for the training model, effectively identifying the fault types of manipulator robots.

Statistical methods are used in manipulator robot fault diagnosis to analyze historical data, identify features related to faults, and use these features to diagnose the faults that occur. For example, statistical indicators such as mean, median, variance, and residuals can be used to describe the operational data of manipulator robots.

3.1.3. Method Based on Mechanism Knowledge-Driven Model

The fault diagnosis method based on a mechanism knowledge-driven model mainly relies on the physical model and accurate mathematical model of manipulator robots. It uses state estimation, and parameter estimation to diagnose faults by comparing and analyzing the difference information between the model output and the actual output [79], as shown in Figure 4.

For example, Piltan et al. [80] used a neural adaptive high-order variable structure observer to accurately estimate the position and torque signals of a manipulator robot and then utilized the characteristics of the position and torque signals of the manipulator robot to improve the robot's fault recognition ability by support vector machines. They designed a modern fuzzy backstepping variable structure controller, used adaptive algorithms to fine-tune controller parameters, and reduced the impact of faults on manipulator robots. Moshgani et al. [81] used a method that combines an unknown input observer with a genetic algorithm to determine the optimal observer parameters and designed an adaptive threshold for the residual signal to achieve fault diagnosis of manipulator robots.

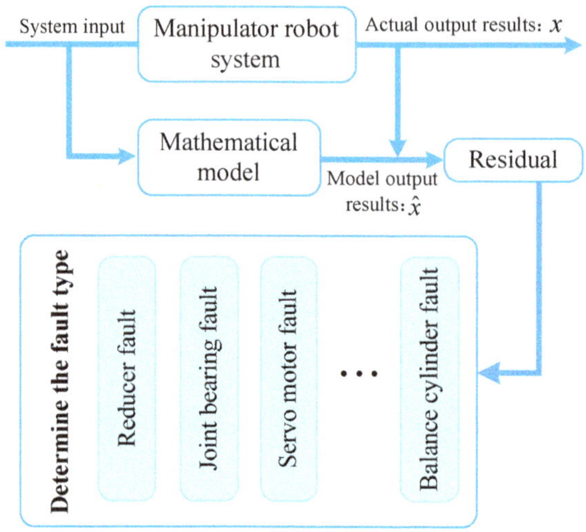

Figure 4. Framework of fault diagnosis method based on mechanism knowledge-driven model.

The Kalman filter is used for fault diagnosis of robots [82], which determines the possible existence of faults by comparing the residual between sensor measurements and Kalman filter predictions [83], as expressed in Equations (3)–(7).

- Prediction

Prediction of the state ahead, and prediction the error covariance

$$x(k+1|k) = Ax(k|k) + Bu(k) \tag{3}$$

$$p(k+1|k) = Ap(k|k)A^T + Q_{k-1} \tag{4}$$

- Correction

Compute the Kalman Gain

$$K_k = \frac{H^T p(k+1|k)}{Hp(k+1|k)H^T + R(k)} \tag{5}$$

Update the estimated value, and update the prediction covariance

$$x(k|k) = x(k|k-1) + K_k[z(k) - Hx(k|k-1)] \tag{6}$$

$$p(k) = (I - K_k H)p(k-1) \tag{7}$$

where $x(k|k)$ is the estimated value of time k, $x(k|k-1)$ is the state estimation of time $k-1$, $x(k+1|k)$ is the predicted value of the next step, $p(k)$ is the covariance matrix corresponding to the state variable, $p(k+1|k)$ is the covariance matrix of the next step of prediction, A is the state transition matrix, B is the matrix, H is the observation matrix, u is the control input, and Q and R are the covariances.

For example, Van et al. [84] proposed a fault diagnosis method based on the Kalman filter to effectively detect actuator faults. This method calculates the estimation error of the Kalman filter (the difference between the actual output based on image features and the estimated output of the Kalman filter) and compares it with a preset threshold to determine whether the system has malfunctioned. Zhang et al. [85] proposed an efficient fault diagnosis and fault-tolerant control method based on adaptive extended Kalman filter and sliding mode control. This method identifies faults based on the control effectiveness coefficient estimated by the adaptive extended Kalman filter and verifies its fault diagnosis and control performance under various fault conditions through simulation.

Mechanism knowledge-driven fault diagnosis methods for manipulator robots focus on the system entity of the diagnosis object, with a solid theoretical foundation, but they demand high accuracy of the mechanism model.

3.1.4. Methods Based on Signal Processing

Signal-based analysis and processing methods involve preprocessing data samples and then extracting time-domain, frequency-domain, and time–frequency-domain fault features from the data samples [86–88]. Signal processing methods diagnose faults by analyzing various signals generated during the operation of robots, such as vibration, current, sound, and temperature [66,89–92]. Common signal processing techniques include Fourier transform, wavelet transform, variational mode decomposition, local mean decomposition, etc. [93–97]. Jiang et al. [95] investigated a method that combines empirical wavelet transform with chaotic oscillators. By using the empirical wavelet transform, the composite fault signal is decomposed into different empirical modes. A fault isolator is then established using an oscillator that includes all individual fault frequencies. Finally, all fault types are identified by observing the irregular motion generated from the mapping of the isolator's output. Wan et al. [96] combined fast spectral kurtosis with variational mode decomposition to process compound fault signals containing weak single components. Zhao et al. [97] investigated a fault signal processing method based on local mean decomposition, considering the local non-stationary characteristics of vibration signals. By defining a non-stationary coefficient to characterize the local non-stationary features of vibration signals, they proposed a composite envelope construction method. This method involves using monotonic segment cubic Hermite interpolation for the non-stationary parts and cubic spline interpolation for the smooth parts. The local mean decomposition algorithm designed based on this envelope method is applied to fault diagnosis.

For the traditional methods, although these methods have shown some effectiveness in the early fault diagnosis of manipulator robots, with the increase in the complexity of industrial scenes, the accuracy and stability of the traditional methods are limited.

3.2. Based on Modern Methods

In recent years, the rapid development of emerging technologies such as artificial intelligence, big data, cloud computing, Internet+, and information fusion has stimulated breakthroughs in manipulator robot fault diagnosis technology [98,99]. By constructing data-friendly signal processing methods to extract features that differ from normal information, fault-related information can be obtained [100]. Then, a fault diagnosis model can be constructed using deep learning or machine learning methods and displayed through a visual interface, as shown in Figure 5.

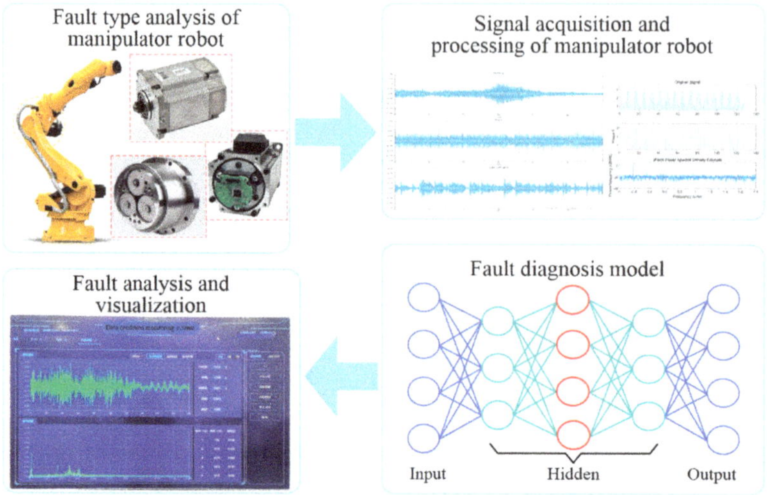

Figure 5. Manipulator robot fault diagnosis based on modern methods.

3.2.1. Method Based on Artificial Intelligence

The resurgence of neural networks and the rise of machine learning have led to breakthroughs in fault diagnosis research [101,102]. Manipulator robot fault diagnosis methods based on neural networks have also become increasingly popular. These methods use historical data that has been collected and subjected to feature selection to train various types of learning algorithms, thereby identifying compound fault patterns. Finally, online data is used as input to the model for fault diagnosis, and the process is illustrated in Figure 6. For example, Bai et al. [103] proposed a method for screening fault data and a fault rate diagnosis framework for manipulator robots. They established a fault rate model for manipulator robots with different subsystems, then used a surrogate model to fit the bathtub curve of the original robot to obtain the early failure time points. The distribution parameters of the original manipulator robot were solved using the maximum likelihood function to quantify the influencing factors of the manipulator robot. Finally, the interval hierarchical analysis method was used to refine the cognitive uncertainty and obtain the correction factor for the fault rate. Although methods based on traditional neural networks offer certain advantages for fault prediction in manipulator robots, limited sample sizes can result in poor generalization performance of these networks.

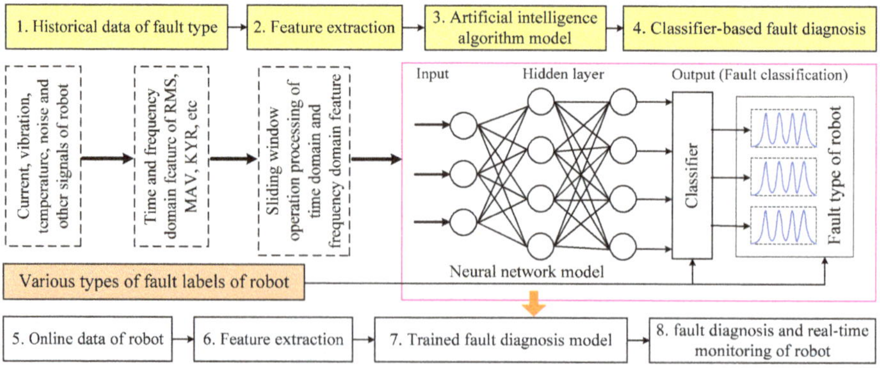

Figure 6. Fault diagnosis process based on artificial intelligence method.

Artificial intelligence methods, such as CNN [104] and GNN [105], have brought significant changes to the field of signal processing and are receiving considerable attention in the field of fault diagnosis. These methods can increasingly accurately fit the nonlinear relationships between samples and label values through continuous sample training. Moreover, by continuously learning from historical and new fault data, it can construct and optimize fault diagnosis models. These methods require minimal human intervention and have advantages in handling complex, nonlinear problems. Lu et al. [11] proposed a fault diagnosis method for manipulator robot reducers using dual module attention CNNs. This method captures different features related to faults by establishing two parallel CNNs with different focuses. Finally, the fault features are fused to obtain the fault diagnosis result (normal or abnormal). Jiao et al. [106] proposed a fault diagnosis method based on deep belief network (DBN) joint information fusion technology to overcome the problems of low accuracy, low efficiency, poor stability, and poor real-time performance of traditional manipulator robot fault diagnosis models in multi-fault state diagnosis. This method, combining DBN and wavelet energy entropy, is used to study the fault diagnosis of manipulator robots. It uses wavelet transform to denoise, decompose, and reconstruct vibration signals of manipulator robot joint bearings, and establishes normalized eigenvectors for reconstructing energy entropy and using them as inputs to DBN. The use of improved D-S evidence theory has solved the problem of eliminating high-conflict evidence and improved the recognition accuracy of fault models. Yang et al. [8] proposed a fault diagnosis method for manipulator robots in dynamic working states. This study utilizes a domain generalization adversarial long short-term memory residual life prediction modeling method to reduce differences between robots. The two-stage health assessment method of principal component analysis prediction error square and p-chart is used to reduce the interference of outliers in the normal operation of robots. The method uses feature smoothing, feature normalization, and feature selection to reduce changes in the working area, thereby effectively reducing errors.

The working state of manipulator robots is affected by external factors, which can lead to a decrease in the accuracy and stability of fault diagnosis models. In addition, the differences between different robots require large-scale adjustments to the fault diagnosis model during deployment. Adaptive learning or transfer learning allows models to dynamically adjust their parameters and structure based on new data, and can quickly improve diagnostic accuracy using previous training experience [107,108]. To address the issue of variable working conditions in manipulator robots, Xia et al. [48] proposed a novel Deep Perceptual Adversarial Domain Adaptation method for fault diagnosis of manipulator robot joint bearings under different operating conditions. This approach employs a novel perception loss based on information entropy to enforce identical distributions between the source and target domains, thereby enhancing the stability of adversarial training. To address the limitations of traditional fault detection methods that require extensive labeled data and exhibit poor generalization capabilities, Kumar et al. [109] proposed a transfer learning-based fault detection method for servo motor bearings, which significantly improves detection efficiency and accuracy in new bearing fault diagnosis tasks by transferring knowledge from pre-trained CNN models. Transfer learning or adaptive learning methods adapt models through prior training experience without requiring full retraining, enabling more efficient deployment across diverse manipulator robot fault diagnosis platforms.

3.2.2. Method Based on Digital Twin Technology

Recently, a new solution has been provided for the fault diagnosis of manipulator robots: digital twin technology. As an emerging cutting-edge technology, digital twin technology utilizes tools such as CAD software(SolidWorks 2022, Dassault Systèmes,

France) [110] to create high-fidelity digital models in virtual space that correspond to physical objects and combines network communication technology to achieve real-time mapping and interaction between the physical world and the digital world [111–113]. Digital twin technology provides new perspectives and methods for the operation and maintenance of manipulator robots, which helps improve the accuracy and real-time performance of fault diagnosis for manipulator robots. The fault diagnosis scheme for manipulator robots based on digital twin technology is shown in Figure 7.

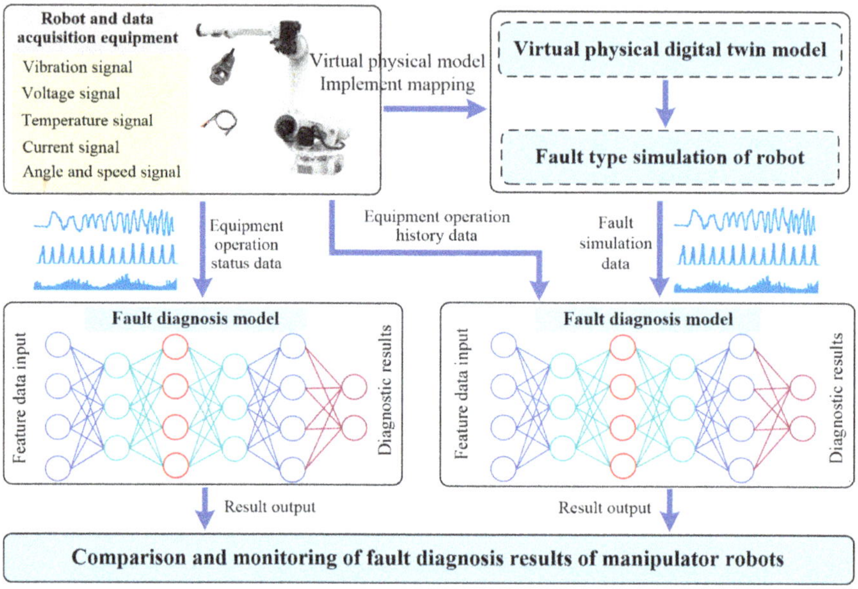

Figure 7. Fault diagnosis architecture for manipulator robots based on digital twins.

Song et al. [114] proposed a manipulator robot joint fault diagnosis system assisted by a twin subsystem. Data generated from the robot dynamics model is utilized as the virtual entity data for the digital twin model. Through the proposed CycleGAN-based digital twin model, a minimal amount of empirical data is employed to map the virtual entity data onto the physical entity. The validity of the simulation data generated by the digital twin model is corroborated using a CNN-ResNet classifier. Ademujimi et al. [115] discussed the application of digital twin technology in fault diagnosis for manipulator robots. The study constructed a co-simulation model that includes network sensors, high-fidelity simulation models, and discrete event simulation models. A structural intervention algorithm is designed to detect directed edges in Bayesian networks and to distinguish between parent nodes and ancestor nodes. The effectiveness of the proposed method was validated through experiments.

3.2.3. Method Based on Multi-Information Fusion

Multi-information fusion is similar to imitating the decision-making process of the brain, perceiving the state information of objective objects through multiple senses, and then conducting joint decision-making and comprehensive analysis through information integration, thereby continuously increasing the cognitive level of objective objects and achieving the goal of recognition. In the diagnostic research of manipulator robots, due to the complexity of the equipment and the harshness of the operating environment, information from single-source sensor devices often fails to extensively extract the global features of faults [116]. This inevitably leads to a reduction in the accuracy of fault diagnosis, and may even result in missed diagnoses and incorrect diagnoses [117]. Multi-information fusion technology provides a new approach to solving the uncertainty problem in mechanical fault diagnosis. In the field of fault diagnosis, multi-information fusion technology comprehensively collects equipment fault status information through multiple sensors and mines the coupling and complementary information between multi-source sensor data. By using multi-feature fusion analysis and joint decision-making as means, it improves the reliability and accuracy of fault diagnosis, overcoming the disadvantages of limited fault information and uncertainty inherent in single sensors [118]. The multi-information fusion method has been widely used in the fault diagnosis of manipulator robots [119–123], as shown in Figure 8.

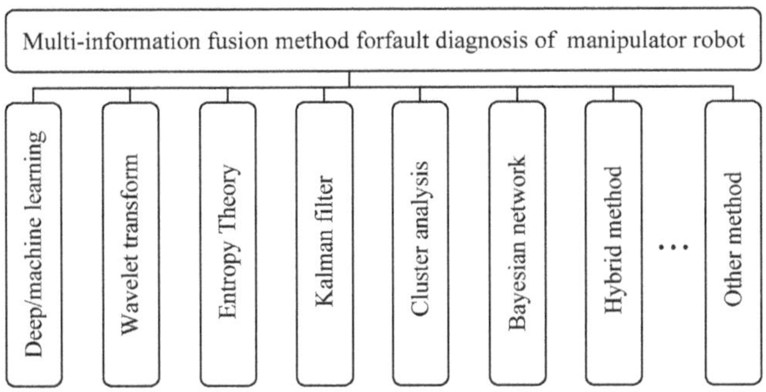

Figure 8. Multi-information fusion methods commonly used in fault diagnosis of manipulator robots.

Gong et al. [117] proposed a hierarchical visual transformer and wavelet time-frequency multi-source information fusion intelligent fault diagnosis method for mechanical components. This method introduces a hierarchical visual transformer to improve feature mapping representation, enriching fault features. It fuses two-dimensional time-frequency diagrams of multi-source signals into the proposed hierarchical visual transformer for a more comprehensive representation of fault features, thereby enhancing the effectiveness of fault diagnosis. Suawa et al. [53] fused vibration and sound data collected by sensors installed on the motor, and analyzed the data through a deep CNN to diagnose faults, thereby addressing the issue of predictive maintenance for motors. Hoang et al. [100] proposed a fault diagnosis method that utilizes deep learning and information fusion techniques. This method primarily collects raw signals from multiple phases of the motor current as direct inputs and extracts features from the current signals of each phase. Then, a CNN is employed to classify each feature set. A novel decision-level information fusion technique is introduced to integrate information from all CNNs, transforming the decision-level information fusion problem into a pattern classification task to obtain accurate fault diagnosis results.

3.2.4. Other Methods for Manipulator Robot Fault Diagnosis

There are also some other combination methods used for fault diagnosis in manipulator robots, which overcome the shortcomings of single methods by combining multiple different forms of methods [124–126]. Composite structure is a proven method. For instance, Yin et al. [125] addressed the issue of integrating prior knowledge with learning networks by proposing a fault diagnosis transfer network driven by both knowledge and data. They extracted prior knowledge from fault signals through envelope analysis and passband selection and used convolutional kernels and designed filters to integrate this prior knowledge. The network employs a deep adaptive network embedding with weight sharing to embed and transfer prior knowledge. Bilal et al. [126] proposed an Internet of

Robotic Things (IoRT) architecture based on transfer learning technology for fault detection of industrial manipulators. This method employs a hybrid 1-D multichannel CNN (MCNN) that combines matrix kernels and RNN to achieve high-precision detection of robot joint states. Furthermore, the study introduced a variable working condition joint fault diagnosis algorithm based on transfer learning with matrix kernels and recurrent neural networks, overcoming the challenges of fault diagnosis for manipulator robots under complex working conditions.

Cross-mode or cross-domain fusion, as a novel method, is applied in the research of fault diagnosis for manipulator robots [127–131]. Ref. [128] proposes a semantic-aware cross-domain industrial process fault diagnosis network based on knowledge mining. The network designs a non-shared attention mechanism to obtain discriminative features for each operating condition. It constructs a fault relationship knowledge graph through a cross-correlation knowledge mining subnetwork to explicitly constrain the local consistency between the source domain, target domain, and cross-domain. A semantic knowledge transfer subnetwork is utilized to impose semantic constraints during the knowledge transfer process. Ultimately, the subnetworks are jointly trained for cross-domain industrial process fault diagnosis. Xu et al. [130] proposed a cross-modal fusion CNN for mechanical fault diagnosis. This network employs two parallel modality feature networks to learn specific modality features of independent signals and a cross-modal knowledge-sharing network to learn shared features from multi-source data. The fault diagnosis performance of the cross-modal fusion convolutional neural network is then enhanced through the online soft-label training strategy. Lee et al. [131] addressed the issue of low prediction confidence in fault diagnosis models in the industrial domain by proposing an interpretable fault diagnosis method that incorporates domain knowledge. This method utilizes the interpretability of generated fuzzy energy pattern image data to infer the causal relationships of faults.

Modern methods have excellent performance in fault diagnosis of manipulator robots, but their application is still limited by factors such as data and computing resources. These limitations will be discussed in the Section 4.

4. Discussion

This paper summarizes the fault types and causes of manipulator robots, as well as the research status of fault diagnosis methods. Tables 2 and 3 respectively list the typical papers of traditional and modern methods for fault diagnosis of manipulator robots.

Table 2. Summary of typical papers on traditional methods.

Form of Method	Sensors	References	Purpose of Research	Key Features of the Research
Expert experience	Robot data and vision sensor	[76]	Avoiding misdiagnosis and missed diagnosis of robot fault	Estimate the motion state of the robot using a Kalman filter and calculate the probability of the corresponding fault mode for the motion state; selecting appropriate knowledge from expert experience systems to diagnostic results
	Vibration measurement sensor	[77]	Strategy and full load working environment, bearing fault diagnosis	Propose a new method of random fuzzy evidence acquisition and intuitionistic fuzzy set fusion by constructing a fuzzy expert system and matching the samples to be tested

Table 2. *Cont.*

Form of Method	Sensors	References	Purpose of Research	Key Features of the Research
Statistic model	Vibration measurement sensor	[78]	A lot of redundant information in the vibration signal, cannot be directly used for fault diagnosis without processing	Principal component analysis, locality preserving projection and isometric feature mapping extract three-dimensional features of vibration signals and form nine-dimensional features, and train naive Bayesian model to identify fault types
Mechanism knowledge-driven	Motor encoder feedback	[80]	Uncertain and nonlinear effects of manipulator robot operating environment fault diagnosis and fault-tolerant control	The neural network adaptive high-order variable structure fault diagnosis observer based on machine learning improves the robustness of fault signal estimation; adaptive modern fuzzy backstepping variable structure controller to reduce the impact of faults on the robot
	Motor encoder feedback	[81]	Fault diagnosis of uncertain and nonlinear effects of manipulator robot operating environment	Using the observer with unknown input, combined with the estimated virtual mode, the residual is generated to detect and diagnose the fault information of the robot
	Image sensor	[84]	Detecting and isolating potential faults in manipulator robots	Kalman filter estimates the system state and calculates the residual between the actual image feature output and the Kalman filter estimation, and determines the fault based on the residual threshold
	Angle sensor and camera	[85]	Improve the safety and robustness of the manipulator when using only position signals	Fault diagnosis and fault-tolerant control methods for adaptive extended Kalman filter and sliding mode, detecting faults and fault-tolerant control of manipulator
Signal Processing	Vibration sensor (accelerometers)	[95]	Different faults occur simultaneously; compound fault decoupling detection is difficult	The empirical wavelet transform uses an adaptive wavelet basis to extract the intrinsic mode of the signal and decomposes the compound fault into single fault with different empirical modes. Each fault is merged into a continuous oscillator, and the fault type is identified by observing the irregular motion generated by the output of the established isolator.
	vibration sensor (accelerometers)	[96]	The faults of bearings are located in different resonance bands and the mutual interference and noise influence between different fault components	The original composite fault signal is decomposed and preprocessed by the variational mode decomposition method, which is decomposed into multiple variational eigenmode function components, and the eigenmode function components are calculated by fast spectral kurtosis
	ICP acceleration sensor	[97]	Vibration signal has nonlinear, non-stationary, multi-component coupling characteristics, and traditional methods cannot effectively extract the characteristics of the vibration signal	Compound interpolation CIE LMD method: nonstationary coefficient is defined to represent the local nonstationarity of vibration signal, monotone piecewise cubic Hermite interpolation is used for the nonstationary part, and cubic spline rubbing is used for the smooth part

Table 3. Summary of typical papers on modern methods.

Form of Method	Sensors	References	Purpose of Research	Key Features of the Research
Artificial intelligence	Motor encoder feedback	[11]	Current methods pay little attention to the correlation and internal differences of test data	Two parallel convolutional neural networks with different attention mechanisms are established to obtain different fault-related features
	vibration measurement sensor	[106]	Low accuracy and efficiency, poor stability and real-time performance of manipulator robot multi-fault state diagnosis	The deep confidence network is combined with wavelet energy entropy; wavelet transform is used to denoise, decompose, and reconstruct the vibration signal, and the normalized eigenvector of reconstructed energy entropy is used as the input of deep confidence network
	Robot controller	[8]	Difficulty in fault diagnosis under dynamic working state	Domain generalization-residual life prediction against long short-term memory, principal component analysis squared prediction error, and p-chart to reduce abnormal interference
Digital twins	Acceleration sensor, Temperature sensor	[114]	The actual joint fault data of the robot is insufficient, which is difficult to obtain in real-time	Based on the digital twin model of CycleGAN, the virtual model is mapped to the physical entity using a small amount of measured data, and the joint state data is obtained in real time
	Vibration measurement sensor; Temperature sensor	[115]	Lack of data and computing power makes it difficult to develop a universal high-fidelity engineering digital twin model for complex systems	Engineering digital twin system: use network physical system to form network sensor and build equipment simulation model; build discrete event simulation model factory; engineering digital twin method injects faults into the virtual system for equipment and plant level fault diagnosis
Multi-information fusion	Accelerometer; Current sensor	[117]	The existing methods cannot effectively capture the time information and global characteristics of the device; Single source fault diagnosis method makes it difficult to accurately extract fault features	Intelligent fault diagnosis method based on multi-source information fusion of hierarchical visual transformer and wavelet time-frequency; the multi-source signals are converted into two-dimensional time-frequency maps and fused into a hierarchical visual transformer
	Accelerometer, Microphone	[53]	The multi-sensor data combination can observe the phenomenon of more abundant system degradation, which is helpful for equipment analysis and decision-making	The influence of data level fusion on the accuracy of different detection tools is studied, and the best combination example of sensor and deep learning fusion algorithm is studied to achieve accurate predictive maintenance; build DCNN, LSTM, and CNN-LSTM models
	Motor polyphase current signal	[100]	Some systems cannot install external sensors; the bearing installed outside the motor cannot effectively measure the current	Taking the multi-phase original signal of motor current as input, the characteristics of each phase current are extracted, and the features are classified by CNN. The decision-making level information fusion technology converts CNN information into simple pattern classification problem

Table 3. *Cont.*

Form of Method	Sensors	References	Purpose of Research	Key Features of the Research
Other method	Accelerometer	[125]	The effective value of prior knowledge and learning network is not mined in the fusion, which reduces the performance of fault diagnosis	Extracts the prior knowledge of the fault signal through envelope analysis and passband selection, fuses the prior knowledge with the convolution kernel and the designed filter, embeds and transmits the prior knowledge and cross-domain knowledge using the weight sharing depth adaptive network, and uses the maximum average deviation of multiple cores for regional adaptation
	Vibration sensor	[126]	The method can self-maintain, diagnose and repair, and carry out reliable and robust online monitoring for robots	The method based on transitive learning makes use of the inherent relationship between the adaptability of transitive learning and different failure modes; hybrid one-dimensional MCNN based on matrix kernel and RNN is used to realize the high-precision detection of robot state; time stamp mapping to overcome time inconsistency of multi-sensor
	Machine monitoring process signal	[128]	Difficulty in obtaining fault samples under changing operating conditions and ineffective application or performance degradation in actual industry; Not utilizing the internal and relevant knowledge under different working conditions	A non-sharing mechanism is designed to obtain the discriminant characteristics of each working condition; the cross-correlation knowledge mining subnet is used to construct the fault relationship knowledge graph to explicitly constrain the local consistency between the source domain, target domain, and cross-domain; semantic knowledge transfer subnet makes output consistent and distinguishable
	Vibration sensors, Sound sensors	[130]	The intrinsic correlation between the distribution gap of CNN and multi-frequency mechanical signals in the learning process is not considered	Two parallel modal-specific networks and a cross-modal knowledge-sharing network are used to explore the independent and shared features of multi-modal mechanical signals; cross-modal fusion is introduced to fuse and transfer the cross-modal features to the next layer
	Accelerometer, Laser sensor, Current sensor, Voltage probe	[131]	Low confidence in robot fault model prediction in the industrial field	Inferring the causal relationship of faults by using the physical interpretability of generating fuzzy energy pattern image data

4.1. Summary and Analysis

The research on fault diagnosis of manipulator robots is a challenging research topic [132]. To achieve the safe and effective operation of manipulator robots, it is urgent to improve the accuracy of manipulator robot fault diagnosis algorithm recognition and explore innovative fault diagnosis technologies. Although some traditional statistical and mechanism knowledge-driven methods can achieve fault diagnosis of manipulator robots, the target objects are mostly single faults. Due to the complex working environment, manipulator robot failures are mostly compound, and the accuracy of detecting compound failures through traditional methods is insufficient. For example, it is difficult to acquire and update the knowledge of expert systems, and the statistical model relies heavily on the assumption of data distribution, which limits its application in complex industrial scenes. In recent years, with the rapid development of artificial intelligence and IoT technology, the use of advanced methods such as deep learning, multi-information fusion, and digital twins for manipulator robot fault diagnosis have attracted attention. Although multi-information fusion methods can significantly improve the accuracy of fault diagnosis in manipulator robots, the time alignment of data collected by multiple sensors, the uniformity of data format, and the computational time cost of deep learning algorithms can affect the overall performance of fault diagnosis. Moreover, the deep learning method has poor interpretability, high requirements for the amount of data, and high model training costs. Although digital twins can realize the fusion of the virtual and the real, they still face the challenges of data synchronization and model construction in practical application. Therefore, further research is needed to enhance the speed and stability of new technologies used for fault diagnosis.

Finally, we summarize the fault diagnosis methods for manipulator robots into two aspects: Firstly, the advantages and disadvantages of using traditional methods such as statistics, mechanism-knowledge-driven, and signal processing for fault diagnosis are analyzed. Secondly, modern methods represented by artificial intelligence and digital twins were analyzed.

(1) Traditional methods for fault diagnosis: Traditional methods for fault diagnosis of manipulator robots have some advantages, but there are still limitations in practical applications.

Advantages: These methods are easy to implement, have strong anti-interference ability, and have good stability in diagnosing a single fault.

Limitations: Insufficient accuracy in compound fault diagnosis of manipulator robot systems.

(2) Modern methods for fault diagnosis: Similarly, modern methods for manipulator robot fault diagnosis also have advantages and limitations.

Advantages: High accuracy in identifying composite faults, and the equipment can be visualized for remote monitoring.

Limitations: The data formats and timing from multi-sensor acquisition are difficult to unify, and deep learning training requires a large number of samples with high training time costs. Signals are easily affected by the acquisition environment, noise, and other factors, leading to insufficient information stability and affecting the accuracy of fault diagnosis. Moreover, the use of digital twin technology for fault diagnosis imposes higher requirements on communication technology, data storage, and computation, and the real-time requirements for visualization and monitoring demand that the hardware system has powerful computational performance.

Traditional and modern methods have advantages and limitations in the fault diagnosis of manipulator robots. For example, in some specific operating conditions, especially for manipulator robot failure modes with periodic signals or known characteristics or single fault types, traditional methods may perform well. However, for the fault diagnosis of manipulator robots in complex nonlinear signals or high-noise environments, traditional methods cannot effectively distinguish fault types and rely on manually designed features. Modern methods have high accuracy in diagnosing complex signals and multiple types of faults and can better handle nonlinear data. However, modern methods such as deep learning are highly dependent on data and have higher computational resource requirements compared to traditional methods, limiting real-time performance. It is worth noting that in modern manufacturing, due to the complexity of tasks and the increasing demand, a single method may not be able to meet actual needs, and it is necessary to choose the appropriate method or a combination of two methods based on reality. The comparison between traditional and modern methods in terms of accuracy, stability, and rapidity is shown in Table 4.

Table 4. Comparison between traditional and modern methods.

Characteristic	Traditional Methods	Modern Methods
Accuracy	Excellent performance in single failure mode or periodic signal, relying on manual feature design;	Excellent performance in complex signals and multiple types of fault modes, but strong data dependence
Stability	Simple method deployment; known working conditions perform well	Strong adaptability to complex working conditions, but requires a large amount of data support
Rapidity	Fast calculation speed, suitable for real-time fault diagnosis	High demand for computing resources, real-time fault diagnosis relies on high-performance hardware devices

Some solutions have the potential to address the aforementioned limitations.

Semi-supervised learning: By integrating a semi-supervised learning mechanism, the model training process effectively mitigates the challenge of limited training data by leveraging scarce labeled samples and abundant unlabeled data.

Multi-sensor data fusion optimization: Improve the efficiency and accuracy of data fusion by improving data preprocessing methods (such as wavelet transform, and PCA dimensionality reduction) and introducing more efficient data fusion algorithms (such as time series analysis based on informer).

Model optimization and improvement: Through transfer learning and meta-learning, to improve the adaptability of the model under different working conditions.

Hardware support and algorithm optimization: Combine edge computing devices (such as high-performance PLC) and optimized algorithm architecture (such as lightweight neural network), reduce computing costs, and improve real-time performance

In summary, both traditional and modern methods have their advantages and disadvantages in practical applications. Next, we will explore the future development direction of manipulator robot fault diagnosis.

4.2. Future Expectations

The fault diagnosis technology of manipulator robots plays an important role in intelligent manufacturing and industrial production. To improve the effectiveness and applicability of its practical application, the development of fault diagnosis for manipulator robots should meet the following requirements:

Intelligentization: Use advanced technologies such as deep learning and cloud computing, combined with high-performance edge computing equipment, to improve the computational power and speed of the model, and improve the intelligent level of fault diagnosis.

Integration: Integrate fault diagnosis technology more closely with the control system and sensor system of the manipulator robot, and combine the mechanism model and artificial intelligence method of the robot to realize online monitoring and real-time early warning.

Servitization: Develop services based on the cloud platform, combine manipulator robot and IoT, and use advanced communication technology to build a high fidelity, real-time, and accurate manipulator robot human networking and visualization system (manipulator robot digital twin system) to provide users with remote fault diagnosis and prediction services.

Quantitative evaluation: According to the application scenarios, application objects, and application environments of different manipulator robots, the corresponding quantitative fault evaluation system should be formulated and improved to promote the wide application of technology.

The first requirement is a prerequisite for achieving accurate fault diagnosis of manipulator robots. By integrating some advanced artificial intelligence algorithms and new IoT technologies, it is possible to effectively combine the operational status data of manipulator robots and extract effective fault feature information. This makes it easier for fault diagnosis models to understand the robot's data, thereby improving the accuracy of fault diagnosis results. The second requirement, building upon the first, involves integrating intelligent diagnostic methods with the inherent systems of manipulator robots. By complementing each other through the robot's mechanism and artificial intelligence methods, it achieves online monitoring of the robot's operational status and fault prediction. The third requirement is a further enhancement, which aims to construct a high-fidelity digital twin for the manipulator robot fault diagnosis system through advanced cloud platforms, industrial internet, and modern communication technologies. This provides technicians with remote visualization of equipment operational status, enabling remote fault diagnosis and predictive services for manipulator robot systems. Furthermore, manipulator robots are applied in a wide range of scenarios with complex application environments, making it difficult to assess the causes and severity of faults under different operating conditions. At present, such assessments are mostly made based on extensive on-site experience or whether the robot can operate normally. Quantitative assessment of manipulator robot fault diagnosis is crucial for establishing and refining fault evaluation systems that correspond to different scenarios and environments. This is of significant importance for utilization in industrial production.

5. Conclusions

Research on fault diagnosis of manipulator robots has attracted increasing attention from both scholars and the industrial community. Establishing a robust fault diagnosis technology system for robots can accurately pinpoint potential faults before they manifest during the production process. This facilitates equipment maintenance and enables early warning and proactive management of potential fault risks, thereby reducing labor costs and enhancing production efficiency. This paper reviews the current research status and major methods of fault diagnosis for manipulator robots. Through comparative analysis, it can be observed that new methods based on artificial intelligence, digital twins, and multi-information fusion have shown great potential and advantages in fault diagnosis. It is concluded that the future development of manipulator robot fault diagnosis may benefit from the following considerations.

(1) Efficient, accurate, and stable fault diagnosis is the foundation for the practical production application of manipulator robots.
(2) The integration of multi-information fusion and artificial intelligence algorithms can compensate for the shortcomings of single signals that fail to effectively reflect fault characteristics of manipulator robots, as well as the challenges posed by traditional algorithms in handling composite faults. However, improvements are still needed in data fusion strategies and the computational time cost of models.
(3) Digital twin technology can achieve remote monitoring of the operational status, fault diagnosis, and predictive maintenance of manipulator robot systems.
(4) Establishing and refining a quantitative assessment system for fault diagnosis of manipulator robots under different operating conditions and environments can enhance production efficiency and reduce production costs.

There is still a long way to go to achieve the aforementioned objectives. However, these methods will become more mature and widespread with the enhancement of computational capabilities and the advancement of big data technologies. Furthermore, although this review systematically discusses the types of faults and fault diagnosis methods for manipulator robots, there are still certain limitations. For fault diagnosis of manipulator robots, the cost of multi-information acquisition and the performance requirements of computational devices for data cleaning and diagnostic models have not been mentioned. These limitations all involve factors that affect the research on fault diagnosis of manipulator robots.

Author Contributions: Conceptualization, Y.Z. and J.W.; methodology, Y.Z. and J.W.; validation, Y.Z. and J.W.; formal analysis, Y.Z., J.W., B.G. and G.C.; investigation, Y.Z.; resources, Y.Z. and H.W.; data curation, Y.Z.; writing—original draft preparation, Y.Z. and J.W.; writing—review and editing, Y.Z.; visualization, Y.Z.; supervision, L.X. and G.C.; project administration, B.G., L.X., C.L. and H.W.; funding acquisition, Y.Z., J.W. and L.X. All authors have read and agreed to the published version of the manuscript.

Funding: This research was funded by the Projects of Shenzhen Institute of Information Technology under Grant SZIIT2025KJ021 and SZIIT2025KJ022, and in part by the National Natural Science Foundation of China under Grant 92467204.

Institutional Review Board Statement: Not applicable.

Informed Consent Statement: Not applicable.

Data Availability Statement: Not applicable.

Conflicts of Interest: The authors declare no conflicts of interest.

Abbreviations

RV	Rotary Vector
CNN	Convolutional neural network
GCN	Graph convolutional network
LSTM	Long short-term memory
DBN	Deep belief network
MCNN	Multichannel convolutional neural network
IoRT	Internet of robotic things
PCA	Principle component analysis
PLC	Programmable logic controller
CAD	Computer aided technology

References

1. Arents, J.; Greitans, M. Smart Industrial Robot Control Trends, Challenges and Opportunities Within Manufacturing. *Appl. Sci.* **2022**, *12*, 937. [CrossRef]
2. Zhong, R.Y.; Xu, X.; Klotz, E.; Newman, S.T. Intelligent Manufacturing in the Context of Industry 4.0: A Review. *Engineering* **2017**, *3*, 616–630. [CrossRef]
3. Wang, B.; Tao, F.; Fang, X.; Liu, C.; Liu, Y.; Freiheit, T. Smart Manufacturing and Intelligent Manufacturing: A Comparative Review. *Engineering* **2021**, *7*, 738–757. [CrossRef]
4. Deng, J.; Wang, T.; Wang, Z.; Zhou, J.; Cheng, L. Research on Event Logic Knowledge Graph Construction Method of Robot Transmission System Fault Diagnosis. *IEEE Access* **2022**, *10*, 17656–17673. [CrossRef]
5. Cheon, E.J.; Schneiders, E.; Skov, M.B. Working with Bounded Collaboration: A Qualitative Study on How Collaboration Is Co-Constructed around Collaborative Robots in Industry. In Proceedings of the ACM on Human-Computer Interaction; Association for Computing Machinery: New York, NY, USA, 2022; Volume 6, pp. 1–34.
6. Hentout, A.; Aouache, M.; Maoudj, A.; Akli, I. Human–Robot Interaction in Industrial Collaborative Robotics: A Literature Review of the Decade 2008–2017. *Adv. Robot.* **2019**, *33*, 764–799. [CrossRef]
7. Villani, V.; Pini, F.; Leali, F.; Secchi, C. Survey on Human–Robot Collaboration in Industrial Settings: Safety, Intuitive Interfaces and Applications. *Mechatronics* **2018**, *55*, 248–266. [CrossRef]

8. Yang, Q.; Li, X.; Cai, H.; Hsu, Y.M.; Lee, J.; Yang, C.H.; Li, Z.L.; Lin, M.Y. Fault Prognosis of Industrial Robots in Dynamic Working Regimes: Find Degradation in Variations. *Meas. J. Int. Meas. Confed.* **2021**, *173*, 108545. [CrossRef]
9. Ruan, D.; Wang, J.; Yan, J.; Gühmann, C. CNN Parameter Design Based on Fault Signal Analysis and Its Application in Bearing Fault Diagnosis. *Adv. Eng. Inform.* **2023**, *55*, 101877. [CrossRef]
10. Wang, X.; Liu, M.; Liu, C.; Ling, L.; Zhang, X. Data-Driven and Knowledge-Based Predictive Maintenance Method for Industrial Robots for the Production Stability of Intelligent Manufacturing. *Expert Syst. Appl.* **2023**, *234*, 121136. [CrossRef]
11. Lu, K.; Chen, C.; Wang, T.; Cheng, L.; Qin, J. Fault Diagnosis of Industrial Robot Based on Dual-Module Attention Convolutional Neural Network. *Auton. Intell. Syst.* **2022**, *2*, 1–12. [CrossRef]
12. Campilho, R.D.S.G.; Silva, F.J.G. Industrial Process Improvement by Automation and Robotics. *Machines* **2023**, *11*, 1011. [CrossRef]
13. Fasuludeen Kunju, F.K.; Naveed, N.; Anwar, M.N.; Ul Haq, M.I. Production and Maintenance in Industries: Impact of Industry 4.0. *Ind. Robot* **2022**, *49*, 461–475. [CrossRef]
14. Nagy, M.; Lăzăroiu, G.; Valaskova, K. Machine Intelligence and Autonomous Robotic Technologies in the Corporate Context of SMEs: Deep Learning and Virtual Simulation Algorithms, Cyber-Physical Production Networks, and Industry 4.0-Based Manufacturing Systems. *Appl. Sci.* **2023**, *13*, 1681. [CrossRef]
15. Xue, S.; Du, C.; Yang, C. Health Management System for the Industrial Robot Control Cabinet. In Proceedings of the 2024 IEEE International Conference on Mechatronics and Automation (ICMA), Tianjin, China, 4–7 August 2024; pp. 142–146.
16. Eang, C.; Lee, S. Predictive Maintenance and Fault Detection for Motor Drive Control Systems in Industrial Robots Using CNN-RNN-Based Observers. *Sensors* **2025**, *25*, 25. [CrossRef]
17. Kumar, P.; Khalid, S.; Kim, H.S. Prognostics and Health Management of Rotating Machinery of Industrial Robot with Deep Learning Applications—A Review. *Mathematics* **2023**, *11*, 3008. [CrossRef]
18. Abid, A.; Khan, M.T.; Iqbal, J. A Review on Fault Detection and Diagnosis Techniques: Basics and Beyond. *Artif. Intell. Rev.* **2021**, *54*, 3639–3664. [CrossRef]
19. Mustafa, F.E.; Khan, A.Q.; Samee, A.; Ahmed, I.; Abid, M.; Alqahtani, M.M.; Khalid, M. Advanced Statistical and Meta-Heuristic Based Optimization Fault Diagnosis Techniques in Complex Industrial Processes: A Comparative Analysis. *IEEE Access* **2023**, *11*, 104373–104391. [CrossRef]
20. Fernandes, M.; Corchado, J.M.; Marreiros, G. Machine Learning Techniques Applied to Mechanical Fault Diagnosis and Fault Prognosis in the Context of Real Industrial Manufacturing Use-Cases: A Systematic Literature Review. *Appl. Intell.* **2022**, *52*, 14246–14280. [CrossRef]
21. Deebak, B.D.; Al-Turjman, F. Digital-Twin Assisted: Fault Diagnosis Using Deep Transfer Learning for Machining Tool Condition. *Int. J. Intell. Syst.* **2022**, *37*, 10289–10316. [CrossRef]
22. Guo, K.; Wan, X.; Liu, L.; Gao, Z.; Yang, M. Fault Diagnosis of Intelligent Production Line Based on Digital Twin and Improved Random Forest. *Appl. Sci.* **2021**, *11*, 7733. [CrossRef]
23. Long, J.; Mou, J.; Zhang, L.; Zhang, S.; Li, C. Attitude Data-Based Deep Hybrid Learning Architecture for Intelligent Fault Diagnosis of Multi-Joint Industrial Robots. *J. Manuf. Syst.* **2021**, *61*, 736–745. [CrossRef]
24. Bo, W.; Jiaxu, W.; Guangwu, Z.; Rongsong, Y.; Hongjun, Z.; Tao, H. Mixed Lubrication Analysis of Modified Cycloidal Gear Used in the RV Reducer. *Proc. Inst. Mech. Eng. Part J J. Eng. Tribol.* **2016**, *230*, 121–134. [CrossRef]
25. Yan, B.; Dong, L.; Yan, K.; Chen, F.; Zhu, Y.; Wang, D. Effects of Oil-Air Lubrication Methods on the Internal Fluid Flow and Heat Dissipation of High-Speed Ball Bearings. *Mech. Syst. Signal Process.* **2021**, *151*, 107409. [CrossRef]
26. Pinto, R.; Cerquitelli, T. Robot Fault Detection and Remaining Life Estimation for Predictive Maintenance. *Procedia Comput. Sci.* **2019**, *151*, 709–716. [CrossRef]
27. Moore, J.; Sawyer, D. Equipment Health Monitoring for Industrial Robotic Arms. In Proceedings of the 2024 IEEE 20th International Conference on Automation Science and Engineering (CASE), Bari, Italy, 28 August–1 September 2024; pp. 1853–1860.
28. Sanfilippo, F.; Hatledal, L.I.; Zhang, H.; Fago, M.; Pettersen, K.Y. Controlling Kuka Industrial Robots: Flexible Communication Interface JOpenShowVar. *IEEE Robot. Autom. Mag.* **2015**, *22*, 96–109. [CrossRef]
29. Chen, L.; Cao, J.; Wu, K.; Zhang, Z. Application of Generalized Frequency Response Functions and Improved Convolutional Neural Network to Fault Diagnosis of Heavy-Duty Industrial Robot. *Robot. Comput. Integr. Manuf.* **2022**, *73*, 102228. [CrossRef]
30. Wu, Y.; Bai, Y.; Yang, S.; Li, C. Extracting Random Forest Features with Improved Adaptive Particle Swarm Optimization for Industrial Robot Fault Diagnosis. *Meas. J. Int. Meas. Confed.* **2024**, *229*, 114451. [CrossRef]
31. Khalastchi, E.; Kalech, M. On Fault Detection and Diagnosis in Robotic Systems. *ACM Comput. Surv.* **2018**, *51*, 1–24. [CrossRef]
32. Wang, J.; Wang, X.; Wang, Y.; Sun, Y.; Sun, G. Intelligent Joint Actuator Fault Diagnosis for Heavy-Duty Industrial Robots. *IEEE Sens. J.* **2024**, *24*, 15292–15301. [CrossRef]
33. Kim, Y.; Park, J.; Na, K.; Yuan, H.; Youn, B.D.; Kang, C. soon Phase-Based Time Domain Averaging (PTDA) for Fault Detection of a Gearbox in an Industrial Robot Using Vibration Signals. *Mech. Syst. Signal Process.* **2020**, *138*, 106544. [CrossRef]
34. Appana, D.K.; Prosvirin, A.; Kim, J.M. Reliable Fault Diagnosis of Bearings with Varying Rotational Speeds Using Envelope Spectrum and Convolution Neural Networks. *Soft Comput.* **2018**, *22*, 6719–6729. [CrossRef]

35. Huang, R.; Xia, J.; Zhang, B.; Chen, Z.; Li, W. Compound Fault Diagnosis for Rotating Machinery: State-of-the-Art, Challenges, and Opportunities. *J. Dyn. Monit. Diagnostics* **2023**, *2*, 13–29. [CrossRef]
36. Nandi, S.; Toliyat, H.A.; Li, X. Condition Monitoring and Fault Diagnosis of Induction Motors: A Review. *IEEE Trans. Energy Convers.* **2005**, *20*, 719–729. [CrossRef]
37. Pan, M.C.; Van Brussel, H.; Sas, P. Intelligent Joint Fault Diagnosis of Industrial Robots. *Mech. Syst. Signal Process.* **1998**, *12*, 571–588. [CrossRef]
38. Hsu, H.K.; Ting, H.Y.; Huang, M.B.; Huang, H.P. Intelligent Fault Detection, Diagnosis and Health Evaluation for Industrial Robots. *Mechanika* **2021**, *27*, 70–79. [CrossRef]
39. Schneider, H.; Frank, P.M. Observer-Based Supervision and Fault Detection in Robots Using Nonlinear and Fuzzy Logic Residual Evaluation. *IEEE Trans. Control Syst. Technol.* **1996**, *4*, 274–282. [CrossRef]
40. Kim, S.; Lee, J.; Kang, C. Analysis of Industrial Accidents Causing through Jamming or Crushing Accidental Deaths in the Manufacturing Industry in South Korea: Focus on Non-Routine Work on Machinery. *Saf. Sci.* **2021**, *133*, 104998. [CrossRef]
41. Han, H.; Xu, Q.; Li, D.; Li, B.; Sun, X. Fault Diagnosis of RV Reducers Used in Industrial Robots Based on Vibration Analysis. In Proceedings of the International conference on the Efficiency and Performance Engineering Network, Baotou, China, 18–21 August 2022; Springer: Berlin/Heidelberg, Germany, 2022; Volume 1, pp. 306–317.
42. Vallachira, S.; Orkisz, M.; Norrlof, M.; Butail, S. Data-Driven Gearbox Failure Detection in Industrial Robots. *IEEE Trans. Ind. Inform.* **2020**, *16*, 193–201. [CrossRef]
43. Rohan, A.; Raouf, I.; Kim, H.S. Rotate Vector (Rv) Reducer Fault Detection and Diagnosis System: Towards Component Level Prognostics and Health Management (Phm). *Sensors* **2020**, *20*, 6845. [CrossRef]
44. Yang, G.; Tao, H.; Du, R.; Zhong, Y. Compound Fault Diagnosis of Harmonic Drives Using Deep Capsule Graph Convolutional Network. *IEEE Trans. Ind. Electron.* **2023**, *70*, 4186–4195. [CrossRef]
45. Yang, G.; Zhong, Y.; Yang, L.; Tao, H.; Li, J.; Du, R. Fault Diagnosis of Harmonic Drive with Imbalanced Data Using Generative Adversarial Network. *IEEE Trans. Instrum. Meas.* **2021**, *70*, 1–11. [CrossRef]
46. Zhou, X.; Zhou, H.C.; He, Y.M.; Huang, S.F.; Zhu, Z.H.; Chen, J.H. Harmonic Reducer In-Situ Fault Diagnosis for Industrial Robots Based on Deep Learning. *Sci. China Technol. Sci.* **2022**, *65*, 2116–2126. [CrossRef]
47. Yazici, B.; Kliman, G.B.; Premerlani, W.J.; Koegl, R.A.; Robinson, G.B.; Abdel-Malek, A. An Adaptive, on-Line, Statistical Method for Bearing Fault Detection Using Stator Current. In Proceedings of the IEEE Transactions on Instrumentation and Measurement, New Orleans, Lousiana, 5–9 October 1997; Volume 70, pp. 213–220.
48. Xia, B.; Wang, K.; Xu, A.; Zeng, P.; Yang, N.; Li, B. Intelligent Fault Diagnosis for Bearings of Industrial Robot Joints Under Varying Working Conditions Based on Deep Adversarial Domain Adaptation. *IEEE Trans. Instrum. Meas.* **2022**, *71*, 1–13. [CrossRef]
49. McInerny, S.A.; Dai, Y. Basic Vibration Signal Processing for Bearing Fault Detection. *IEEE Trans. Educ.* **2003**, *46*, 149–156. [CrossRef]
50. Raouf, I.; Kumar, P.; Kim, H.S. Deep Learning-Based Fault Diagnosis of Servo Motor Bearing Using the Attention-Guided Feature Aggregation Network. *Expert Syst. Appl.* **2024**, *258*, 125137. [CrossRef]
51. Zhang, X.; Yan, C.; Liu, Y.; Yan, P.; Wang, Y.; Wu, L. Dynamic Modeling and Analysis of Rolling Bearing with Compound Fault on Raceway and Rolling Element. *Shock Vib.* **2020**, *2020*, 8861899. [CrossRef]
52. Junior, R.F.R.; Areias, I.A.d.S.; Campos, M.M.; Teixeira, C.E.; da Silva, L.E.B.; Gomes, G.F. Fault Detection and Diagnosis in Electric Motors Using 1d Convolutional Neural Networks with Multi-Channel Vibration Signals. *Meas. J. Int. Meas. Confed.* **2022**, *190*, 12–14. [CrossRef]
53. Suawa, P.; Meisel, T.; Jongmanns, M.; Huebner, M.; Reichenbach, M. Modeling and Fault Detection of Brushless Direct Current Motor by Deep Learning Sensor Data Fusion. *Sensors* **2022**, *22*, 3516. [CrossRef]
54. Kim, H.; Lee, H.; Kim, S.W. Current Only-Based Fault Diagnosis Method for Industrial Robot Control Cables. *Sensors* **2022**, *22*, 1917. [CrossRef]
55. Kim, H.; Lee, H.; Kim, S.; Kim, S.W. Attention Recurrent Neural Network-Based Severity Estimation Method for Early-Stage Fault Diagnosis in Robot Harness Cable. *Sensors* **2023**, *23*, 5299. [CrossRef]
56. Yang, Q.; Li, X.; Wang, Y.; Ainapure, A.; Lee, J. Fault Diagnosis of Ball Screw in Industrial Robots Using Non-Stationary Motor Current Signals. *Procedia Manuf.* **2020**, *48*, 1102–1108. [CrossRef] [PubMed]
57. Gawde, S.; Patil, S.; Kumar, S.; Kamat, P.; Kotecha, K.; Abraham, A. Multi-Fault Diagnosis of Industrial Rotating Machines Using Data-Driven Approach: A Review of Two Decades of Research. *Eng. Appl. Artif. Intell.* **2023**, *123*, 106139. [CrossRef]
58. He, S.; Chen, J.; Zhou, Z.; Zi, Y.; Wang, Y.; Wang, X. Multifractal Entropy Based Adaptive Multiwavelet Construction and Its Application for Mechanical Compound-Fault Diagnosis. *Mech. Syst. Signal Process.* **2016**, *76–77*, 742–758. [CrossRef]
59. Zhou, X.; Zeng, H.; Chen, C.; Xiao, H.; Xiang, Z. An Attention-Enhanced Multi-Modal Deep Learning Algorithm for Robotic Compound Fault Diagnosis. *Meas. Sci. Technol.* **2023**, *34*, 014007. [CrossRef]
60. He, Y.; Zhao, C.; Zhou, X.; Shen, W. MJAR: A Novel Joint Generalization-Based Diagnosis Method for Industrial Robots with Compound Faults. *Robot. Comput. Integr. Manuf.* **2024**, *86*, 102668. [CrossRef]

61. Chen, C.; Wang, T.; Liu, C.; Liu, Y.; Cheng, L. Lightweight Convolutional Transformers Enhanced Meta-Learning for Compound Fault Diagnosis of Industrial Robot. *IEEE Trans. Instrum. Meas.* **2023**, *72*, 1–12. [CrossRef]
62. Cheng, F.; Raghavan, A.; Jung, D.; Sasaki, Y.; Tajika, Y. High-Accuracy Unsupervised Fault Detection of Industrial Robots Using Current Signal Analysis. In Proceedings of the 2019 IEEE International Conference on Prognostics and Health Management (ICPHM), San Francisco, CA, USA, 17–20 June 2019; pp. 1–8.
63. Goyal, D.; Choudhary, A.; Pabla, B.S.; Dhami, S.S. Support Vector Machines Based Non-Contact Fault Diagnosis System for Bearings. *J. Intell. Manuf.* **2020**, *31*, 1275–1289. [CrossRef]
64. Gawde, S.; Patil, S.; Kumar, S.; Kotecha, K. A Scoping Review on Multi-Fault Diagnosis of Industrial Rotating Machines Using Multi-Sensor Data Fusion. *Artif. Intell. Rev.* **2023**, *56*, 4711–4764. [CrossRef]
65. Hassan, I.U.; Panduru, K.; Walsh, J. An In-Depth Study of Vibration Sensors for Condition Monitoring. *Sensors* **2024**, *24*, 740. [CrossRef]
66. Emre, Y.; Gürsel, N.; Gümüs, L. Intelligent Worm Gearbox Fault Diagnosis under Various Working Conditions Using Vibration, Sound and Thermal Features. *Appl. Acoust.* **2022**, *186*, 108463.
67. Yun, H.; Kim, H.; Jeong, Y.H.; Jun, M.B.G. Autoencoder-Based Anomaly Detection of Industrial Robot Arm Using Stethoscope Based Internal Sound Sensor. *J. Intell. Manuf.* **2023**, *34*, 1427–1444. [CrossRef]
68. Fedorova, D.; Tlach, V.; Kuric, I.; Dodok, T.; Zajačko, I.; Tucki, K. Technical Diagnostics of Industrial Robots Using Vibration Signals: Case Study on Detecting Base Unfastening. *Appl. Sci.* **2025**, *15*, 270. [CrossRef]
69. Algburi, R.N.A.; Gao, H. Health Assessment and Fault Detection System for an Industrial Robot Using the Rotary Encoder Signal. *Energies* **2019**, *12*, 2816. [CrossRef]
70. Islam, S.; Kim, M.; Ku, K.; Kim, H.; Kim, K. Study on Fault Diagnosis and Data Processing Techniques for Substrate Transfer Robots Using Vibration Sensor Data. *J. Microelectron. Packag. Soc.* **2024**, *31*, 45–53.
71. Schoen, R.R.; Habetler, T.G.; Kamran, F.; Bartheld, R.G. Motor Bearing Damage Detection Using Stator Current Monitoring. *IEEE Trans. Ind. Appl.* **1995**, *31*, 1274–1279. [CrossRef]
72. Long, J.; Qin, Y.; Yang, Z.; Huang, Y.; Li, C. Discriminative Feature Learning Using a Multiscale Convolutional Capsule Network from Attitude Data for Fault Diagnosis of Industrial Robots. *Mech. Syst. Signal Process.* **2023**, *182*, 109569. [CrossRef]
73. Zhang, J.; Luo, Y.; Wang, B.; Lu, C.; Si, J.; Song, J. Deep Reinforcement Learning for Load Shedding Against Short-Term Voltage Instability in Large Power Systems. *IEEE Trans. Neural Networks Learn. Syst.* **2023**, *34*, 4249–4260. [CrossRef]
74. Yu, L.; Yao, X.; Yang, J.; Li, C. Gear Fault Diagnosis through Vibration and Acoustic Signal Combination Based on Convolutional Neural Network. *Information* **2020**, *11*, 266. [CrossRef]
75. Visinsky, M.L.; Cavallaro, J.R.; Walker, I.D. Expert System Framework for Fault Detection and Fault Tolerance in Robotics. *Comput. Electr. Eng.* **1994**, *20*, 421–435. [CrossRef]
76. Liu, Y.; Chen, J. Integrated Fault Diagnosis Method of Mobile Robot. *Commun. Comput. Inf. Sci.* **2011**, *164*, 372–379.
77. Sun, X.; Jia, X. A Fault Diagnosis Method of Industrial Robot Rolling Bearing Based on Data Driven and Random Intuitive Fuzzy Decision. *IEEE Access* **2019**, *7*, 148764–148770. [CrossRef]
78. Wu, Y.; Fu, Z.; Fei, J. Fault Diagnosis for Industrial Robots Based on a Combined Approach of Manifold Learning, Treelet Transform and Naive Bayes. *Rev. Sci. Instrum.* **2020**, *91*, 015116. [CrossRef]
79. Xu, Y.; Sun, Y.; Wan, J.; Liu, X.; Song, Z. Industrial Big Data for Fault Diagnosis: Taxonomy, Review, and Applications. *IEEE Access* **2017**, *5*, 17368–17380. [CrossRef]
80. Piltan, F.; Prosvirin, A.E.; Sohaib, M.; Saldivar, B.; Kim, J.M. An SVM-Based Neural Adaptive Variable Structure Observer for Fault Diagnosis and Fault-Tolerant Control of a Robot Manipulator. *Appl. Sci.* **2020**, *10*, 1344. [CrossRef]
81. Moshgani, S.M.H.; Fakharian, A. Fault Detection and Isolation for Manipulator Robot Using Optimal Unknown Input Observer. *Control Eng. Appl. Inform.* **2019**, *21*, 71–79.
82. Sadeghzadeh-Nokhodberiz, N.; Poshtan, J. Distributed Interacting Multiple Filters for Fault Diagnosis of Navigation Sensors in a Robotic System. *IEEE Trans. Syst. Man Cybern. Syst.* **2017**, *47*, 1383–1393. [CrossRef]
83. Zhang, Y.; Cao, G.; Li, L.; Diao, D. AEKF-Based Trajectory-Error Compensation of Knee Exoskeleton for Human–Exoskeleton Interaction Control. *Robot. Intell. Autom.* **2024**, *44*, 84–95. [CrossRef]
84. Van, M.; Wu, D.; Ge, S.S.; Ren, H. Fault Diagnosis in Image-Based Visual Servoing with Eye-in-Hand Configurations Using Kalman Filter. *IEEE Trans. Ind. Inform.* **2016**, *12*, 1998–2007. [CrossRef]
85. Zhang, T.; Shi, P.; Li, W.; Yue, X. Adaptive EKF Enhanced Fault Diagnosis and Fault Tolerant Control for Space Manipulators with Position Measurements Only. *J. Franklin Inst.* **2024**, *361*, 106824. [CrossRef]
86. Betta, G.; Liguori, C.; Paolillo, A.; Pietrosanto, A. A DSP-Based FFT-Analyzer for the Fault Diagnosis of Rotating Machine Based on Vibration Analysis. *IEEE Trans. Instrum. Meas.* **2002**, *51*, 1316–1321. [CrossRef]
87. Chen, L.; Wen, S.; Wang, H.; Hu, H. Fault Mechanism Analysis and Diagnosis for Closed-Loop Drive System of Industrial Robot Based on Nonlinear Spectrum. *Sci. Rep.* **2022**, *12*, 18350. [CrossRef] [PubMed]

88. Pappachan, B.K.; Caesarendra, W.; Tjahjowidodo, T.; Wijaya, T. Frequency Domain Analysis of Sensor Data for Event Classification in Real-Time Robot Assisted Deburring. *Sensors* **2017**, *17*, 1247. [CrossRef] [PubMed]
89. Qiao, W.; Lu, D. A Survey on Wind Turbine Condition Monitoring and Fault Diagnosis—Part II: Signals and Signal Processing Methods. *IEEE Trans. Ind. Electron.* **2015**, *62*, 6546–6557. [CrossRef]
90. Ding, K.; Chen, X.; Jiang, M.; Yang, H.; Chen, X.; Zhang, J.; Gao, R.; Cui, L. Feature Extraction and Fault Diagnosis of Photovoltaic Array Based on Current–Voltage Conversion. *Appl. Energy* **2024**, *353*, 122135. [CrossRef]
91. Banerjee, T.P.; Das, S. Multi-Sensor Data Fusion Using Support Vector Machine for Motor Fault Detection. *Inf. Sci.* **2012**, *217*, 96–107. [CrossRef]
92. Aherwar, A. An Investigation on Gearbox Fault Detection Using Vibration Analysis Techniques: A Review. *Aust. J. Mech. Eng.* **2012**, *10*, 169–183. [CrossRef]
93. Al Kazzaz, S.A.S.; Singh, G.K. Experimental Investigations on Induction Machine Condition Monitoring and Fault Diagnosis Using Digital Signal Processing Techniques. *Electr. Power Syst. Res.* **2003**, *65*, 197–221. [CrossRef]
94. Chen, K.; Huang, C.; He, J. Fault Detection, Classification and Location for Transmission Lines and Distribution Systems: A Review on the Methods. *High Volt.* **2016**, *1*, 25–33. [CrossRef]
95. Jiang, Y.; Zhu, H.; Li, Z. A New Compound Faults Detection Method for Rolling Bearings Based on Empirical Wavelet Transform and Chaotic Oscillator. *Chaos Solitons Fractals* **2016**, *89*, 8–19. [CrossRef]
96. Wan, S.; Zhang, X.; Dou, L. Compound Fault Diagnosis of Bearings Using Improved Fast Spectral Kurtosis with VMD. *J. Mech. Sci. Technol.* **2018**, *32*, 5189–5199. [CrossRef]
97. Haiyang, Z.; Jindong, W.; Lee, J.; Ying, L. A Compound Interpolation Envelope Local Mean Decomposition and Its Application for Fault Diagnosis of Reciprocating Compressors. *Mech. Syst. Signal Process.* **2018**, *110*, 273–295. [CrossRef]
98. Chen, H.; Liu, Z.; Alippi, C.; Huang, B.; Liu, D. Explainable Intelligent Fault Diagnosis for Nonlinear Dynamic Systems: From Unsupervised to Supervised Learning. *IEEE Trans. Neural Netw. Learn. Syst.* **2024**, *35*, 6166–6179. [CrossRef]
99. Maurya, M.; Panigrahi, I.; Dash, D.; Malla, C. Intelligent Fault Diagnostic System for Rotating Machinery Based on IoT with Cloud Computing and Artificial Intelligence Techniques: A Review. *Soft Comput.* **2024**, *28*, 477–494. [CrossRef]
100. Hoang, D.T.; Kang, H.J. A Motor Current Signal-Based Bearing Fault Diagnosis Using Deep Learning and Information Fusion. *IEEE Trans. Instrum. Meas.* **2020**, *69*, 3325–3333. [CrossRef]
101. Lei, Y.; Yang, B.; Jiang, X.; Jia, F.; Li, N.; Nandi, A.K. Applications of Machine Learning to Machine Fault Diagnosis: A Review and Roadmap. *Mech. Syst. Signal Process.* **2020**, *138*, 106587. [CrossRef]
102. Zhu, Z.; Lei, Y.; Qi, G.; Chai, Y.; Mazur, N.; An, Y.; Huang, X. A Review of the Application of Deep Learning in Intelligent Fault Diagnosis of Rotating Machinery. *Meas. J. Int. Meas. Confed.* **2023**, *206*, 112346. [CrossRef]
103. Bai, B.; Li, Z.; Wu, Q.; Zhou, C.; Zhang, J. Fault Data Screening and Failure Rate Prediction Framework-Based Bathtub Curve on Industrial Robots. *Ind. Robot* **2020**, *47*, 867–880. [CrossRef]
104. He, F.; Ye, Q. A Bearing Fault Diagnosis Method Based on Wavelet Packet Transform and Convolutional Neural Network Optimized by Simulated Annealing Algorithm. *Sensors* **2022**, *22*, 1410. [CrossRef] [PubMed]
105. Li, T.; Zhao, Z.; Sun, C.; Yan, R.; Chen, X. Multireceptive Field Graph Convolutional Networks for Machine Fault Diagnosis. *IEEE Trans. Ind. Electron.* **2021**, *68*, 12739–12749. [CrossRef]
106. Jiao, J.; Zheng, X.J. Fault Diagnosis Method for Industrial Robots Based on DBN Joint Information Fusion Technology. *Comput. Intell. Neurosci.* **2022**, *2022*, 4340817. [CrossRef]
107. Chen, Z.; Wu, K.; Wu, J.; Deng, C.; Wang, Y. Residual Shrinkage Transformer Relation Network for Intelligent Fault Detection of Industrial Robot with Zero-Fault Samples. *Knowl. Based Syst.* **2023**, *268*, 110452. [CrossRef]
108. Yang, B.; Lei, Y.; Li, X.; Li, N. Targeted Transfer Learning through Distribution Barycenter Medium for Intelligent Fault Diagnosis of Machines with Data Decentralization. *Expert Syst. Appl.* **2024**, *244*, 122997. [CrossRef]
109. Kumar, P.; Raouf, I.; Kim, H.S. Transfer Learning for Servomotor Bearing Fault Detection in the Industrial Robot. *Adv. Eng. Softw.* **2024**, *194*, 103672. [CrossRef]
110. Sarajchi, M.; Sirlantzis, K. Evaluating the Interaction between Human and Paediatric Robotic Lower-Limb Exoskeleton: A Model-Based Method. *Int. J. Intell. Robot. Appl.* **2025**, 1–15. [CrossRef]
111. Hu, C.; Zhang, Z.; Li, C.; Leng, M.; Wang, Z.; Wan, X.; Chen, C. A State of the Art in Digital Twin for Intelligent Fault Diagnosis. *Adv. Eng. Inform.* **2025**, *63*, 102963. [CrossRef]
112. Mihai, S.; Yaqoob, M.; Hung, D.V.; Davis, W.; Towakel, P.; Raza, M.; Karamanoglu, M.; Barn, B.; Shetve, D.; Prasad, R.V.; et al. Digital Twins: A Survey on Enabling Technologies, Challenges, Trends and Future Prospects. *IEEE Commun. Surv. Tutor.* **2022**, *24*, 2255–2291. [CrossRef]
113. Mykoniatis, K.; Harris, G.A. A Digital Twin Emulator of a Modular Production System Using a Data-Driven Hybrid Modeling and Simulation Approach. *J. Intell. Manuf.* **2021**, *32*, 1899–1911. [CrossRef]
114. Song, Z.; Shi, H.; Bai, X.; Li, G. Digital Twin-Assisted Fault Diagnosis System for Robot Joints with Insufficient Data. *J. F. Robot.* **2023**, *40*, 258–271. [CrossRef]

115. Ademujimi, T.; Prabhu, V. Digital Twin for Training Bayesian Networks for Fault Diagnostics of Manufacturing Systems. *Sensors* **2022**, *22*, 1430. [CrossRef]
116. Ye, M.; Yan, X.; Hua, X.; Jiang, D.; Xiang, L.; Chen, N. MRCFN: A Multi-Sensor Residual Convolutional Fusion Network for Intelligent Fault Diagnosis of Bearings in Noisy and Small Sample Scenarios. *Expert Syst. Appl.* **2025**, *259*, 125214. [CrossRef]
117. Gong, C.; Peng, R. A Novel Hierarchical Vision Transformer and Wavelet Time–Frequency Based on Multi-Source Information Fusion for Intelligent Fault Diagnosis. *Sensors* **2024**, *24*, 1799. [CrossRef] [PubMed]
118. Zhang, Y.; Zang, Z.; Zhang, X.; Song, L.; Yu, Z.; Wang, Y.; Gao, Y.; Wang, L. Fault Diagnosis of Industrial Robot Based on Multi-Source Data Fusion and Channel Attention Convolutional Neural Networks. *IEEE Access* **2024**, *12*, 82247–82260. [CrossRef]
119. Li, Z.; Liu, K.; Wang, X.; Yuan, X.; Xie, H.; Wang, Y. A Signal-to-Image Fault Classification Method Based on Multi-Sensor Data for Robotic Grinding Monitoring. *J. Intell. Manuf.* **2023**, *36*, 537–550. [CrossRef]
120. Xue, H.; Li, Z.; Li, Y.; Jiang, H.; Chen, P. A Fuzzy Diagnosis of Multi-Fault State Based on Information Fusion from Multiple Sensors. *J. Vibroengineering* **2016**, *18*, 2135–2148. [CrossRef]
121. Kheirandish, M.; Yazdi, E.A.; Mohammadi, H.; Mohammadi, M. A Fault-Tolerant Sensor Fusion in Mobile Robots Using Multiple Model Kalman Filters. *Robot Auton. Syst.* **2023**, *161*, 104343. [CrossRef]
122. Wang, Z.; Liang, H.; Chen, C.; Wang, T.; Chen, Z. Graph Fusion and Propagation for Fault Diagnosis in Industrial Robots with Limited Labeled Data. *IEEE Sens. J.* **2024**, 1–14. [CrossRef]
123. Liu, B.; Bi, X.; Gu, L.; Wei, J.; Liu, B. Application of a Bayesian Network Based on Multi-Source Information Fusion in the Fault Diagnosis of a Radar Receiver. *Sensors* **2022**, *22*, 6396. [CrossRef]
124. Chen, J.; Wen, K.; Xia, J.; Huang, R.; Chen, Z.; Li, W. Knowledge Embedded Autoencoder Network for Harmonic Drive Fault Diagnosis Under Few-Shot Industrial Scenarios. *IEEE Internet Things J.* **2024**, *11*, 22915–22925. [CrossRef]
125. Yin, T.; Lu, N.; Guo, G.; Lei, Y.; Wang, S.; Guan, X. Knowledge and Data Dual-Driven Transfer Network for Industrial Robot Fault Diagnosis. *Mech. Syst. Signal Process.* **2023**, *182*, 109597. [CrossRef]
126. Bilal, H.; Obaidat, M.S.; Aslam, M.S.; Zhang, J.; Yin, B.; Mahmood, K. Online Fault Diagnosis of Industrial Robot Using IoRT and Hybrid Deep Learning Techniques: An Experimental Approach. *IEEE Internet Things J.* **2024**, *11*, 31422–31437. [CrossRef]
127. Zhao, C.; Shen, W. Dual Adversarial Network for Cross-Domain Open Set Fault Diagnosis. *Reliab. Eng. Syst. Saf.* **2022**, *221*, 108358. [CrossRef]
128. Wang, L.; Liu, J.; Zhang, H.; Zuo, F. KMSA-Net: A Knowledge-Mining-Based Semantic-Aware Network for Cross-Domain Industrial Process Fault Diagnosis. *IEEE Trans. Ind. Inform.* **2024**, *20*, 2738–2750. [CrossRef]
129. Zhao, K.; Liu, Z.; Zhao, B.; Shao, H. Class-Aware Adversarial Multiwavelet Convolutional Neural Network for Cross-Domain Fault Diagnosis. *IEEE Trans. Ind. Inform.* **2024**, *20*, 4492–4503. [CrossRef]
130. Xu, Y.; Feng, K.; Yan, X.; Sheng, X.; Sun, B.; Liu, Z.; Yan, R. Cross-Modal Fusion Convolutional Neural Networks With Online Soft-Label Training Strategy for Mechanical Fault Diagnosis. *IEEE Trans. Ind. Inform.* **2024**, *20*, 73–84. [CrossRef]
131. Lee, J.; Noh, I.; Lee, J.; Lee, S.W. Development of an Explainable Fault Diagnosis Framework Based on Sensor Data Imagification: A Case Study of the Robotic Spot-Welding Process. *IEEE Trans. Ind. Inform.* **2022**, *18*, 6895–6904. [CrossRef]
132. Dev Anand, M.; Selvaraj, T.; Kumanan, S. Fault Detection and Fault Tolerance Methods for Industrial Robot Manipulators Based on Hybrid Intelligent Approach. *Adv. Prod. Eng. Manag.* **2012**, *7*, 225–236. [CrossRef]

sensors

MDPI

Systematic Review

Proactive Maintenance of Pump Systems Operating in the Mining Industry—A Systematic Review

Sylwia Werbinska-Wojciechowska * and Rafal Rogowski

Faculty of Mechanical Engineering, Wroclaw University of Science and Technology, Wyspianskiego 27, 50-370 Wroclaw, Poland; rafal.rogowski@pwr.edu.pl

* Correspondence: sylwia.werbinska@pwr.edu.pl; Tel.: +48-71-3203427

Abstract: Recently, there has been a growing interest in issues related to mining equipment maintenance, with particular focus on pumping systems' continuous operation. However, despite wide applications of pump system maintenance in a wide range of industries, such as water and wastewater, aviation, petrochemical, building (HVAC system), and nuclear power plant industries, the literature on maintenance of pump systems operating in the mining industry still needs development. This study aims to review the existing literature to present an up-to-date analysis of maintenance strategies for mining pumps, with a particular focus on proactive maintenance approaches. Key aspects considered include predictive diagnostics and prognosis, health status monitoring, maintenance management, and the integration of intelligent mining systems to enhance operational reliability and efficiency in harsh mining environments. The proposed methodology includes a systematic literature review with the use of the Primo multi-search tool, adhering to the Preferred Reporting Items for Systematic Reviews and Meta-Analyses (PRISMA) guidelines. The selection criteria focused on English studies published between 2005 and 2024, resulting in 88 highly relevant papers. These papers were categorized into six groups: (a) condition/health status monitoring, (b) dewatering system operation and maintenance, (c) health diagnosis and prognosis, (d) intelligent mining (modern technologies), (e) maintenance management, and (f) operational efficiency and reliability optimization. A notable strength of this study is its use of diverse scientific databases facilitated by the multi-search tool. Additionally, a bibliometric analysis was performed, showcasing the evolution of research on pump maintenance in the mining sector over the past decade and identifying key areas such as predictive diagnostics, dewatering system optimization, and intelligent maintenance management. This study highlights the varied levels of research and practical implementation across industries, emphasizing the mining sector's unique challenges and opportunities. Significant research gaps were identified, including the need for tailored diagnostic tools, real-time monitoring systems, and cost-effective maintenance strategies specific to harsh mining environments. Future research directions are proposed, focusing on advancing predictive maintenance technologies, integrating intelligent systems, and enhancing operational efficiency and reliability. The study concludes with a detailed discussion of the findings and their implications, offering a roadmap for innovations in pump maintenance within the mining industry.

Keywords: predictive maintenance; condition-based monitoring; pump diagnosis and prognosis; pump systems; the mining industry; systematic review

1. Introduction

The mining industry faces significant challenges driven by dynamic market demands, technological advancements, and evolving safety and environmental regulations. Effective machinery maintenance is a critical component in sustaining reliable and safe mining operations. Among the various systems used in mining, pumping stations play a pivotal role in water management and dewatering processes, particularly in underground mines where water intrusion poses substantial risks to operational safety and productivity. The ability to maintain the durability and reliability of these systems is vital for ensuring continuous mining operations and preventing costly downtime [1]. In addition, maintenance costs constitute a significant portion of total mining expenditures. According to industry reports, including data from the Empowering Pumps & Equipment portal [2], maintenance expenses can account for 35% to 50% of total mining project costs. Unplanned downtime of critical systems, such as pumps, can lead to severe financial consequences due to production losses and increased repair expenses. Therefore, adopting effective maintenance strategies, particularly transitioning from reactive to proactive maintenance, is essential for reducing operational risks and improving cost efficiency.

Mining operations must navigate a variety of factors that impact the performance of pumping systems. Harsh operating conditions, including abrasive slurries, corrosive substances, and fluctuating mechanical loads, accelerate wear and degradation of pumps and associated equipment. Maintenance strategies must, therefore, address these challenges through a combination of predictive and preventive practices. Predictive maintenance, leveraging technologies such as vibration analysis, thermography, and real-time data analytics, has become increasingly important in identifying potential failures before they result in breakdowns. This proactive approach minimizes unplanned downtime, reduces repair costs, and enhances overall system reliability [3].

Preventive maintenance remains a cornerstone of pumping system upkeep, focusing on routine inspections, scheduled servicing, and the timely replacement of components. For instance, tasks such as inspecting seals, lubricating bearings, and cleaning filtration systems help mitigate common wear-related issues. However, the complexity of maintaining large-scale and interconnected pumping systems in mining requires careful planning and effective resource management. Companies must develop comprehensive maintenance schedules tailored to the specific demands of their equipment and operating environment [4,5].

Design choices and material selection also influence the reliability of mining pumping systems. Pumps exposed to abrasive media benefit from wear-resistant materials, while corrosive environments necessitate corrosion-resistant coatings and components. Optimizing these design elements alongside robust maintenance practices improves the equipment's resilience, ensuring better performance and extended service life [4]. Moreover, advancements in digital technologies, including the Internet of Things (IoT) and data-driven diagnostics, are transforming maintenance management by enabling remote monitoring and predictive analytics [6].

Another significant factor in maintaining mining systems is the competency and training of maintenance personnel. A skilled workforce equipped with the knowledge and tools to perform diagnostics, repairs, and proactive interventions is essential. Developing a strong maintenance culture, reinforced by regular training and adherence to safety protocols, enhances both the effectiveness and safety of maintenance activities. Organizations must prioritize continuous improvement to remain adaptive to new technologies and changing operational requirements [3,7].

Additionally, the necessity for a quick response in case of failures and unforeseen situations poses a significant challenge in mining operations. Pump failures and unplanned downtime can lead to severe operational disruptions, increased costs, and heightened safety risks. Therefore, mining companies must implement robust measures to minimize the likelihood of system breakdowns and ensure continuous operation. Employing modern maintenance technologies, such as predictive analytics, real-time condition monitoring, and automated diagnostics, significantly improves the responsiveness and efficiency of maintenance processes. These innovations enhance the reliability and availability of pumping systems while reducing maintenance costs and operational risks. Investments in advanced maintenance strategies and technologies strengthen the overall resilience of mining operations, ensuring sustainable performance, safety, and competitiveness in a demanding industry. In conclusion, addressing the challenges of mining system maintenance, particularly for pumping stations, requires a holistic approach that integrates predictive and preventive strategies, advanced technological solutions, and well-trained personnel. As mining operations evolve and the complexity of machinery increases, companies must invest in innovative maintenance practices to enhance reliability, reduce costs, and maintain a competitive edge in a demanding global market.

Recently, numerous studies and publications have emerged related to the mining sector, focusing on maintenance management and diagnostic strategies aimed at improving the reliability and operational efficiency of critical machinery (for a comprehensive review, see, for example, [8–12]).

Given the extensive range of maintenance strategies discussed in the literature—spanning from reactive maintenance to predictive maintenance—one of the key decision-making challenges is selecting the most appropriate strategy that aligns with mining companies' specific needs and operational capabilities [13,14]. Several studies have addressed this issue, including analyses of preventive maintenance and inspection policies [15–17]. The feasibility of implementing condition-based maintenance in the mining sector is explored in works such as [18–21], while predictive maintenance strategies and their implementation are reviewed in [6,22,23]. Moreover, numerous studies focus on predictive maintenance applications and diagnostic or early warning methods, including fan failure prediction (e.g., [24]), gearboxes maintenance (e.g., [19]), bearings diagnostics (e.g., [25]), or mobile mining equipment maintenance (e.g., [26–28]). Additionally, research addressing the integration of Industry 4.0 technologies and artificial intelligence in mining equipment maintenance can be found in, e.g., [29–31].

When it comes to the problem of pump maintenance, the aspects of reliability and diagnostics are extensively explored, with examples provided in [32], where case studies on centrifugal and reciprocating pumps are presented. A review of predictive maintenance strategies for pump systems is available in works such as [33,34], although these studies primarily analyze heat pumps in combined heat and power plants. Additionally, the maintenance challenges of large-scale heat pumps are comprehensively summarized in [35].

Taking one step further, a search for English-language review publications in the Scopus database, using the keywords *"mining industry OR mine"* AND *"pump maintenance"* AND *"review* OR *state of the art* OR *current state"*, resulted in the identification of eight relevant records published between 2019 and 2023.

In work [10], the authors focus on the role of artificial intelligence (AI) and machine learning (ML) in improving predictive maintenance (PdM) strategies for machinery and equipment in the mining industry, which is crucial for continuous production but involves high costs and complexity. It provides a systematic review of current research, examining PdM methodologies, architectures, and models, as well as potential applications and challenges specific to the mining sector. The issues in the PdM area also examined in work [36]. The authors focus on the PdM in the context of Industry 4.0, highlighting its potential to reduce downtimes, lower costs, and enhance productivity through data-driven solutions in smart manufacturing. They provide a systematic review of PdM applications across various manufacturing sectors (including the mining sector), offering a comparative decision support map, insights into technology readiness, and a framework to guide the development of PdM strategies while addressing existing challenges. Another work, [37], reviews the current state of digital twin (DT) technology and its applications in the minerals industry, addressing challenges such as geological unpredictability, legacy system integration, and cybersecurity. It emphasizes the potential of combining immersive visualization with real-time spatial graphics to enhance the usability and acceptance of DTs, particularly in operational scenarios, and highlights the need for further research in this area. In addition, there are distinguished reviews that focus on the specific areas on maintenance (e.g., [38]) or type of objects—(e.g., [39,40]). The article [38] reviews the importance of condition-based monitoring for induction motors (IMs) to reduce operational and maintenance costs through early fault detection, minimizing downtime and unexpected failures. It presents an overview of IM faults, diagnostic schemes, and monitoring techniques, emphasizing the potential of non-invasive methods for automating maintenance scheduling and predicting failures in industrial applications. Article [40] focuses on wear issues in hydraulic components of hydrostatic transmission systems, exploring solutions like thermal coating and surface treatments, including laser beam and plasma coatings, to extend the lifespan of pumps, valves, and cylinders. Meanwhile, article [39] reviews the use of vibration signal analysis for monitoring rotating components, discussing advancements in signal processing, AI-based diagnostics, and prognostics and identifying the need for better interpretability, experimental validation, and data reproducibility to facilitate real-time industrial applications. Together, these reviews provide a comprehensive perspective on modern approaches to equipment maintenance and reliability in industrial environments. However, despite the increasing recognition of the role of pumps in ensuring continuous material flow and water management in mining operations, there is a notable gap in comprehensive review articles specifically addressing maintenance strategies, diagnostic technologies, and automation potential for mining pumps. Only a few of the identified studies touch upon general trends in pump maintenance without a detailed examination of the mining sector's unique operational constraints and evolving technological landscape. Some examples of mine-specific dewatering solutions can be found in [41], where a comparative analysis of basic dewatering systems used in coal mines in India highlights efficiency-related issues in these systems. This lack of focus is particularly striking in the context of Maintenance 4.0, where advanced data-driven techniques, predictive analytics, and real-time monitoring are revolutionizing maintenance practices.

Consequently, a systematic review of existing approaches, technological advancements, and research gaps in mining pump maintenance is necessary to provide a holistic understanding of proactive maintenance strategies. This includes an assessment of predictive diagnostic methods, cost-effective maintenance planning, and the application of real-time monitoring technologies. The findings aim to guide both future research and industrial practices in enhancing the reliability and efficiency of pump systems in the mining sector.

Following the above considerations, this study provides a comprehensive overview of academic research on the issues of pump system maintenance, with particular emphasis on the mining industry application field. The primary goal is to identify key research trends in this field and suggest potential future research directions. Additionally, based on the literature review, a framework for proactive maintenance in pump systems from the mining industry is developed. This framework outlines a systematic approach for improving the reliability and efficiency of pumping systems by integrating predictive maintenance techniques, real-time monitoring, and the application of advanced technologies such as IoT and AI. Consequently, this paper contributes to the existing knowledge on pump systems maintenance in three ways: (1) identifying the major research trends related to mining industry applications in the maintenance of pump systems; (2) outlining future research directions for the study of pump maintenance in the mining industry sector; and (3) developing a framework for proactive maintenance of pump systems in the mining industry.

Based on these objectives, the research questions are as follows:

RQ1: What is the state of the literature on mining equipment maintenance, with a particular focus on pump systems, between 2005 and 2024?

RQ2: What are the main research and knowledge gaps in pump systems maintenance, especially in the mining industry sector?

RQ3: What trends can be identified in proactive maintenance approaches, and how have they evolved over recent years in the mining sector?

RQ4: What scope should the framework for proactive maintenance of pump systems in the mining industry have?

This paper addresses the research questions posed above by employing bibliometric performance analysis and systematic analysis using the PRISMA method (Preferred Reporting Items for Systematic Reviews and Meta-Analyses) [42]. This approach is designed to summarize and pinpoint the key research areas within the identified application fields.

In summary, the article is structured into seven sections. Following the Introduction (Section 1), the Theoretical Background (Section 2) outlines the issues of pump systems in the mining industry with a particular view on the maintenance of pump systems. The Review Methodology (Section 3) details the primary methods used for the review, including the strategy for the literature search and the criteria used to assess the relevance of the analyzed documents. Section 4 presents the main findings of the systematic literature review for the selected papers within the six identified application fields. Section 5 then discusses the results related to these application fields, identifying gaps in the existing literature and knowledge. Section 6 introduces a framework for proactive maintenance in pump systems from the mining industry. The final section, Conclusions (Section 7), provides a summary of contributions, outlines limitations, and offers recommendations for future research.

2. Theoretical Background

The analysis of the literature review would be incomplete without a presentation of the theoretical background related to the issue under study. Within the framework of their research work, the authors focus on pump systems operating in the mining industry. Therefore, in the next subsections, issues related to the specificity of such systems, focusing on mine dewatering systems in the investigated industry and currently used maintenance strategies for mining pumps, will be discussed.

2.1. Pump Systems in the Mining Industry

The mining industry involves discovering and extracting naturally occurring minerals from the earth [43]. Materials such as coal, metals, oil shale, rock salt, chalk, stone, and gravel are all extracted in mining. The recovery of these raw materials is based on the economic viability of investing in labor, mining tools, and energy to run mining operations, refining, and transport [44].

The exploitation of mineral deposits varies in terms of the way the work is carried out. Indeed, we may identify two main types of mining: surface mining and underground mining [45]:

- Surface mining activity means that all the work related to the extraction of raw materials takes place on the surface, and the entire process of extracting minerals is based on the discovery of successive layers of the deposit of raw material.

- Underground mining, also known as subsurface mining, consists of digging or blasting tunnels and shafts into the earth's crust to reach buried orebodies. Ore deposits and tailings are brought to the surface to be refined by onsite processing plants.

In addition, both types of mining activities use different methods for raw materials extraction (Figure 1). Their use results in various operational and maintenance problems that occur in mining companies.

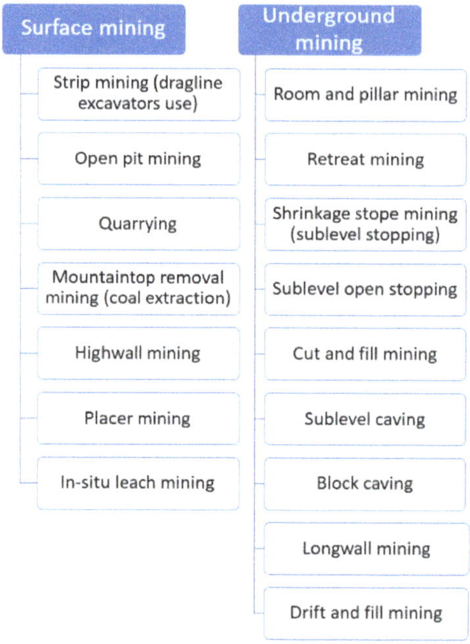

Figure 1. Surface mining and underground mining methods. Source: own contribution based on [46].

Going one step further, it is evident that underground as well as surface mining of various mineral deposits involve the presence of water [41]:

- Water that is needed for technological processes performance;
- Water that remains after technological processes performance;
- Water as an inflow to mine working areas caused by ore exploitation.

Water can also enter the pits due to rainfall in the surface mines.

The demand for water needed for operations performance and the removal of surplus water is one of the main problems in the mining industry. This fact enforces the maintenance of dewatering systems at a high level to ensure the possibility of conducting mining operations. In addition, the problem becomes more complex due to the presence of different dewatering systems in the mine:

- Horizontal dewatering system;
- Main (vertical) dewatering system;
- Technological water system;
- Other water systems, such as a fire water system or an air-conditioning system.

The authors present an example of a mine shaft dewatering system's operation and maintenance in [47], introducing the main challenges in this area.

The proper performance of water systems in the mine is based on using various pump systems. Pumps belong to a group of working machines called fluid conveyors. Their task is to pump (move) liquids (or mixtures of liquids with solids) through the system (overcoming local and linear flow resistance). In addition, the pump's tasks are also connected with lifting the liquid from a lower level to a higher level and pumping the liquid from a vacuum tank with a lower pressure to a compression tank with a higher pressure to create a high working pressure of the liquid (e.g., hydraulic drives). In the pumping process, the pump takes mechanical energy from the motor and transfers it to the liquid flowing through it via the working unit (piston, rotor, impeller, etc.) [48]. Indeed, following this short introduction, the first classification of pump systems operating in mine dewatering systems is presented in Figure 2. The classification focuses on the application area in the mining industry—mine dewatering.

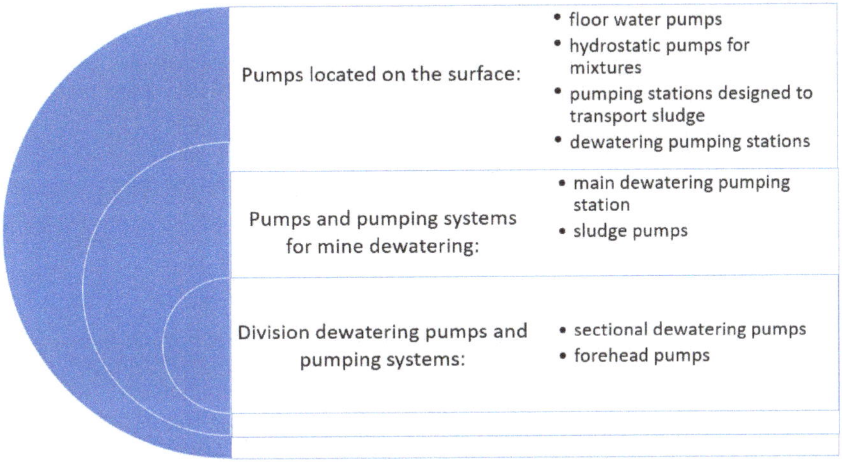

Figure 2. Pumps and pumping systems used in the mining industry. Source: own contribution based on [47,49,50].

Water flowing out of the underground, along with technological water used, for example, in the process of drilling boreholes, is accumulated in the pits of forehead pumping stations. These are small pumping stations located in the mine workings, near the mining fields. Each mining field has at least one forehead pumping station. Water from the forehead pumping stations is transported to local pumping stations built at the main mine roadways. Local pumping stations have small settling tanks where water is collected and pre-cleaned of solids by sedimentation. Divisional pumping stations collect water from local pumping stations from a given mining district. They have settling tanks to ensure adequate water retention and sedimentation. The final stage of the dewatering process is the main pumping stations located at the shafts. Their task is to collect and drain water from the entire mine. Their tanks have retention, allowing them to collect water, purify it thoroughly, and pump out the daily inflow [51].

In addition, main dewatering systems for underground mine workings can be divided into the following categories [47]:

- Single stage;
- Multi-level;
- Indirect;
- Direct.

Moreover, according to [52–56], we may define three types of pumps used in the mining industry with their applicability:

- Slurry pumps, widely used in mining to transport abrasive mixtures of solids and liquids. They are prone to wear and tear due to the abrasive nature of the materials they handle.
- Submersible pumps, commonly used for mine dewatering. They are preferred for their ease of installation and maintenance.
- Centrifugal pumps, used for various applications, including dewatering and slurry transport. They require careful selection and maintenance to ensure longevity and efficiency.

Going step further, we can distinguish four single-stage dewatering systems (Figure 3). Systems (a) and (d) are characterized by the fact that the pumps are located above the liquid mirror. Pumps in such systems must be characterized by good cavitation properties. This condition forces the use of pumps with no more than 1500 rpm speeds. Pump systems of (b) and (c) are inflow systems that allow using pumps with lower cavitation properties. They may use pumps that reach 3000 rpm.

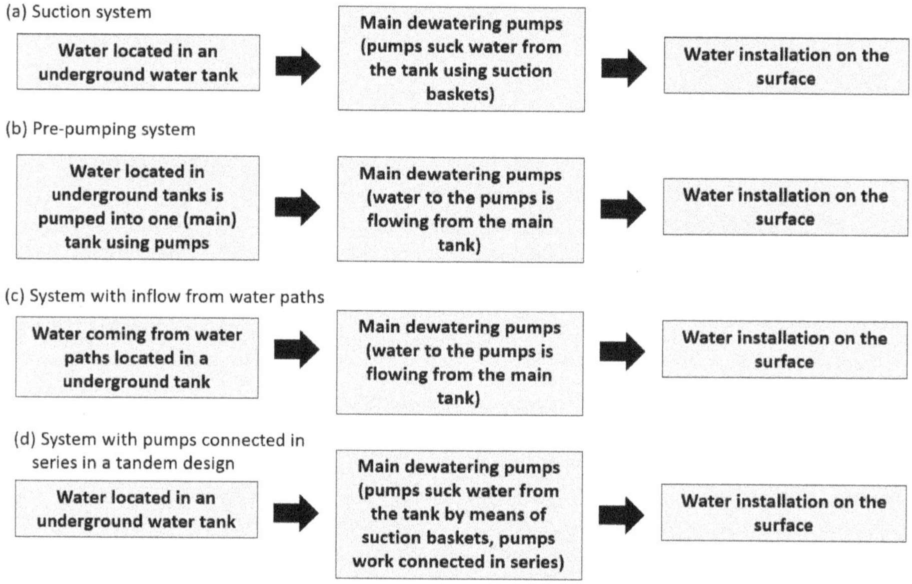

Figure 3. Single-stage dewatering systems. Source: own contribution based on [47].

Multi-level dewatering systems are characterized in Figure 4.

The selection of the right dewatering system in which the drainage pumps operate depends on a number of factors [57,58]:

- The number of levels on which mining is executed;
- The distance from the mining decks;
- Water inflow on each level;
- The irregularity of the inflows;
- The possibility of collecting water in the pits.

Going one step further, pumps can be classified into one of two main groups: dynamic (non-positive displacement) pumps and positive displacement pumps [59]. The first group of pumps uses the fluid velocity and the resulting momentum to generate the pumping power and move the fluid through the system. The second type of pump can operate by forcing a fixed volume of fluid from the inlet pressure section of the pump into the discharge zone of the pump [59,60]. In another work [61], the authors provide a more detailed classification of pumps based on three groups' definitions: rotodynamic pump, positive displacement pump, and other pumps (impulse pump, gravity pump, steam pump, and valveless pump).

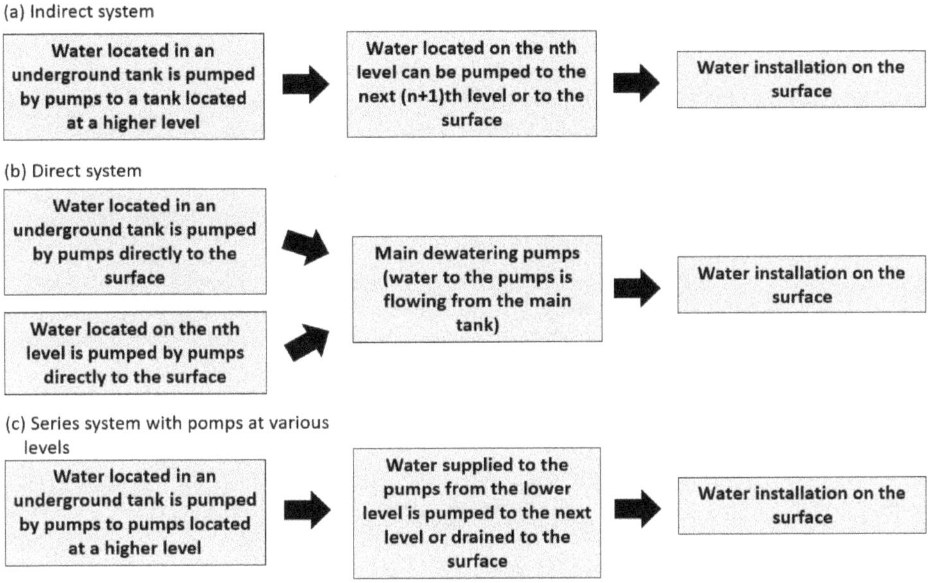

Figure 4. Multi-level dewatering systems. Source: own contribution based on [47].

Pump systems are the second most important systems for the operation of a mine, following ventilation systems. They ensure the safety of workers and allow the mine to function. Therefore, these systems must be maintained at the highest possible efficiency to ensure operational continuity. Indeed, in the next subsection, pump maintenance issues are investigated.

2.2. Maintenance of Pump Systems

In mining, there are, among other things, mineral sludge, wastewater, kaolin slurries, and compacted sludge, which have to be dealt with by suitable pumping systems and equipment. Therefore, many types of pumps in the mining industry are designed for pumping clean and contaminated water and sludge with various muds. These pumps have to meet the requirements of each substance, and the safety of workers and the efficiency of their work depend on the quality of the pump used and the correctness of the maintenance processes carried out [62,63]. At the same time, machinery and equipment operating in the mining industry are also exposed to many factors related to the working environment, e.g., salts, dust, humidity, and high temperatures. In order to meet these requirements, these systems are designed in such a way that their sensitivity to these factors is as low as possible. Therefore, due to the difficult operating conditions, maintenance issues become particularly important in the case of such systems [41].

Pump systems, second only to ventilation systems in importance, are critical to mine operations. These systems ensure worker safety and enable continuous mine functionality. Consequently, maintaining pump systems at peak performance is essential for uninterrupted operations. Main dewatering pump stations must continuously remove water from mining excavations, regardless of failure conditions or production halts. Two independent energy sources must power such systems. Control systems for these pumps should allow both automatic and manual switching between units. Depending on system design and operational requirements, various parameters are monitored, including flow rate, water temperature, bearing temperature, supply voltage, motor current, lift height, and operating pressure. These measurements guide pump regulation and inform diagnostic actions [57].

2.2.1. Pump Systems Reliability Issues

Enhancing the reliability of machinery operating in the mining industry requires a comprehensive understanding and identification of the root causes of failures, which form the basis for implementing predictive maintenance systems. Despite advancements in diagnostic methods, the current analysis and response to early failure symptoms often remain delayed, preventing the full realization of benefits offered by predictive maintenance strategies. Reactive maintenance, instead of proactive measures, contributes to increased operational costs and a lower utilization rate of machinery [63].

The reliability of pump systems largely depends on evaluating their electrical, mechanical, and electromechanical parameters. Condition assessment can be performed through two approaches: a limited-scope evaluation using diagnostic parameters from machine-installed sensors and a comprehensive, periodic global assessment. Local diagnostics involve monitoring technical conditions, operational parameters, and protective systems. Continuous online monitoring of parameters is essential to obtain a complete diagnostic picture. Data acquisition from operating equipment is facilitated by sensors and transducers selected based on the monitored circuit and parameters, such as pressure, temperature, and current draw. All collected data are archived in a central supervisory computer.

The mining industry employs a wide range of pump types, and accurately diagnosing failure sources requires a thorough analysis of the factors contributing to the malfunction. A critical aspect of pump diagnostics is determining the time of the first failure symptoms and tracking their progression over time. In facilities adhering to predictive maintenance principles, equipment condition forecasting is based on information obtained through diagnostic monitoring. An early step in this process involves defining threshold values that trigger preventive actions to avoid unplanned downtime [64].

Effective prevention and prediction of pump failures require a thorough understanding of both the pump's internal construction and its operational environment. The primary components susceptible to failure include impellers, seals, bearings, shafts, and casings. Failures in these elements are often accompanied by identifiable signs of damage, such as the following [4,65,66]:

- Excessive temperature generated at seals;
- Rotor looseness;
- Corrosion of the casing or impeller due to turbulence or cavitation;
- Damage to rolling elements (bearings);
- Failures caused by improper use of the pump outside its design parameters.

Before the pump's performance is visibly impaired due to component failure, early warning symptoms may appear within the system, signaling operational anomalies. Although these signals do not yet indicate a complete loss of functionality, their timely detection enables preventive maintenance actions. Diagnostic efforts should focus on observing physical phenomena produced by the system, emphasizing periodic or intermittent inspections to capture the moments when corrective or repair actions are necessary.

The key measurable parameters during pump operation include the following [47]:

- Temperature: excessive heating in areas experiencing high friction;
- Pressure: a drop in pump efficiency;
- Power consumption: increased power draw, indicating inefficiencies;

- Vibrations: an increase in overall vibration levels, with bearing failures often preceded by changes in vibration patterns. Excessive vibrations may result from pump misalignment or cavitation resonance between the pump, its base, or valves located in the suction or discharge lines;
- Leakage: monitoring seal integrity and bolted connections.

An increase in any of these parameters typically signals a deterioration in the pump's technical condition.

Depending on the criticality and diagnostic capabilities of the system, condition monitoring can be performed manually by diagnostics specialists or through automated systems using information networks that collect data from the entire machinery fleet. All operational data are archived, and deviations from normal conditions are flagged to prompt maintenance interventions.

2.2.2. Pump Systems Maintenance Issues

The available literature offers numerous definitions of the term "maintenance" [67]. For the purposes of this research, the definition provided in the European standard PN-EN 13306:2010 [68] has been adopted. According to this standard, maintenance encompasses all technical, organizational, and managerial actions performed throughout the lifecycle of an asset aimed at retaining or restoring it to a condition in which it can fulfill its intended function. A comprehensive literature review on the maintenance of systems and equipment can be found in works such as [69–71]. A general classification of fundamental maintenance strategies is presented in Figure 5. A comprehensive review of the topic of maintenance modeling is provided in [67].

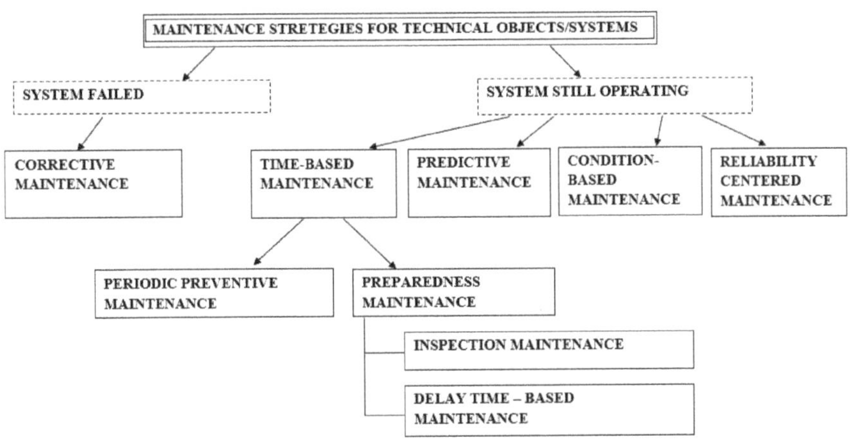

Figure 5. Technical system maintenance strategies. Source: own contribution based on [67].

Effective pump maintenance involves a combination of traditional scheduled maintenance, condition monitoring, and predictive maintenance strategies. Leveraging technological advancements and best practices can significantly enhance pump reliability and operational efficiency, ultimately reducing costs and improving productivity.

Pump maintenance can be broadly categorized into preventive, corrective, and predictive maintenance strategies. Preventive maintenance refers to scheduled tasks performed to prevent potential issues, such as checking for leaks, cleaning the pump components, and replacing seals and bearings before they fail. On the other hand, corrective maintenance is carried out after a malfunction or failure occurs, addressing the issue at hand to restore the pump's functionality. Predictive maintenance, the most advanced of the three, relies on monitoring systems and sensors to predict potential failures based on real-time data, such as vibration, temperature, and pressure readings. This type of maintenance is designed to minimize unplanned downtime by identifying issues before they lead to significant failures.

A significant body of literature is dedicated to pump maintenance, with numerous review papers addressing various aspects of the field. A preliminary literature analysis on pump maintenance was conducted using the Scopus database. A search for the term "pump maintenance" in article titles, abstracts, and keywords yielded 7252 results. To ensure relevance, the search was limited to subject areas such as engineering, energy, environmental science, computer science, mathematics, and business management and accounting. A keyword co-occurrence analysis based on using VOSviewer software ver. 1.6.18 of the obtained results allowed defining six main thematic clusters (Figure 6).

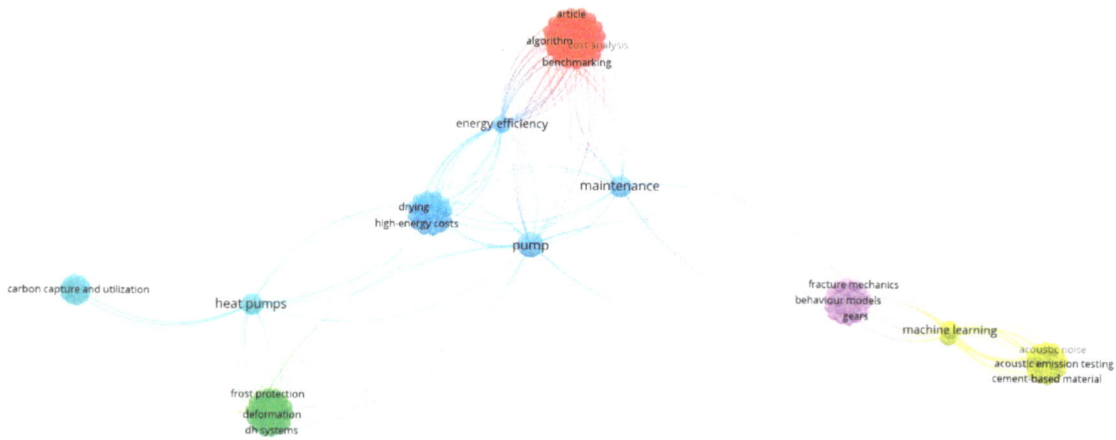

Figure 6. Mapping of the keywords that have occurred in the selected publications at least once—the Scopus database. Source: own development using VOSviewer software [72].

The first cluster—the largest cluster (red one, 31 keywords)—mostly focuses on the aspects of pumping scheduling and operational optimization. The second cluster (green one, 24 keywords) strictly investigates interactions between infrastructure, particularly railways, and thermal conditions. The third cluster (blue one, 23 keywords) focuses on improving the energy performance of pumps. The yellow cluster (22 keywords) develops the possible usability of AI-based methods in health monitoring. The next cluster (violet one, 22 keywords) is connected with the aspects of assessing and mitigating risks associated with failures of different types of pumps, and the last cluster (light blue, 15 keywords) is mostly focused on performance analysis in relation to typical faults of heat pumps.

The investigation of the reviewing literature in this area may extend this initial analysis. These works often focus on methods for improving maintenance strategies, such as the development of more accurate diagnostic tools, the integration of predictive maintenance technologies, and the analysis of failure modes and their causes. Many reviews also examine case studies where pump maintenance practices have been optimized, offering valuable insights into real-world applications and best practices. The growing interest in sustainability and the need for energy-efficient solutions has further fueled research in this area, as maintaining pumps in an optimal state can directly impact an organization's environmental footprint.

The literature also highlights the importance of condition-monitoring techniques and the role of digitalization in modern pump maintenance. With advancements in sensor technologies and data analytics, pumps can now be monitored in real time, allowing for immediate corrective action when necessary. These innovations have led to intelligent pump systems that can self-diagnose and predict future maintenance needs, significantly reducing the likelihood of unexpected failures. Such systems are increasingly used in industries that rely on high uptime, such as power generation, water treatment, and mining. A summary of recent reviewing articles that focus on pump maintenance issues is presented in Table 1.

Table 1. A summary of recent papers focusing on providing a literature overview in the area of pump maintenance.

Ref.	Publication Year	Research Objectives	Methodology Used	Databases Analyzed	Papers Analyzed	Analyzed Sector or Type of Pumps
[73]	2011	Review on major fault modes in centrifugal pumps, especially in the context of the water and sewage industry applications.	n/a	n/a	n/a	Water and sewage industry/ Centrifugal pumps
[74]	2014	Review on diagnostics and prognostics of centrifugal pumps, focusing on the application area, detection of fault modes, and test results.	n/a	n/a	n/a	Centrifugal pumps
[33]	2014	A comprehensive review of the development, applications, and performance characteristics of pumps operating in turbine mode (PAT) for small-scale hydroelectric power systems, with a focus on their feasibility in remote communities lacking grid connections.	n/a	n/a	n/a	Pump as turbine (PAT)
[34]	2020	Review on prognostics and health management for pump systems operating in thermal power plants.	Structured overview	n/a	n/a	Energy sector
[35]	2021	Analysis and characteristics of faults in large-scale heat pumps, with particular attention to recurrent faults in critical components, to assess existing fault detection and diagnosis methods, identify their limitations, and propose future research directions.	n/a	n/a	n/a	Large-scale heat pumps
[75]	2022	Critical analysis of machine learning's most current advances in the field of artificial intelligence-based system health management.	Systematic literature review	Web of Science, Science Direct, Scopus	88	Oil and gas sector
[61]	2022	Review various types of faults and their identification possibilities by various traditional and ML-based techniques.	n/a	n/a	n/a	Centrifugal pumps

Table 1. *Cont.*

Ref.	Publication Year	Research Objectives	Methodology Used	Databases Analyzed	Papers Analyzed	Analyzed Sector or Type of Pumps
[76]	2022	Investigation of the potentials and challenges of integrating emerging technologies (digital twin, building information modeling, machine learning) in the PdM of pumps.	Systematic literature review	Web of Science, Scopus	118	Civil infrastructure applications
[77]	2022	Review on condition monitoring techniques for pumps with a particular focus on the sensing technologies.	n/a	n/a	n/a	Oil and gas sector
[78]	2022	Review on machine learning-based fault diagnosis of centrifugal pumps.	n/a	n/a	n/a	Centrifugal pumps
[79]	2022	Review on recent developments in hydraulic pumps' fault diagnosis and health management.	Systematic literature review	n/a	n/a	Hydraulic pumps
[80]	2022	Review on health monitoring methods suitable for airframe fuel pumps under flight conditions.	n/a	n/a	n/a	Aviation sector
[81]	2023	Review on the artificial intelligence (AI)-based models for PdM of water injection pumps.	Systematic literature review	Scopus, IEEE Xplore, ScienceDirect, and ACM Digital Library	16	Oil and gas sector
[82]	2024	General overview of control systems used to operate pump units.	n/a	n/a	n/a	Offshore industry
[83]	2024	Review of the centrifugal pump cavitation monitoring methods (flow-head method, high-speed photography, pressure pulsation method, acoustic emission method, vibration method).	n/a	n/a	n/a	Centrifugal pumps

n/a—not available.

The articles presented in the table, focusing on reviewing pump maintenance issues, demonstrate that the topic has been extensively analyzed from various perspectives, including preventive strategies, predictive techniques, and technological advancements. These reviews highlight the importance of pump maintenance in ensuring operational efficiency and reducing downtime across multiple industries. However, there is a notable gap in the literature regarding review articles specifically focused on the mining industry sector. Given the critical role of pumps in mining operations, a comprehensive review tailored to this sector could provide valuable insights into industry-specific challenges, maintenance practices, and opportunities for improvement in this demanding environment.

Despite the lack of review articles, numerous studies are dedicated to pump maintenance in the mining sector, addressing key challenges such as abrasive wear, high-pressure demands, and energy efficiency. These works provide valuable insights into maintenance practices and solutions tailored to the unique conditions of mining operations. In the maintenance of the pump systems within the mining industry, two primary approaches are currently employed: the "run-to-failure" or corrective maintenance (CM) strategy and the preventive maintenance (PM) strategy [74]. The CM approach involves repairing or replacing components only after a failure occurs, leading to significant financial losses, downtime, and potential safety risks. In contrast, the PM strategy, particularly time-based preventive maintenance, may result in the premature replacement of components simply because their scheduled service life has ended, even if they are still functional. This method can be inefficient due to parts, labor, and service time costs. Additionally, unexpected failures can still occur between scheduled maintenance intervals, potentially leading to severe consequences [74].

An alternative approach gaining popularity in pump maintenance is condition-based maintenance (CBM). This strategy involves continuously monitoring key parameters that reflect the technical condition of a pump and its components using diagnostic tools [84,85]. CBM enables maintenance to be performed just before a failure occurs, improving efficiency by focusing on fault detection, degradation tracking, and failure prediction. Unlike CM or traditional PM, CBM emphasizes proactive monitoring of wear and tear to optimize maintenance activities. Studies such as [86–88] highlight its effectiveness in enhancing reliability.

Another increasingly applied strategy is predictive maintenance [89,90], which uses advanced diagnostic technologies to analyze the operating conditions of pumps and predict future failures. Continuous condition monitoring detects early signs of wear, allowing for timely interventions such as adjusting operating settings or improving lubrication [32]. PdM techniques commonly include thermal imaging, vibration analysis, and acoustic measurements [91]. Research on predictive maintenance of pumps, including fault detection and diagnosis, is presented in works like [22,92], while comprehensive reviews on fault diagnosis methods for pumping systems are discussed in [61,79]. Despite this, limited research has been conducted specifically to address maintenance challenges for mining pumps [65].

In summary, interest in pump maintenance in the mining industry is steadily growing, as confirmed by the increasing number of studies addressing the unique challenges of this sector. While traditional approaches like corrective maintenance (CM) and preventive maintenance (PM) remain widely used, their limitations in terms of efficiency, cost, and reliability underscore the need for alternative strategies. Condition-based maintenance (CBM) and predictive maintenance are emerging as promising solutions, leveraging advanced diagnostic tools and continuous monitoring to enhance operational efficiency, minimize downtime, and predict failures. Despite the growing body of research on pump maintenance across various industries, there is a notable lack of comprehensive reviews specifically focused on the mining sector.

This article aims to fill this gap by offering a detailed synthesis of current knowledge on maintenance practices for pumps in mining operations. It addresses two critical gaps in the literature:

- Lack of comprehensive reviews on mining-specific pump maintenance: While numerous studies broadly explore pump maintenance, few focus on the unique environmental and operational challenges the mining industry faces. This review provides a structured and in-depth analysis tailored to these conditions.
- Limited exploration of advanced maintenance strategies in mining: The application of CBM and PdM in mining pump systems remains underexplored. This article highlights the potential of these approaches to improve reliability and efficiency in harsh mining environments and proposes future research directions to optimize their implementation.

By addressing these gaps, this review aims to advance the understanding of pump maintenance in the mining industry and provide a foundation for developing more effective and sustainable maintenance practices. Additionally, this article will propose a framework for proactive maintenance in pump systems tailored to the unique conditions of mining operations, offering practical guidelines for improving maintenance strategies and minimizing downtime. This framework will be valuable for enhancing reliability, efficiency, and cost-effectiveness in mining pump systems.

3. Review Methodology

This section presents the key assumptions and steps undertaken in the systematic literature review (SLR) conducted for this study. An SLR aims to systematically identify, evaluate, and synthesize all relevant research pertaining to a specific research question, topic, or phenomenon of interest [93]. It is widely recognized as a rigorous scientific methodology that adheres to standardized procedures for designing, conducting, and reporting research processes [94,95].

The primary objective of the review was to explore the main research trends and identify gaps in pump maintenance in the mining sector. The review process followed established guidelines outlined in [94,96,97] and was conducted in alignment with the PRISMA framework [98,99]. The SLR methodology employed by the research team consisted of nine steps, grouped into three phases: planning (three steps), conducting (three steps), and documenting (three steps), as shown in Figure 7.

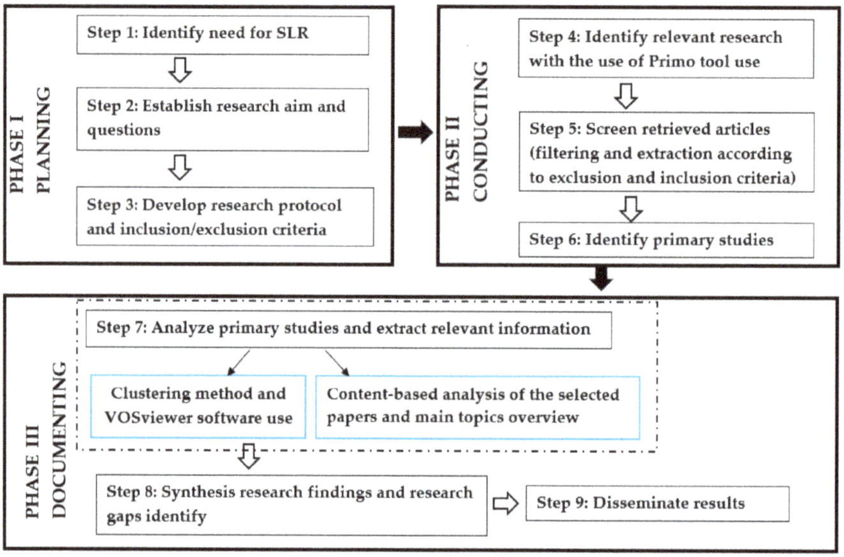

Figure 7. Research framework and methods/tools used for systematic literature review. Source: own contribution.

The following subsections provide a detailed discussion of the research activities carried out during each of these three phases.

3.1. Planning of the SLR Study

This section outlines the key steps in this SLR study's planning phase. The primary aim of the research is to review, classify, and synthesize the existing literature on pump maintenance in the mining industry based on a thematic analysis of relevant studies. The planning process's first step involved defining the review's main objectives. These objectives include (a) establishing a comprehensive understanding of maintenance strategies for pumps used in mining operations, (b) identifying key knowledge gaps and research opportunities in this field, and (c) highlighting trends and challenges to guide future research efforts. To achieve these objectives, a set of research questions was formulated to structure the review (step 2). These questions, presented in the Introduction section, were designed based on an in-depth analysis of existing studies on pump maintenance strategies, diagnostic and prognostic approaches, and the unique challenges associated with mining operations presented in Section 2. The preliminary review also emphasized the need to address the lack of a comprehensive framework tailored to the proactive maintenance of pumps in the mining sector.

Additionally, the research questions and the initial literature analysis informed the development of the study's research framework, the selection of appropriate tools and methods, and the establishment of inclusion and exclusion criteria for the review process. This systematic approach ensures a focused and rigorous analysis of the relevant literature while addressing the identified research gaps (step 3).

3.2. Conducting the SLR Study

The second phase of the performed methodology includes steps 4, 5, and 6, which are discussed in detail below.

3.2.1. Collection of Publications for Review

The first step involved conducting the literature search process, utilizing the Primo multi-search tool [100]. Primo allows users to explore a library's collection of resources through keyword searches, with various filters available to refine the results. This approach enables the development of search strategies that incorporate resources from multiple scientific databases, including Scopus, Web of Science, and ScienceDirect, as well as databases from publishers such as Elsevier, Wiley, and Springer. The literature search was carried out over a one-month period, specifically from 20 December to 30 December 2024.

The search strategy utilized English terms combined with the Boolean operators AND and OR to form a comprehensive search query. Keywords related to digital twins, maintenance, and transportation aspects were selected for this purpose. Maintenance-related terms were identified based on existing review articles [10,34,79,101]. To ensure thoroughness, the keyword selection process incorporated both contemporary and historical terminology pertinent to pump maintenance in relation to the mining sector. This approach ensured relevance to both the research scope and historical developments in the field.

An initial broad search strategy was implemented, which included examining related terms (e.g., "dewatering system", "maintenance AND mining industry", "proactive maintenance AND pump systems") before finalizing the keyword list. Ultimately, the selected keywords were chosen to achieve the broadest possible coverage of pump maintenance research within the selected industry sector context. The final search engine configuration includes the following keywords:

(ALL = (pump maintenance)) AND (ALL = (pump condition OR pump monitoring OR pump prognosis OR pump diagnosis OR pump fault)) AND (ALL = (mining industry))

Based on the presented criteria, the initial search allowed for the identification of 1142 scientific papers, which were later analyzed in the screening process (Section 3.2.2).

3.2.2. Screening of Collected Publications

The screening process was crucial for identifying studies relevant to full-text analysis. Initially, the search was narrowed to include papers published between 2005 and 2024, ensuring a focus on recent advancements in the field. To ensure the quality of the selected papers, additional evaluation criteria were applied, including an analysis of the research methodology, the diagnostic tool quality, and the consistency of results with industrial requirements. The quality assessment was conducted based on the guidelines of the systematic review protocol (PRISMA), which also involved evaluating the reliability of sources and the transparency of results. Papers that did not meet these criteria were excluded from further analysis. Specifically, studies that lacked clear results or employed methodologies unsuitable for pump diagnostics in mining were excluded.

To thoroughly evaluate each study's relevance to the topic of pump maintenance in relation to the mining industry, two primary inclusion criteria were applied: full-text availability and publication in English. Full-text access was essential for a detailed review of each study's methodologies, findings, and alignment with the research focus. Limiting the selection to English-language publications ensured consistent interpretation and a comprehensive understanding of the studies. Based on these criteria, 375 papers were excluded from further consideration.

The screening process aimed to filter out studies unrelated to the research topic and was carried out in two steps. In the first step, all authors reviewed the titles and abstracts of the identified records. Papers were included if their abstracts focused on DT applications in transportation systems, operations, and maintenance. The team then discussed their assessments in a research meeting, and in cases of disagreement, articles were retained for full-text analysis. Following this step, 390 papers remained for further evaluation, with duplicates and review articles subsequently removed.

In the second step, a full-text review was conducted to determine the paper's relevance to the thematic scope. Studies were evaluated based on their contributions to understanding pump maintenance applications in mine dewatering systems, the maintenance methods and models applied, and condition monitoring, diagnosis, and prognosis. As in the first step, team members performed evaluations individually, followed by group discussions during research meetings. When disagreements arose, the team conducted a more detailed document analysis. This step led to the exclusion of 730 papers for various reasons, such as studies focused on maintenance issues in unrelated fields like medical applications (pump fusions).

In the second step of the screening process, the authors examined the papers in the full-text research. The main aim was to assess their relevance to the investigated thematic area. Papers were evaluated based on their contributions to understanding DT applications, methodologies used, and their implications for maintenance and operational practices in transportation systems. As in the first step, the research teams made the evaluation individually. Later, at research team meetings, we compared team members' opinions. In case of discrepancies in assessing the paper's suitability, the team members focused on a more detailed analysis of the full document. As a result, 302 papers were excluded for specific reasons. For example, the studies that describe maintenance issues, e.g., medicine applications, were excluded.

3.2.3. Primary Studies Identification

Following the screening process, 88 publications were chosen for subsequent qualitative and quantitative analysis. A cross-sectional review of the selected papers was also performed to ensure a comprehensive and representative collection of the relevant literature [102]. Papers were excluded if they did not meet the inclusion criteria or did not contribute adequately to the topic of pump maintenance in mining operations. Additionally, studies lacking methodological rigor or applied diagnostic tools irrelevant to mining pump systems were excluded. These exclusions were based on an assessment of the study's research design, the proposed diagnostic methods' applicability, and the findings' relevance to the mining industry. This approach confirmed that the identified studies collectively offer a thorough overview of the current state of research in the analyzed area.

The flow diagram illustrating the study selection process, aligned with PRISMA guidelines, is shown in Figure 8. Additionally, the PRISMA checklist detailing the methodology is provided in the Supplementary Materials (Table S1).

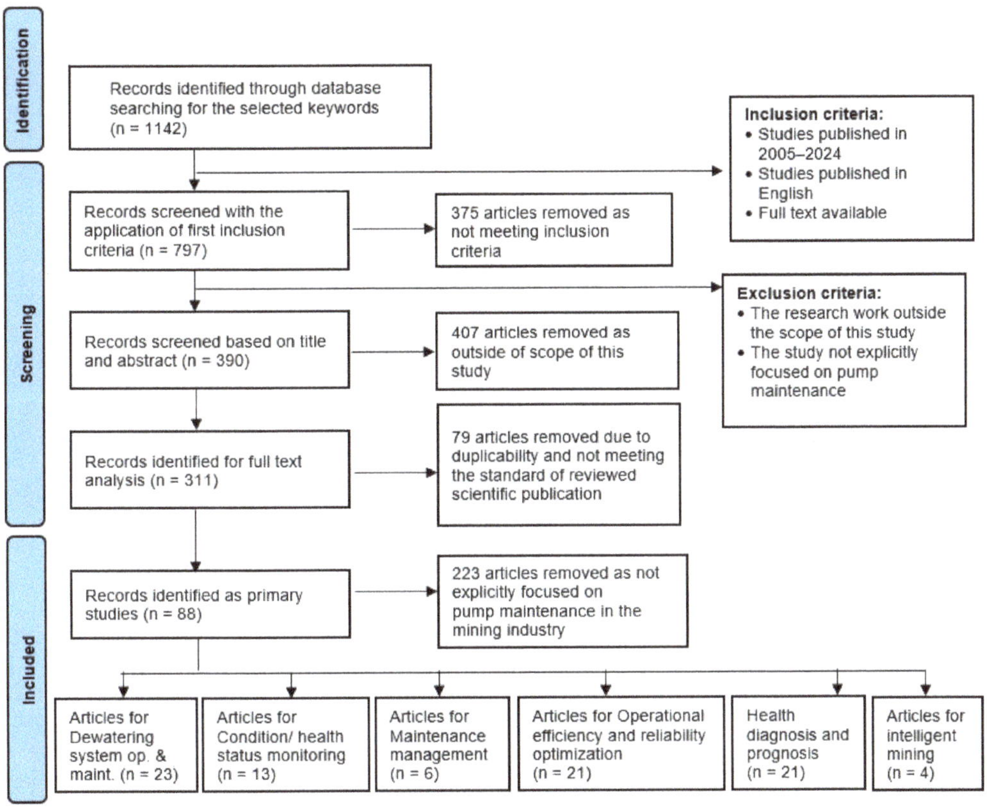

Figure 8. PRISMA-based flowchart of systematically selecting relevant studies in the analyzed research area. Source: own contribution based on [42].

3.3. Documenting of the SLR Study

This phase corresponds to steps 7, 8, and 9 of the SLR process. As part of step 7, a bibliometric analysis was conducted.

Bibliometrics, a subfield of scientometrics, applies mathematical and statistical techniques to evaluate scientific activities. This analysis facilitates the exploration of networks formed around key representative keywords and highlights how citations, authors, affiliations, countries, and publications reflect the significance of specific topics in the research field. Notably, there has been a growing interest in bibliometric studies within the scientific community (see, e.g., [103–107]).

Following the analysis, the selected articles were organized, documented, and categorized using the Mendeley reference manager [108,109]. For primary content-based analysis,

MS Excel Professional Plus 2019 and VOSviewer ver. 1.6.18 software were utilized [72]. The main findings were summarized using variables such as authors' geographic locations or publication years.

VOSviewer, as described in [110], is a specialized tool for creating and analyzing bibliometric maps, allowing for a detailed examination of the results. The quantitative summary includes trend analyses and distributions by publication year and source. Using the functionalities of VOSviewer, bibliometric maps were created, focusing on keyword co-occurrence and relationships between leading authors. The findings are detailed in Section 4.

Step 8 involved synthesizing the research findings and aligning the results with the four defined research questions.

Finally, step 9 outlined the study's limitations and suggested directions for future research. The outcomes of steps 8 and 9 are presented in Section 5 and form the foundation for developing the framework for proactive maintenance in pump systems from the mining industry (Section 6).

4. Results

This section includes the results of the conducted systematic review according to the defined research methodology (Figure 8). As a result, in the next subsections, bibliometric analysis and content-based analysis are presented.

4.1. Bibliometric Analysis

In the first step, a bibliometric analysis of already-selected publications for further research on the topic of pump maintenance in the mining industry was carried out. A total of 88 publications from the six subject areas analyzed were accepted for detailed analysis. The largest number of articles (23 papers) was in the area of dewatering system operation and maintenance (O&M). The number of analyzed publications in the other areas is as follows: health diagnosis and prognosis—21 papers; operational efficiency and reliability optimization—21 papers; condition/health status monitoring—13 publications; maintenance management—6 publications; and intelligent mining—4 publications.

The geographic distribution of authors and scientific institutions was also examined as part of the bibliometric analysis. The results reveal significant disparities in research activity worldwide. The United States and China lead in the number of publications, contributing approximately 23% and 21% of the analyzed papers, respectively. Poland and India also demonstrate strong research engagement, with 10 and 6 publications, respectively. Other countries show varying levels of contribution, including Africa and Hong Kong (5 papers each), as well as Germany, South Korea, and Australia (3 papers each).

When viewed from a continental perspective, Asia emerges as the most active region, accounting for nearly 40% of all analyzed publications. Europe and North America also exhibit substantial research output in this domain, representing 26% and 24% of the total publications, respectively. In contrast, Africa contributes 5%, Australia 3%, and South America 2% of the total analyzed works. The regional distribution of authorship is illustrated in Figure 9.

The distribution of publications on pump maintenance across different countries and continents can be attributed to several key factors, including industrial activity, research funding, and technological development.

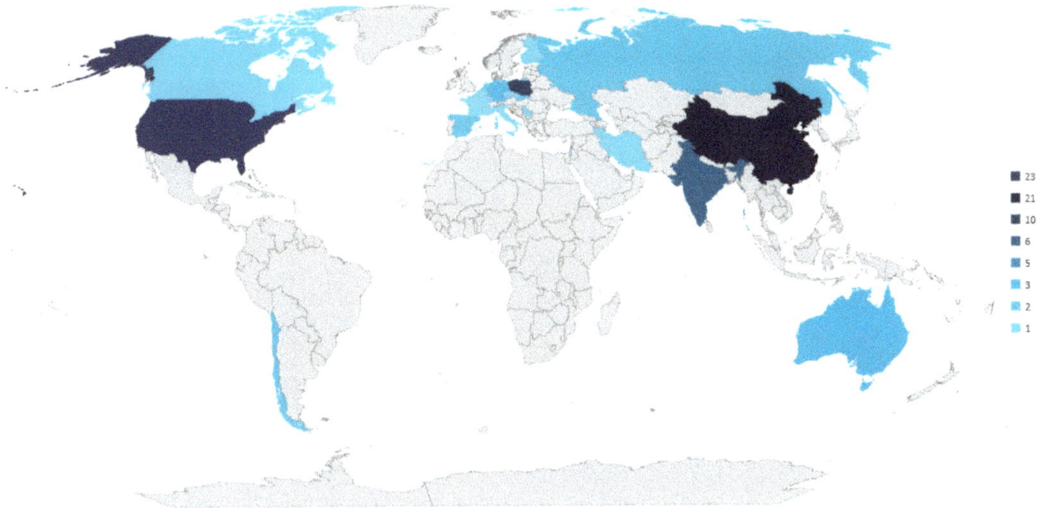

Figure 9. The number of papers by the location where the investigated study took place.

Countries like the United States and China lead in the number of publications due to their strong industrial bases, significant research and development investments, and major academic institutions collaborating with the industry. Both nations have extensive mining, energy, and manufacturing sectors, where pump maintenance is critical for operational efficiency and cost reduction. Additionally, government-backed initiatives and corporate research efforts drive innovation in predictive and proactive maintenance strategies.

Poland and India also show substantial contributions, which can be linked to their strong engineering and technical research traditions and the importance of heavy industry and energy production in their economies. Poland's engagement in mining and manufacturing, along with its growing interest in Industry 4.0 technologies, supports active research in maintenance strategies. Conversely, India has a rapidly expanding industrial sector and is increasing investment in smart maintenance solutions, driving academic interest in the field.

These trends are further confirmed by an analysis of publications in the Scopus database, where the keyword *"pump maintenance"* via a title/abstracts/keywords search yields 9439 publications for the period 2000–2024, showing a clear dominance of the United States and China in this research area (Figure 10).

The dominance of Asia as a region, with nearly 40% of all analyzed publications, aligns with its rapid industrialization, large-scale infrastructure projects, and focus on digital transformation in maintenance practices. Europe and North America also have significant research activity due to their well-established engineering sectors and advanced technological solutions in industrial maintenance.

The lower publication activity in regions such as Africa, South America, and Australia may be due to fewer research institutions focusing on pump maintenance, limited funding for maintenance-specific research, or a greater reliance on industry-led rather than academic-led innovation. However, given the importance of mining and energy industries in these regions, the potential for increased research in pump maintenance remains significant.

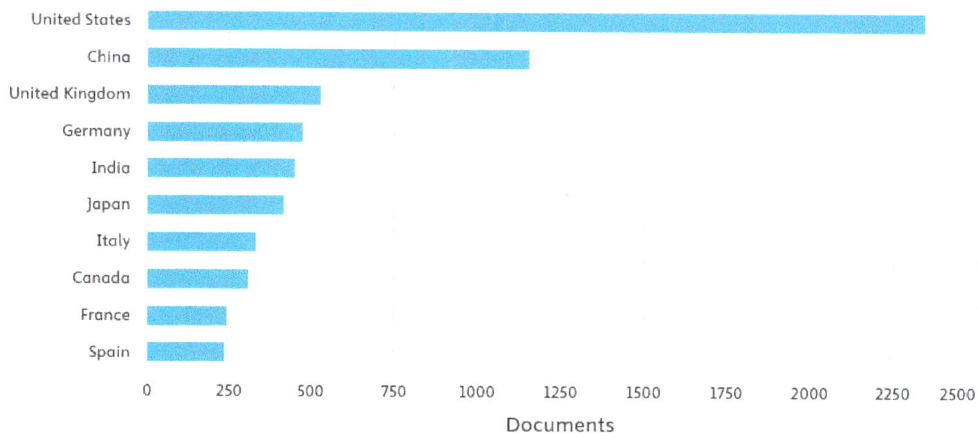

Figure 10. According to the Scopus database search via the keyword "pump maintenance" for the period 2000–2024, the number of documents by the location where the investigated study took place.

In addition, this review brings together 88 publications that were published between 2005 and 2024. Figure 11 illustrates the distribution of the publications according to their publication year. As we can see, the distribution of publications by year is highly diverse. The highest number of publications falls within the period 2015–2023, with 59 publications accounting for 67% of all publications selected for analysis. Notably, the peak years were 2010 and 2021, with the most publications related to the studied area.

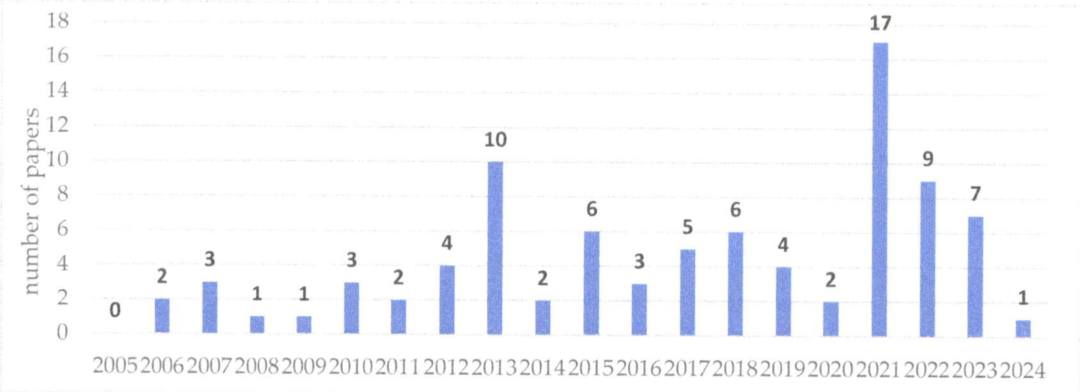

Figure 11. Distribution of publications by year.

Interestingly, since 2021, the number of these publications has been declining, reaching just a single entry in 2024. This downward trend may be attributed to a shift in research focus, where topics related to the predictive maintenance of pumps are being explored in a broader context without specifically emphasizing solutions dedicated to pumps within the mining industry. This conclusion is supported by the previously presented analysis of publications in the Scopus database, where the keyword *"pump maintenance"* via a title/abstract/keywords search for the period 2000–2024 was mentioned. According to the obtained results from the Scopus database, we can see a clear year-over-year upward trend (Figure 12).

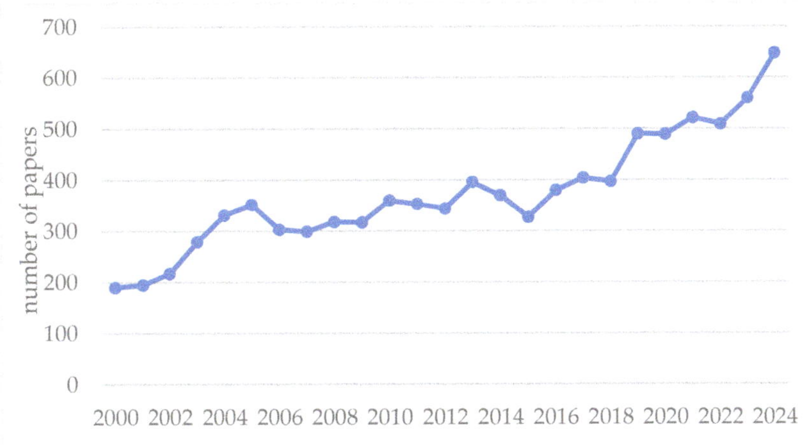

Figure 12. Distribution of publications by year, according to the Scopus database search via the keyword "pump maintenance" for the period 2000–2024.

Moreover, the analyzed articles were published across 44 different journals, reflecting the broad interest in the subject within various research communities. Figure 13 presents a graphical representation of journals that featured at least two publications related to the studied area. In total, 38 articles from 13 journals were considered in the analysis.

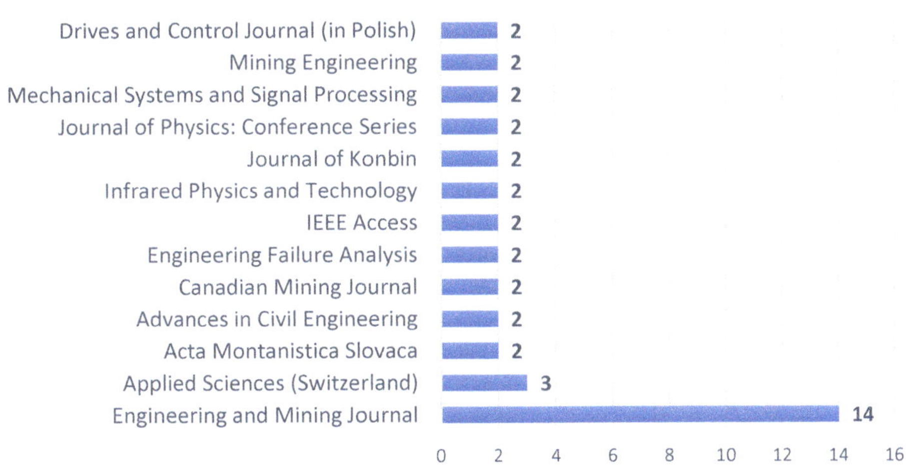

Figure 13. Number of publications with journal sources (for journals with at least two published articles out of the 88 articles analyzed).

The *Engineering and Mining Journal* emerged as the most prominent publishing outlet, with 14 articles dedicated to the topic. *Applied Sciences* ranked second, featuring three relevant publications. The remaining journals each contributed two articles, indicating a more distributed but still notable engagement with the subject.

This distribution of publications highlights the interdisciplinary nature of research on pump maintenance and predictive strategies, spanning both industry-focused and broader scientific journals. The dominance of the *Engineering and Mining Journal* suggests a strong connection between pump maintenance research and the mining sector, where reliable equipment operation is crucial. Meanwhile, publications in *Applied Sciences* and other journals reflect the increasing academic interest in integrating new technologies, such as artificial intelligence and IoT, into predictive maintenance frameworks.

The many publications in these journals may also indicate their prestige and broad visibility within the scientific community in the analyzed research field. Authors may prefer to publish their work in widely read and respected journals, which enhances the impact of their research on the development of mining technologies.

To supplement the conducted analysis, a co-occurrence of authors was investigated using VOSviewer software and Excel software. A total of 222 authors were identified from the selected papers. Figure 14 shows the results for the top 15 authors with the highest number of co-authorship connections. The author with the most links (16) is Guangjie Peng from China, who co-authored three papers [111–113]. Two other authors, Qiang Chen [111,112] and Hao Chang [112,113], also from China, each have a link strength of 11.

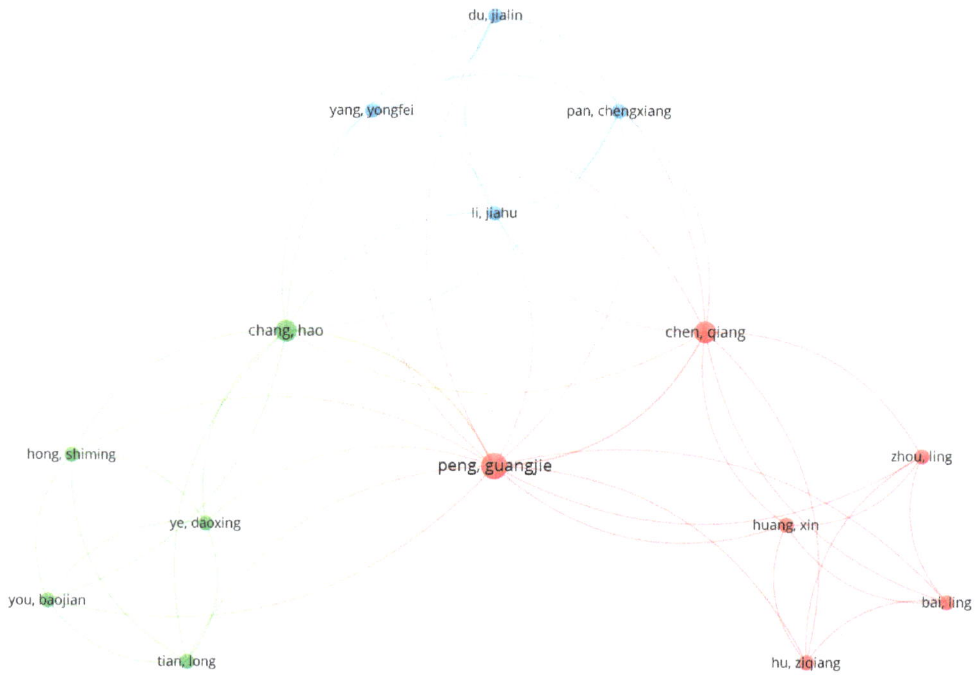

Figure 14. The largest set of connected items based on co-authorship links. Source: own development using VOSviewer software [72].

Furthermore, Figure 15 presents the distribution of publications based on the number of authors per paper. The findings reveal that single-author papers are the most common (30 publications). In contrast, papers with larger research teams (more than five authors) are less frequent in the analyzed sample.

The last part of the bibliometric analysis was a keyword co-occurrence analysis based on using VOSviewer software. The initial study focused on the keywords that occurred in the publications at least once. As a result, 220 keywords were identified for the selected papers (Figure 16). The most used words were maintenance (17 links), diagnostics (16 links), centrifugal pump (15 links), and fault diagnosis (15 links). The results present the used keywords in 28 clusters.

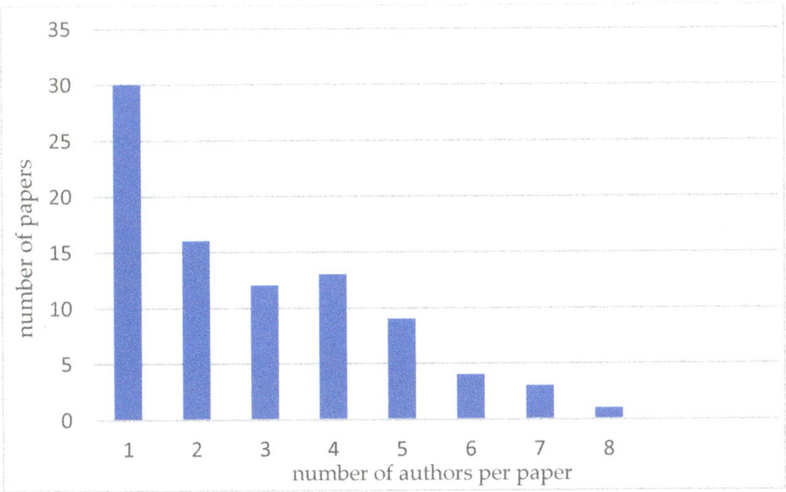

Figure 15. Distribution of publications per number of authors.

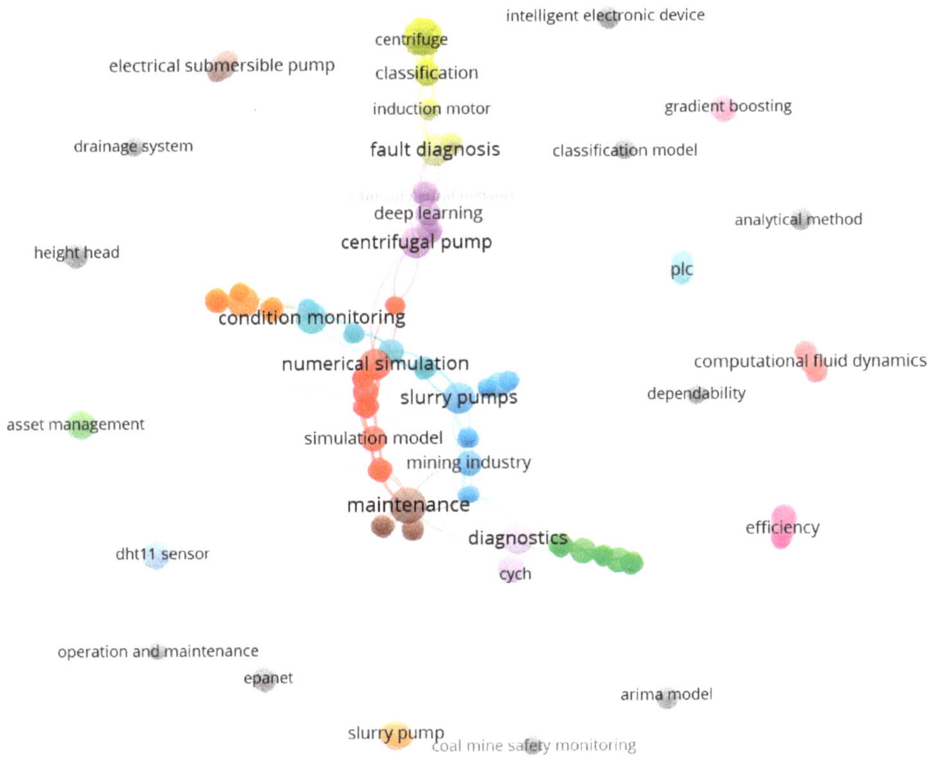

Figure 16. Mapping of the keywords that have occurred in the selected publications at least once. Source: own development using VOSviewer software [72].

A more detailed analysis was focused on developing a co-occurrence map based on the text from the titles and abstracts of the selected papers. The analysis focused on the 52 terms that have occurred at least five times in the papers (Figure 17). The obtained results are classified into five clusters.

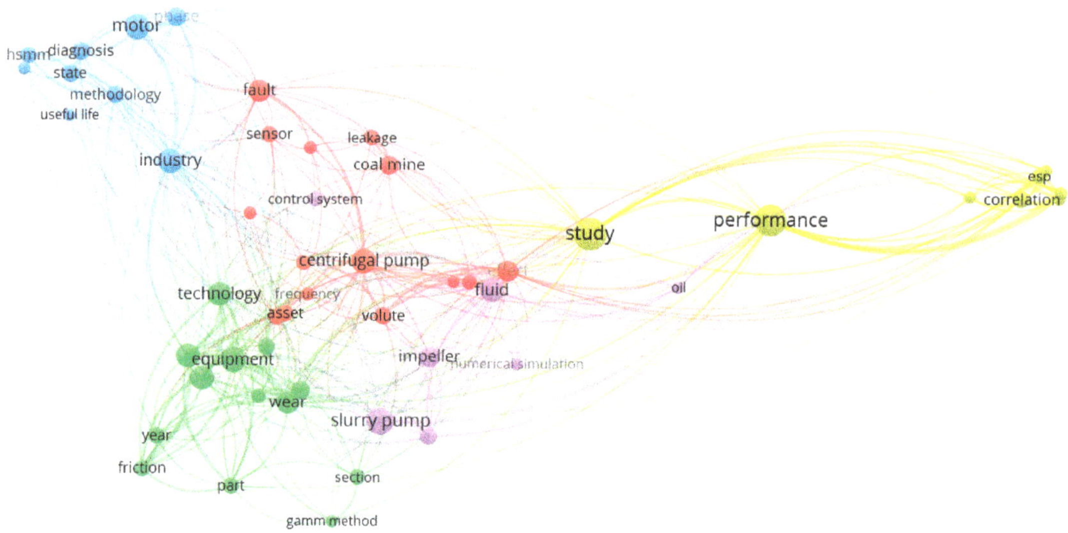

Figure 17. Mapping of the keywords with the largest occurrence. Source: own development using VOSviewer software [72].

The largest cluster (fifteen items, red one in Figure 17) is strongly associated with terms related to mining operations and machinery, such as centrifugal pump, coal mine, and open pit mine. This cluster represents a body of research focused on the operational performance, failure modes, and monitoring of mining environments, where equipment reliability and efficiency are critical.

The second cluster of publications (thirteen items, green one in Figure 16) reflects research focused on the technological aspects of equipment maintenance, emphasizing wear and friction in components such as pump parts.

The third cluster (nine items, blue one in Figure 17) is centered around assessing and extending equipment lifespan in industrial settings, particularly in the context of pump maintenance in the mining industry. This cluster likely represents research focused on developing and applying diagnostic methodologies to evaluate equipment conditions, predict failures, and optimize maintenance strategies.

The next cluster (eight items, yellow one in Figure 17) is strongly related to evaluating pump performance in real-world mining conditions, particularly for submersible pumps used in underground and deep-well applications. Indeed, this cluster represents research that combines experimental and field-based approaches to improve the understanding of submersible pump behavior, assess their performance in varying downhole conditions, and develop strategies for enhancing their operational lifespan in mining applications.

In the end, the fifth cluster (seven items, violet one in Figure 17) is closely related to the design, optimization, and performance analysis of pumps, particularly slurry pumps used in the mining industry. This cluster represents research focused on both theoretical and practical aspects of pump design, particularly for heavy-duty applications like slurry transport in mining. The studies within this group likely explore ways to enhance impeller efficiency, reduce wear, and implement advanced control strategies to improve the longevity and performance of pumps operating in harsh conditions.

The performed bibliometric analysis introduces the comprehensive content-based analysis, which is carried out in the next section.

4.2. Content-Based Analysis

As a result of the conducted research, following the methodology adopted (Figure 7), we focus on the content-based analysis.

The identification of the main problems and issues raised in the context of pump maintenance in the mining industry was based on an extensive review of the available literature. The prepared literature analysis was also supplemented by review publications in the area of pump maintenance (e.g., works [34,74,76]). As a result of the research carried out, six core research areas were defined, which have been most extensively developed over the last twenty years (Figure 18). These are discussed in detail in the next subsections.

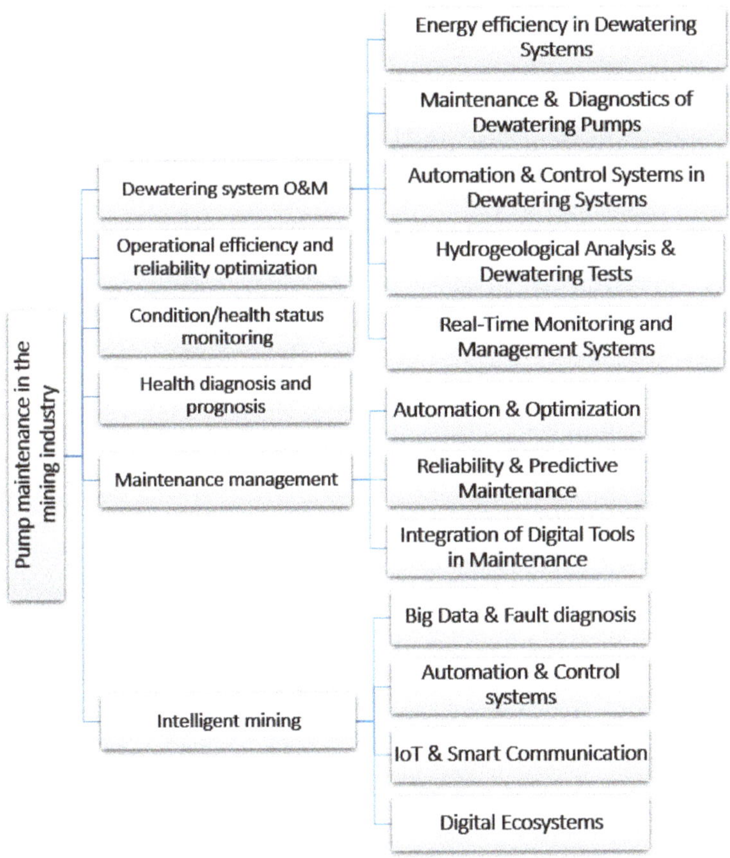

Figure 18. The main areas in the context of pump maintenance in the mining sector. Source: own contribution.

4.2.1. Dewatering System Operation and Maintenance

The first research area under consideration is dewatering system operation and maintenance. According to the theory background presented in Section 2, dewatering systems play a critical role in the mining industry, particularly in operations where the removal of excess water from excavated areas or slurry is essential to maintaining operational efficiency and safety. Effective dewatering ensures that mining operations continue without interruptions caused by water accumulation, which can lead to flooding, equipment damage, and hazardous working conditions. These systems are primarily used to control water levels in open-pit and underground mines, manage water from tailings ponds, and handle the water content in slurry during processing.

The operation and maintenance of dewatering systems involve a variety of technologies and processes designed to remove water efficiently, whether it is groundwater, rainwater, or process water. Common methods include pumping, filtration, and evaporation, which rely on different types of pumps, drainage networks, and treatment processes. Maintaining the systems in peak working conditions is vital, given the critical nature of dewatering in preventing operational disruptions. Regular maintenance ensures that the pumps and other components perform optimally, reducing the risk of system failure and associated downtime, which can be costly in terms of both time and resources.

The twenty-three articles in this field provide a wide range of perspectives on dewatering system operation and maintenance, covering aspects such as system design, operational challenges, performance optimization, and the application of new technologies. These papers offer valuable insights into best practices for managing dewatering systems, including integrating automated monitoring systems, predictive maintenance strategies, and optimizing energy usage in pump operations. By improving the efficiency and reliability of dewatering systems, mining companies can enhance productivity, reduce environmental impact, and ensure safer working conditions for their personnel. In addition, the operation and maintenance of dewatering systems involve a variety of technologies and processes designed to remove water efficiently, whether it is groundwater, rainwater, or process water. Common methods include pumping, filtration, and evaporation, which rely on different types of pumps, drainage networks, and treatment processes. Maintaining the systems in peak working conditions is vital given the critical nature of dewatering in preventing operational disruptions. Regular maintenance ensures that the pumps and other components perform optimally, reducing the risk of system failure and associated downtime, which can be costly in terms of both time and resources.

One of the works in this area [114] offers a historical overview of dewatering pump technology, from steam-powered pumps in the 18th century to the electric pumps of the late 19th century. It highlights the evolution of pump technology and its importance in the mining industry's dewatering systems. In another work [115], the authors discuss the safety issues related to damaged manholes in urban drainage systems. It presents a system for monitoring drainage conditions using sensors and the Internet of Things (IoT), which could provide useful insights for improving water management systems in mining sites, particularly in urban mining operations. Additionally, the paper [116] presents various dewatering equipment used in mining, including centrifugal pumps and filter presses. It focuses on energy-efficient, low-maintenance solutions, such as hybrid pumping systems and continuous steam filtration. The paper discusses the cost-effectiveness and performance improvements these technologies bring to mining dewatering processes. In addition, the technological advancements are discussed in [117,118]. The possibility of hose pump application in abrasive mining applications is presented in [119].

Another research area is connected with dewatering system reliability issues. In work [120], centrifugal pump connections are explored in series and parallel for dewatering processes in mining. The article investigates how these configurations affect pump efficiency, energy consumption, and reliability at different operational conditions (e.g., varying rotation speeds). The research concludes that the parallel connection is optimal for higher flow rates, while the series connection is suited for higher discharge heads. The problem of selecting a reliable dewatering pump for mining applications is investigated in [56,57] and continued in [121]. Another study [122] investigated the importance of pump availability in improving load-shifting performance in South African mining dewatering systems. The research introduces a methodology for investigating how reduced pump availability negatively impacts performance, highlighting significant cost increases related to pump unavailability. It emphasizes the critical link between operational efficiency and pump availability, especially in peak electricity demand periods. The next paper is focused on the dewatering pumps' operational effectiveness [123]. This study focuses on the importance of effective operation and maintenance of dewatering pumps in underground mining, particularly the challenges posed by energy consumption and pump reliability. The study underscores the need for regular maintenance to minimize energy consumption and reduce operational costs. In addition, a mining automatic drainage monitoring and control system based on Siemens PLC (Siemens 314C-2DP) and KingView 6.53 software in the upper computer is presented. Energy efficiency and cost-effectiveness are discussed in [124]. This paper examines the energy efficiency of main dewatering pumps in mining operations, noting that excessive operational time between maintenance cycles can significantly increase energy consumption. It emphasizes the importance of monitoring and optimizing pump usage to reduce operational costs. The analysis was performed for the case of an underground mine in Poland. The energy efficiency problem in mining companies due to pumps and pipework operation is continued in, e.g., [125].

Energy consumption prediction is also under investigation by the author of [126]. The paper investigates machine learning models, such as random forest and k-nearest neighbors, to predict the energy consumption of underground dewatering pumps in gold mines. The results show that random forest outperforms other models, presenting an opportunity for predictive maintenance and energy optimization in mining operations. Another aspect of pump energy efficiency is investigated in [127]. This paper explores the impact of dewatering pump inefficiency in mining due to oversized pumps that are initially required for high pumping rates but often operate inefficiently during maintenance pumping. It discusses the importance of reviewing pump operation regularly to reduce electricity consumption, lower greenhouse gas emissions, and cut operating costs. The focus is on the lifecycle costs of pumps, with energy usage being the most significant cost. The results suggest that substantial savings could be achieved by replacing or optimizing pumps, especially considering future carbon taxes. The last work in the area of energy management is [58]. This paper discusses the monitoring and management of a dewatering system at the Drmno lignite open pit mine in Serbia. The system includes approximately 200 drainage wells with a total pump capacity of 2 MW, which removes 15 million cubic meters of water annually. The paper critiques the existing monitoring system and proposes a computer-assisted monitoring and management system to optimize dewatering processes, improve energy efficiency, and track operational costs.

Fault diagnosis for mining pumps in dewatering systems is the next topic under study. In work [128], a method combining minimum entropy deconvolution (MED), wavelet packet decomposition (WPD), and radial basis function neural networks (RBFNN) is proposed for diagnosing faults in mine drainage pumps. The approach effectively reduces noise interference and enhances fault recognition, boasting a high diagnostic accuracy of 94.44%.

The problem of mine dewatering systems monitoring and diagnostics continues to be addressed, e.g., in work [129]. This study highlights the importance of regular maintenance and monitoring of pumps in mine dewatering systems. It presents the preliminary results of diagnostic tests on pumps in a mine shaft dewatering system, using vibration analysis, thermal imaging, and acoustic testing to assess performance. The study emphasizes the need to choose the appropriate maintenance methods to ensure operational safety and reduce downtime, along with an initial operational test plan for further analysis. This problem is also investigated in [130]. This paper focuses on diagnosing the main pump drive systems in mines, with an emphasis on the supply, control, and drive systems. It describes the use of a PLC-based automatic pump control system, focusing on monitoring pump parameters and identifying failure conditions. The paper presents the results of diagnostic investigations, including real-time monitoring and visualization of pump performance in industrial plants.

The issues of cost-effective monitoring and control systems for underground drainage pumps in coal mines are investigated in works [131,132]. The authors present a solution in the first work utilizing a 32-bit ARM7 microcontroller and a touchscreen interface. It emphasizes the importance of user-friendly human-machine interfaces (HMIs) for efficient system control and monitoring. The system incorporates a graphical LCD to display pump parameters, which can be analyzed and adjusted based on real-time data [131]. The second paper outlines the design of a system based on Siemens S7-200 PLC and a Delta DOP-B touchscreen. The system monitors pump parameters, controls startup and shutdown actions, and gives water-level alarms when necessary. The system is designed to be reliable, cost-effective, and suitable for promotion in similar mining operations.

Another type of research problem being investigated recently regards hydrogeological conditions in coal mines. The authors in [133] present a three-step drawdown dewatering test to estimate water yields and hydrogeological conditions in coal mines, particularly focusing on the Siwan coal mine. The test combines steady and unsteady flow methods to assess aquifer storage and water yield predictions. The results offer insights into water availability and dewatering challenges, helping inform mine water management strategies.

In conclusion, the operation and maintenance of dewatering systems in mining operations involve a variety of technologies and processes aimed at efficiently removing water, such as centrifugal pumps, filtration, and evaporation. However, the literature presents contradictory views regarding the optimal selection of technologies. For instance, some studies advocate for low-maintenance solutions like hybrid pumping systems, as seen in [116], while others emphasize the costs associated with longer maintenance intervals [124]. Moreover, discussions on energy efficiency in dewatering systems show a lack of consensus on the best approaches. Papers like [124,127] stress the importance of optimizing pump operation to reduce energy consumption, but the results are not universally applicable, particularly in mines with specific energy needs. Predictive tools, such as those discussed in [126], highlight the potential for energy optimization through machine learning, but concerns about their reliability under unstable conditions persist. In terms of system reliability, research on pump configurations (e.g., [120,121]) suggests that parallel connections are better for higher flow rates, while series connections are more suitable for higher discharge heads. However, there is no clear guidance on when and where each configuration should be applied, leaving room for interpretation depending on the mine's operational specifics. Furthermore, studies on pump monitoring and fault diagnosis (e.g., [128,129]) emphasize the use of technologies like vibration analysis and thermal imaging, but the effectiveness of these methods can vary depending on the operating conditions and the diagnostic equipment used. This lack of standardization across diagnostic approaches often leads to differing results. The application of advanced technologies, such as IoT and automated monitoring systems, is widely discussed, with works like [115,116] noting their potential to improve efficiency. However, the literature also indicates that these technologies may not be universally applicable, especially in regions with specific hydrogeological conditions, as noted in [133]. Overall, the literature review reveals a growing trend of incorporating modern technologies to enhance the performance of dewatering systems, but it also underscores the need for clearer, more unified guidelines for selecting appropriate technologies. This will help address the contradictions and inconsistencies found in existing research, ensuring more consistent and effective recommendations for practice.

4.2.2. Operational Efficiency and Reliability Optimization

Mining operations rely heavily on pumping systems for dewatering, slurry transport, and process water management. The efficiency and reliability of these pumps directly impact operational costs, energy consumption, and overall mine productivity. However, harsh working conditions, abrasive materials, and variable operating demands pose significant challenges to pump performance and maintenance strategies.

To enhance operational efficiency and ensure long-term reliability, research in this field focuses on optimizing pump selection, control strategies, predictive maintenance techniques, and failure prevention methods. Various approaches, including computational fluid dynamics (CFD), machine learning, and experimental studies, have improved pump performance, reduced downtime, and extended service life.

According to our study, we have identified 21 relevant studies that contribute to optimizing operational efficiency and reliability in mining pump maintenance. These works address key aspects such as energy consumption analysis, predictive maintenance using data-driven methods, impeller design improvements, and numerical simulations of pump behavior under different conditions. By synthesizing these research contributions, we aim to provide a comprehensive overview of current advancements and highlight future directions for optimizing mining pump performance. From analyzing the identified research works, we define six key research topics (Figure 19).

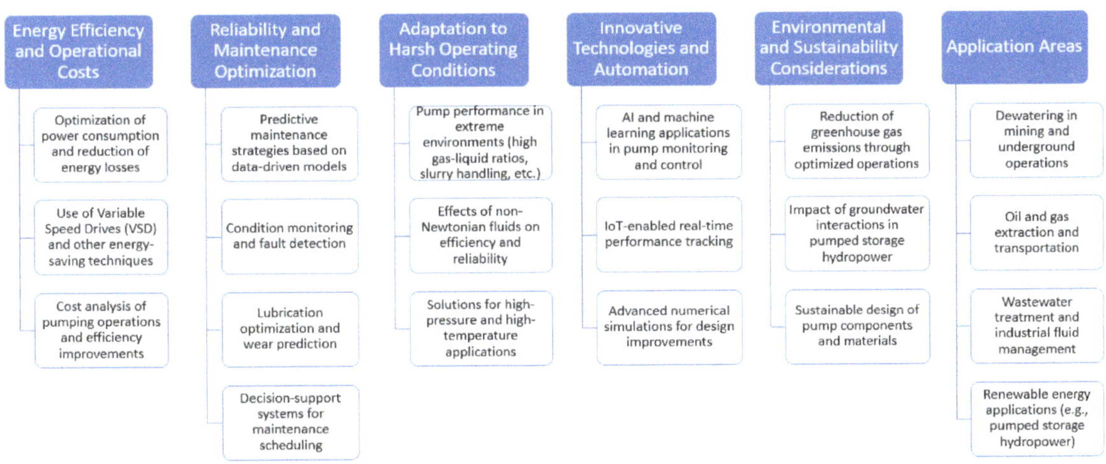

Figure 19. Main aspects of operational efficiency and reliability optimization in pump maintenance. Source: own contribution.

The first research aspect is focused on energy efficiency and operational cost optimization. Three identified studies [120,134,135] focus on optimizing pump operations to reduce energy consumption. Article [134] identifies inefficiencies in mine drainage systems. It examines the operating costs of electric pumps used in dewatering the Artois uranium mine. An energy balance was conducted over three years, revealing high specific energy consumption and inefficiencies in the current drainage system. The findings highlight areas for potential optimization. Next, ref. [135] develops an optimization model to enhance the scheduling process. Indeed, this research presents a modified EPANET2 toolkit (ETTAR) that enables rule-based control optimization for pump scheduling. The results demonstrate that optimizing pump control based on multiple conditions (e.g., tank levels and time of day) reduces operational costs while ensuring robust performance. The third article [120] explores centrifugal pump configurations, highlighting the advantages of parallel and serial connections in energy-efficient dewatering.

The second key research topic is focused on reliability and maintenance optimization issues. Research works [63,136,137] emphasize predictive maintenance strategies. The first work [63] applies data mining to predict lubricant service life in water injection pumps, ensuring reliability. Various algorithms, including gradient boosting and random forest, were tested, with results indicating that data-driven predictive models can effectively support pump reliability management. The authors in [136] explore flow characteristics in pump as turbine (PAT) systems, focusing on vortex formation, pressure pulsations, and rotor-stator interactions. The findings help improve PAT stability and efficiency by optimizing flow guidance and reducing pressure fluctuations. Later, in [137], a data-driven model integrating neural networks and Markov processes is proposed for optimizing pump scheduling in wastewater treatment plants. The model minimizes operational and maintenance costs through hierarchical particle swarm optimization, demonstrated in two case studies.

The third key research area is connected with adaptation to harsh operating conditions. Studies like [112,138,139] address pump performance in extreme conditions.

The first study [138] examines the impact of free gas presence on ESP performance under downhole conditions. Correlations predicting two-phase gas-liquid pump performance were developed using both field and laboratory data. The findings help optimize ESP operation but are recommended mainly for conditions similar to those in studied wells. The next article [139] focuses on handling non-Newtonian fluids in drilling applications. The study simulates the performance of centrifugal pumps transporting drilling fluids, showing that they exhibit lower head and efficiency but higher shaft power consumption than water pumps. It provides insights into flow patterns, pressure pulsations, and energy losses, contributing to pump design improvements. In another article [112], the authors introduce an impeller with splitter blades to reduce energy loss and wear in slurry pumps. By analyzing vortex structures and entropy generation, they demonstrate that the new design improves efficiency and durability, providing guidelines for pump optimization in mining and energy industries.

The next key research topic is connected with innovative technologies and automation introduction in order to improve pump operations. Article [140] evaluates LiDAR for mine inspections, demonstrating its feasibility for non-critical applications. The research assesses the accuracy and applicability of the iPhone 13 Pro LiDAR scanner for mining shaft inspections. Results indicate that while it lacks high precision, it is suitable for rapid inspections, offering a cost-effective alternative for preliminary 3D mapping. In addition, ref. [141] optimizes vacuum pump performance through adaptive discharge area control. This study examines how varying discharge areas impact liquid-ring vacuum pumps' inspiratory capacity and power consumption. Experimental and numerical results confirm the importance of optimizing discharge areas to enhance compression efficiency and practical applications.

The last research aspect regards environmental and sustainability improvement in pump operations. In this area, ref. [134] highlights the high energy cost of mine drainage, emphasizing the need for efficiency improvements. Additionally, research work [142] explores UPSH as an alternative for renewable energy storage, balancing efficiency with environmental impact. In more depth, this numerical study evaluates the influence of groundwater interactions on UPSH system efficiency. The results show that groundwater exchanges enhance efficiency but also increase environmental impact, requiring a balance between energy performance and sustainability.

Additionally, we have identified another research area focused on implementation challenges and case studies related to mining pump maintenance. This category includes publications that examine real-world applications [57,119], practical deployment issues [143–145], and case-specific solutions for improving pump performance and reliability [146–149]. These studies provide valuable insights into the practical constraints, operational limitations, and industry best practices necessary to successfully implement advanced maintenance strategies in mining operations.

In conclusion, the reviewed literature offers various strategies for optimizing mining pump performance, focusing on energy efficiency, reliability, adaptability, and sustainability. However, differences in methodologies and results highlight the need for a more unified approach. Studies on energy efficiency and predictive maintenance, such as those by [63,134], show contrasting methods, suggesting that no single approach is universally applicable. Similarly, research on harsh conditions and pump design improvements, like [112,138], reveals inconsistencies in findings across different fluid types and pump configurations.

Innovative technologies, including [140] LiDAR for inspections and ref. [141] adaptive discharge control, show promise but face practical limitations. Environmental concerns raised by [134,142] emphasize the need to balance energy efficiency with sustainability.

Case studies, such as [57,119] or [146–149], offer valuable insights but lack broader applicability across diverse mining environments. While progress is evident, future research should focus on integrating these approaches and addressing existing contradictions to develop standardized, universally applicable solutions.

4.2.3. Condition/Health Status Monitoring

Condition monitoring (CM) and health status monitoring (HSM) are essential techniques used to ensure the optimal operation and longevity of pumps in the mining industry. These methodologies involve continuously assessing the condition of equipment—such as pumps—by collecting and analyzing real-time data to detect any deviations or signs of wear that may lead to failure. The goal is identifying potential issues before they lead to significant breakdowns or inefficiencies, minimizing unplanned downtime and optimizing maintenance schedules.

The thirteen selected articles on condition/health status monitoring in mining pumps focus on various aspects of monitoring pump performance and detecting potential faults that may affect their operation. These studies employ diverse methodologies, technologies, and models to analyze the health of pumps and related systems. The first article [150] proposes a model for monitoring the leakage and diffusion of gas drainage pipelines in underground coal mines. The use of field data to validate the model ensures that gas sensors downstream of leakage points can effectively detect gas leaks. While not focused directly on pumps, it contributes to the broader context of safety and equipment health, highlighting the importance of monitoring systems in mining operations. It shows how a real-time monitoring system can help prevent accidents related to gas leaks.

The second work [53] focuses on the problem of wear monitoring in slurry pumps. This study investigates the wear status of slurry pumps using unsupervised clustering ensemble methods to assess and measure the wear of wetted components, like impellers. The article emphasizes the importance of condition-based monitoring to identify pump wear before it results in unplanned outages, showcasing how predictive techniques can optimize maintenance schedules and prevent costly downtime.

The next four works are also dedicated to the problems of slurry pump wear analyses [111,113,151,152]. In the first work [111], the authors focus on the wear patterns of slurry pump impellers, utilizing a multiphase flow model, and investigate how operational conditions such as flow rate and particle concentration affect wear. The findings suggest that wear-prone areas can be identified, which helps improve the design and maintenance strategies for slurry pumps, thus reducing unplanned downtimes and enhancing pump efficiency. Similar to this study, ref. [113] analyzes the wear characteristics of slurry pumps under varying operational conditions. The study emphasizes how different particle sizes and concentrations contribute to the wear of impellers. These insights can be used to improve slurry pump designs and monitoring systems to optimize their performance and lifespan. The work [151] discusses the main challenges and future pump-related technologies in relation to their possible development. The last work [152] presents Metso's plant control monitoring software with an example of its use in slurry pump operation.

The next work [153] introduces a model to predict abnormal ESP operations when the pump is compromised by free gas. The research provides a method for monitoring ESP health and anticipating failures using nodal analysis and field case studies. The detection of abnormal operation parameters allows for early interventions to prevent premature pump failure and production losses.

Intelligent electronic devices (IEDs) are under investigation in another work [154]. The paper discusses the application of IEDs in monitoring motor control centers and drives used in mining operations. By using the IEC 61850 standard, these devices can detect potential faults and allow for real-time monitoring of motor health. This highlights the importance of adopting intelligent systems for preventive maintenance in mining machinery.

In another work [155], the authors investigate hydraulic transmission system health monitoring. In this study, the health of hydraulic transmission systems, including axial piston pumps, is monitored by analyzing leakage flow characteristics. The simulation-based approach focuses on detecting performance degradation due to increased leakage and wear, which indicates pump health. This is a preventive measure aimed at minimizing system failures and improving reliability.

The problem of axial piston pump condition monitoring is later continued in work [156]. This article develops a mathematical model for axial piston pumps, which uses pressure and flow measurements to identify changes in pump health, particularly related to wear and leakage. The research emphasizes pressure control's significance in maintaining optimal performance and preventing system breakdowns. This approach underlines the importance of continuously monitoring operational parameters for early fault detection.

The next work is worth investigating [157], focusing on centrifugal pump health monitoring using fluid pressure. This study focuses on detecting faults like blockages and cavitation in centrifugal pumps by analyzing fluid pressure signatures. The deep learning-based binary classification method allows for the identification of fault types and severity, offering a robust way to monitor pump health and take timely corrective actions.

The next work [158] presents a cost-effective solution for condition monitoring by using a central power measurement device to analyze the power consumption patterns of industrial devices, including pumps. Applying spectral analysis and data mining models demonstrates how monitoring power usage can be a predictive tool for identifying real-time equipment failures.

Safety issues and real-time monitoring in coal mining are investigated in work [159]. The article highlights the use of multi-sensor systems to monitor various equipment, including pumps, in coal mining operations. By collecting and analyzing data from sensors that monitor gas concentration, temperature, and pressure, this system provides insights into the health status of the equipment and ensures the safety of mining operations.

In summary, the articles on condition/health status monitoring in pump maintenance highlight several key insights and future trends. The research focuses on advanced modeling, simulation, and real-time monitoring technologies used to predict failures, improve maintenance strategies, and ensure operational safety (Figure 20).

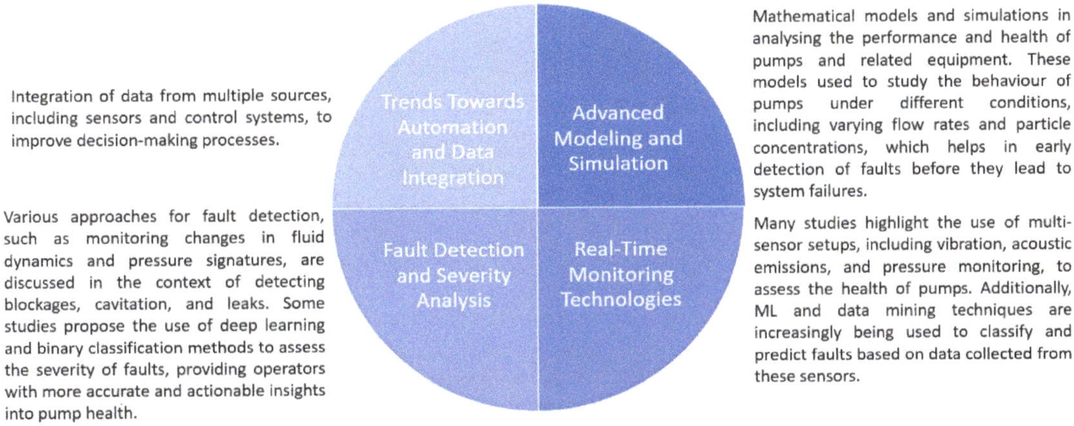

Figure 20. Main aspects of condition/health status monitoring in pump maintenance. Source: own contribution.

In conclusion, the reviewed literature on condition and health status monitoring (CM/HSM) in mining pumps provides valuable insights into optimizing pump performance and preventing failures. While the studies contribute to understanding pump health using various methods, such as wear monitoring, simulation-based analysis, and real-time data collection, their approaches have notable differences. Some studies, like [53,111], highlight the use of predictive models to monitor wear and operational conditions, which can effectively prevent unplanned downtimes. However, methodologies differ, with some focusing on specific pump types or operating conditions, raising concerns about their findings' generalizability across different mining environments.

For instance, the application of intelligent electronic devices (IEDs) [154] for real-time monitoring contrasts with other predictive maintenance techniques, showing diverse approaches in addressing similar issues of fault detection. Similarly, studies on slurry pump wear [111,113] and ESP operation [153] point to the importance of condition-based monitoring, yet their results sometimes conflict with the effectiveness of specific sensor-based models. Moreover, while several studies emphasize real-time monitoring, such as [159], others focus on simulation-based approaches, like [156], which may not always reflect actual operating conditions.

Future research should aim to integrate these varied approaches into more unified systems, balancing automation, predictive maintenance, and real-time data collection. Addressing contradictions in methodologies and results will be crucial for developing more reliable, broadly applicable monitoring systems for mining pumps.

4.2.4. Health Diagnosis and Prognosis

In the context of mining pump maintenance, health diagnosis and prognosis play crucial roles in ensuring the continuous and efficient operation of pumps, which are integral to various mining processes such as dewatering, slurry transport, and fluid handling. Health diagnosis refers to the process of detecting, identifying, and assessing the current condition or "health" of the pump and its components. This can involve identifying faults, wear, and degradation in real time, often using sensors, condition monitoring systems, and diagnostic algorithms.

On the other hand, prognosis is the prediction of the future health of the pump, particularly in terms of its remaining useful life (RUL) and potential failures. Prognostic models utilize data from diagnostic tools, historical performance, and environmental conditions to forecast when a pump might fail or require maintenance. These predictions allow for proactive measures to be taken before failures occur, preventing costly downtimes and unscheduled maintenance.

Research in this area is advancing with the use of data-driven approaches, such as machine learning, predictive analytics, and condition-based monitoring systems. These technologies help enhance the accuracy of diagnostic assessments and prognostic predictions. Furthermore, this research field is becoming increasingly important as mining operations strive for higher levels of automation and cost-effective operations, where predictive maintenance models are becoming indispensable tools for ensuring continuous production without sacrificing equipment reliability.

By focusing on health diagnosis and prognosis, mining pump maintenance strategies can move from traditional, reactive methods to a more proactive and predictive approach, ultimately optimizing the performance and reliability of pumps in harsh mining environments. This research area is vital for improving the overall sustainability and efficiency of the mining industry, contributing to smarter, data-driven decision-making processes in pump management.

In our study of this research area, we have identified 21 selected articles that are highly relevant and contribute to the advancement of health diagnosis and prognosis models in mining pump maintenance. These works focus on innovative methods, technologies, and case studies that improve diagnostic accuracy and prediction capabilities, further enhancing maintenance strategies and operational sustainability. The selected research works can be classified into four main research areas (Figure 21).

The first group focuses on methods and techniques for predicting the remaining useful life (RUL) of machinery and equipment, which is a crucial aspect of condition-based maintenance (CBM). Predicting RUL is vital for determining the optimal time for maintenance actions, preventing unexpected failures, and extending the operational life of systems. The techniques often involve advanced data analytics, machine learning algorithms, and statistical models to estimate how long a system or component will continue to function before failure.

Few papers focus on prediction models and algorithms by discussing the application of sophisticated models like segmental hidden semi-Markov models (HSMMs) and relevance vector machines (RVMs) to predict RUL. These models are built to handle the complexities and temporal patterns in real-world machinery operations, particularly in hydraulic and industrial pumps. One paper [160] presents an HSMM-based methodology for both diagnosis and prognosis in a unified framework capable of predicting the RUL of hydraulic pumps. It improves the recognition of pump states and helps predict remaining useful life (RUL), with testing showing improved recognition rates and prognosis capabilities over traditional models. Another paper on this topic [161] highlights the use of RVMs to predict the RUL of pump impellers, comparing their effectiveness to traditional exponential fitting methods. The model outperforms traditional methods by predicting RUL more accurately, demonstrating its effectiveness with field data. In addition, the methodologies are often applied to real-world scenarios to test their practical utility. The health monitoring of hydraulic pumps and oil sand pump impellers is used as a case study, where real operational data are analyzed to validate the models' accuracy in predicting RUL.

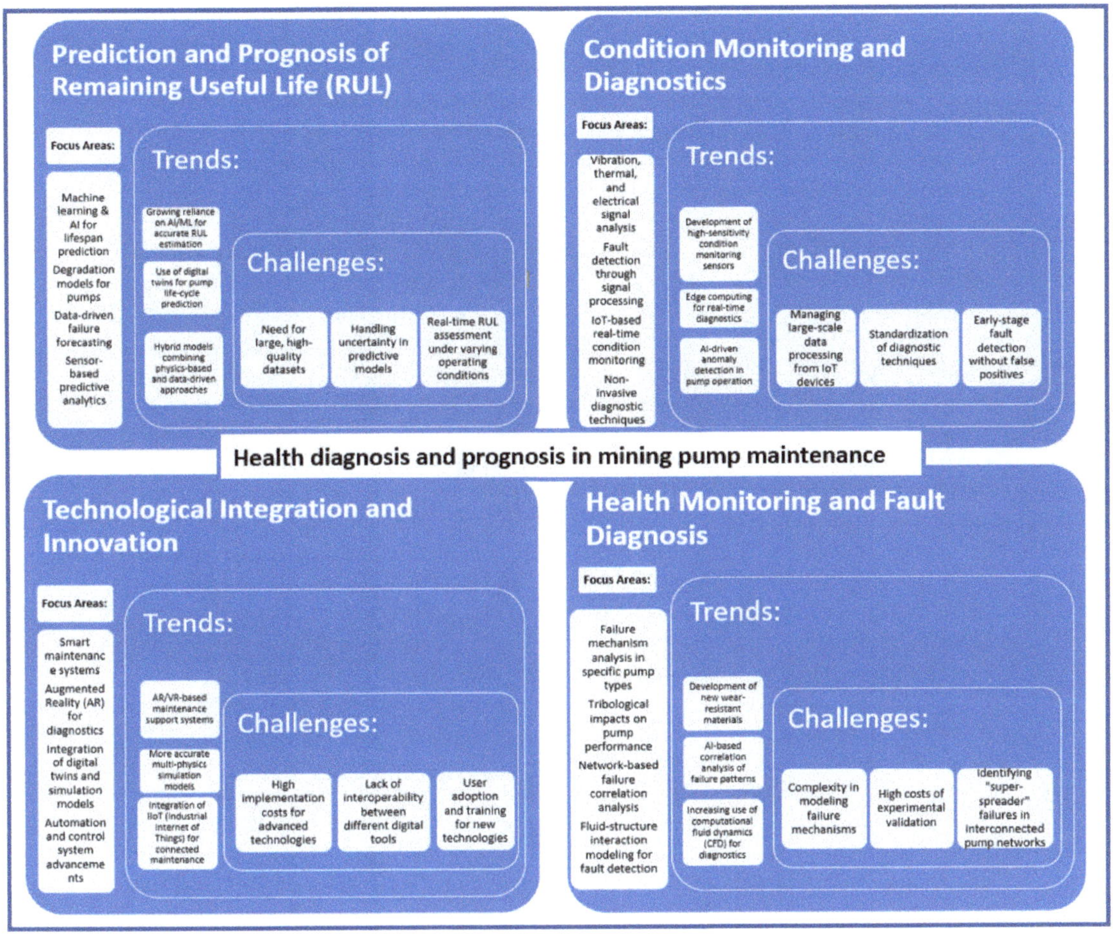

Figure 21. Main aspects of health diagnosis and prognosis in mining pump maintenance. Source: own contribution.

The next work [162] presents a framework for multi-sensor-based health monitoring and prognosis of equipment using an adaptive hidden semi-Markov model (AHSMM). AHSMMs enhance the traditional hidden semi-Markov models (HSMM) by addressing computational challenges, making them more efficient for large-scale applications. The approach integrates sensor data to predict RUL accurately. This model is applied to monitor hydraulic pumps from Caterpillar Inc., demonstrating its effectiveness in real-world conditions. The study concludes that this method provides valuable insights into the degradation process and offers improved prediction accuracy compared to conventional models.

Another problem is defined in [163], where the authors use the auto-regressive integrated moving average (ARIMA) model to forecast the RUL of axial piston pumps used in hydraulic systems. The ARIMA model is especially useful for time-series forecasting, where it analyzes historical data, including parameters like leakage volume, to predict future performance. The study focuses on finding the optimal ARIMA model configuration, with the ARIMA (1, 1, 2) model showing the lowest forecasting error. The predicted RUL of the pump is 28 months, providing a solid foundation for maintenance planning and failure prevention in applications like mining and agriculture.

The application of relevance vector machines for predicting the RUL of slurry pump impellers in the oil-mining industry is presented in [54]. The study introduces a novel feature extraction process to tackle the non-stationary nature of slurry pump data, which is often characterized by fluctuating operational conditions. The RVM model, combined with a two-summed exponential function, successfully captures the degradation trend of the impellers, enabling the prediction of their RUL and helping avoid unplanned breakdowns. The model's accuracy in predicting pump degradation is validated with real-world field data.

The last work in this area is [52]. This work explores the use of support vector machines (SVMs) to estimate the RUL of slurry pumps, which are critical in industries such as mining, waste treatment, and oil sands. Given the harsh, abrasive operating conditions, slurry pumps experience continuous wear, especially on components like impellers. Traditional maintenance strategies often fail to anticipate failures in time, leading to unnecessary part replacements or unplanned breakdowns. The study proposes a data-driven technique using SVMs to analyze the pumps' sensor data, identifying the components' health status and providing a reliable estimate of the RUL. The effectiveness of SVM is demonstrated through sensor data analysis, emphasizing the importance of sensor placement and frequency band selection for optimal results.

The second research field focuses on condition monitoring and diagnostics for continuously assessing the health of industrial machinery and systems in order to identify potential faults and prevent unexpected breakdowns. The main goal is to improve maintenance practices by enabling proactive actions based on real-time performance data, which helps in extending the lifespan of equipment, improving reliability, and minimizing unplanned downtime. The first work in this area [164] focuses on vibration-based monitoring for detecting faults in reciprocating compressors used on an oil rig in the Baltic Sea. Due to the strong impacts and disturbances in the vibration signals, the system must handle significant signal distortion. The paper introduces the use of enhanced resolution envelope analysis to address the challenges posed by these strong vibrations. The methodology is applied specifically to detect bearing faults in the compressor, highlighting the importance of selecting the appropriate diagnostic method based on the nature of the fault and the machinery's operating conditions. This paper demonstrates the role of vibration analysis in machinery diagnostics, especially in harsh operating environments like offshore oil rigs.

Thermal imaging for fault diagnosis in three-phase induction motors is introduced in [165]. The study presents a fault diagnostic technique that analyzes thermal images to identify issues such as broken bars or faulty rings in the motor's squirrel-cage rotor. The MoASoID (method of areas selection of image differences) technique is used to extract significant features from thermal images, which are then classified using machine learning techniques like nearest neighbor (NN), k-means, and backpropagation neural network (BNN). This approach provides a non-invasive, efficient method for identifying faults, thereby protecting motors from damage that could lead to significant downtime or economic loss.

Automated pump diagnostics using PLC control is investigated in the next work [130]. This work introduces an automated diagnostics system for pump drive systems, focusing on the supply system, control system, and drive part of the pumps. The diagnostics are integrated with a PLC controller, enabling real-time monitoring of various parameters to detect failure states. The paper discusses the effectiveness of the diagnostic system in industrial plants, where it was successfully verified. The diagnostic method helps in identifying operational issues in the pump system, such as power loss or mechanical failure, by continuously tracking and visualizing key parameters. This system supports online diagnostics and can be used for preventive maintenance, ensuring that pumps continue operating efficiently.

The last work [166] focuses on slurry pumps used in oil sand pumping operations, where the impellers are subject to continuous wear and erosion due to the abrasive nature of the liquid-solid mixtures they pump. The paper introduces two wear degradation indices—the moving-average mean and moving-average deviation wear indices—to track the health of the impellers over time. These indices are based on vibration signals collected from the pump, and their effectiveness is validated through real-world data. By assessing the degradation of impeller performance, this approach helps predict impeller failure and improve the reliability of slurry pumps, reducing costly breakdowns and extending their operational lifespan.

The third research field—technical integration and innovation—focuses on the incorporation of advanced technologies to enhance the efficiency, reliability, and predictive capabilities of industrial systems. This includes leveraging sensor-based diagnostics, dynamic modeling, augmented reality (AR), and machine learning algorithms to improve maintenance strategies and optimize system performance. In this area, ref. [167] proposes a novel approach to pump diagnostics that involves using the motor itself as a sensor, analyzing current and voltage signals to detect faults such as cavitation, hydraulic blockage, and dry running. This extends traditional motor diagnostics techniques to assess pump health, offering a more integrated and cost-effective solution for predictive maintenance. The research focuses on testing different classification methods to determine the most effective diagnostic approach. The next work [153] proposes advanced nodal analysis software (e.g., Schlumberger PIPESIM) that is used to monitor and predict failures in electric submersible pumps (ESPs) operating under high gas conditions. By analyzing pump intake pressure (PIP) and pump discharge pressure (Pdis), abnormalities in ESPs related to gas lock are diagnosed.

The oil and gas sector, especially in remote and extreme environments (e.g., Arctic regions), requires efficient maintenance strategies. As a result, in work [168], dynamic modeling is integrated with augmented reality (AR) (via Microsoft HoloLens 2) to enhance diagnostics and reduce maintenance time. AR allows technicians to access real-time system data, interactive maintenance guides, and predictive analytics, streamlining repair and servicing processes. The combination of AR with simulation-based diagnostics represents a breakthrough in remote maintenance assistance and digital twin applications.

The next work in this research field is focused on data mining and AI for predictive maintenance in centrifugal pumps [169]. This work applies big data analytics and machine learning (ML) techniques to process vast amounts of data generated during pump operations. The C5.0 Decision Tree (DeT) algorithm analyzes vibration signals (horizontal, vertical, and axial planes) and classifies pump health status.

The last work focuses on the application of fault diagnosis and fault tolerance in the areas of electrical, mechanical, and chemical engineering as well as computer science [170].

The next research field is focused on health monitoring and fault diagnosis in specific pump types. Health monitoring and fault diagnosis of pumps are critical for ensuring reliability and minimizing operational costs, particularly in industries like mining, power generation, and heavy manufacturing. The reviewed articles provide insights into various failure mechanisms, diagnostic methods, and predictive maintenance strategies across different pump types. Ref. [161] focuses on hydraulic plunger pumps used in coal mines. It challenges prior assumptions about suction valve failure, revealing that rotational wear and material loss, rather than impact wear, are the main causes of degradation. Using computational fluid dynamics (CFD) and nonlinear finite element analysis, researchers identified uneven flow fields and low-speed rotation as key contributors to premature wear. Improved valve materials based on these findings led to increased durability in field tests. The next study [171] presents a global analysis of energy consumption due to friction and wear in mining operations, where pumps play a significant role. The study found that 40% of energy in mineral mining is used to overcome friction, and additional energy is expended on manufacturing replacement parts. By implementing advanced materials, coatings, and lubrication technologies, energy losses could be reduced by up to 30% over 20 years, resulting in major economic and environmental benefits. In another work [172], the authors highlight a failure rate monitoring initiative for large industrial pumps in the power and mining sectors. Over four years, data were collected on 17 different pump types across multiple locations in the Czech Republic. The study presents baseline failure rate evaluations and identifies unexpected fluctuations in failure rates, providing potential solutions to improve pump reliability. The next work [173] explores cascading failures in pump networks, where failures in one pump can trigger additional failures in connected assets. Using complex network analysis on data from 5655 pumps and 50,000 work orders in a mining company, researchers identified "super-spreader" pumps whose malfunctions led to increased maintenance demands. The study suggests that preventive maintenance measures can be implemented to mitigate cascading effects by recognizing these patterns. The last paper in this research field [174] investigates piston diaphragm pumps used for transporting abrasive and aggressive slurries in high-pressure applications. The study proposes a fluid-structure interaction (FSI) simulation to analyze pressure pulsation behavior, allowing for more accurate predictions of deformation and fluctuations. This model helps optimize pump design and improve operational stability.

In the field of pump health diagnosis and prognosis, various methods and approaches have been proposed by researchers. These methods often differ in terms of the underlying technology, accuracy, computational requirements, and application in real-world systems. A critical analysis and comparison of these methods reveal both strengths and weaknesses, as well as contradictions in the findings of different authors. To facilitate a clearer understanding, the following Table 2 summarizes several key approaches, highlighting their respective advantages and disadvantages and the inconsistencies found in the conclusions across studies.

The comparison of different pump condition monitoring and RUL prediction methods reveals a range of strengths and limitations. HSMM offers high accuracy, but its complex computational requirements make it challenging for large systems, and its performance may degrade in unstable conditions. RVM, while accurate with lower data needs, is sensitive to data errors, leading to inconsistent results. ARIMA is simple and effective for time series data but struggles with irregular data and varying pump systems, showing inconsistent results across cases.

Vibration-based monitoring is cost-effective and provides early fault detection but may miss certain failure types and is affected by environmental noise. Augmented reality is useful for real-time diagnostics but comes with high costs and infrastructure demands, limiting its applicability. SVM is effective for sensor data analysis but requires large datasets and precise tuning and performs poorly with low-frequency measurements.

In conclusion, while these methods each have their merits, they also exhibit contradictions and limitations, underlining the need for further refinement and possibly integrating approaches to enhance their reliability in diverse real-world conditions.

Table 2. Comparison of approaches in pump health diagnosis and prognosis.

Approach/Method	Ref.	Technology/Model	Advantages	Disadvantages	Contradictions/Inconsistencies in Findings
Support Vector Machines (SVMs)	[52]	SVM for RUL assessment	Effective with sensor data analysis, good for damage identification	Requires large amounts of data and well-tuned parameters	Does not perform well with low-frequency measurements
Segmental Hidden Semi-Markov Models (HSMMs)	[160]	Hidden semi-Markov model for RUL prediction	Improved accuracy in RUL prediction, better state detection in pumps	High computational requirements, challenging implementation in large systems	Effective on field data, but not always reliable in unstable operational conditions
Relevance Vector Machines (RVMs)	[161]	RVM for RUL prediction	More accurate than traditional methods, lower data requirements	Needs precise feature selection, sensitive to data errors	Results can vary depending on operational conditions
ARIMA	[163]	ARIMA model for RUL prediction	Simple to apply, effective for time series data	Low effectiveness with irregular data, parameter tuning required	Inconsistent results across different pump systems
Vibration-based Monitoring	[164]	Vibration monitoring for failure detection	Early damage detection, cost-effective	May not detect all types of failure, needs specialized sensors	Results vary depending on background vibration in different environments
Augmented Reality (AR) for Diagnostics	[168]	AR technology for diagnostics	Enables real-time diagnosis, supports technicians	High implementation cost, requires advanced infrastructure	Sometimes insufficient data for full analysis in field conditions

4.2.5. Maintenance Management

Maintenance management in the mining industry, particularly for pump systems, refers to the systematic planning, monitoring, and optimization of maintenance activities to ensure the reliable, efficient, and cost-effective operation of pumps used in water management, slurry transport, and hydraulic systems. It integrates preventive, predictive, and condition-based maintenance strategies to minimize unplanned failures, optimize energy consumption, and extend equipment lifespan [34].

Following this, we have identified six main research papers in this research area. The reviewed articles focus on various aspects of pump maintenance in mining operations, emphasizing automation, data-driven decision-making, and advanced maintenance strategies. Their central theme revolves around improving efficiency, reliability, and cost-effectiveness in mining pump systems through different technological approaches.

One key aspect explored in the articles is the automation of water pumping systems. The first article [175] discusses the importance of automating underground clear water pump stations to optimize energy consumption, prevent flooding, and reduce operational costs. Similarly, another article [176] presents a control method for pump speed regulation using a high-voltage synchronous electric motor drive and frequency converters, ensuring efficient water supply management. These approaches highlight the role of automation and real-time monitoring in optimizing pump performance.

Another significant theme in the revised area is data-driven maintenance and failure analysis. The GAMM method, discussed in two of the articles, provides a graphical analysis of equipment reliability and identifies anomalies affecting pump operations. By analyzing historical data, GAMM enhances decision-making in maintenance management and helps detect failures stemming from operational misuse, design flaws, or maintenance inefficiencies. This aligns with broader trends in predictive maintenance, where statistical modeling and reliability estimation are used to improve equipment longevity [177,178].

The role of computer-aided maintenance management is also highlighted in the discussion of computerized maintenance management systems (CMMS) applied to the

Polish copper mining industry [179]. By leveraging CMMS data, mining companies can analyze failure intensities of critical components, such as hydraulic pumps in haul trucks, and refine their maintenance strategies accordingly. This case study emphasizes the shift from time-based maintenance (MTTR, MTTF, MTBF) to data-driven, condition-based maintenance, improving efficiency and reducing unplanned downtimes.

The last identified research work [180] presents a case study that discusses the successful implementation of a total asset management plan by Weir Minerals at the Tronox mineral sands mine in Cooljarloo, Western Australia. The key objectives of the asset management plan include enhancing asset reliability, minimizing maintenance costs, and optimizing operational efficiency. By leveraging Weir's expertise, Tronox benefits from a structured maintenance strategy, reducing unplanned downtime and ensuring the longevity of critical equipment. This case study demonstrates how comprehensive asset management, strategic maintenance practices, and innovative equipment upgrades contribute to sustainable and cost-efficient mining operations.

To sum up, effective maintenance management for mining pumps is essential for ensuring operational reliability, cost efficiency, and sustainability in mining operations. The six analyzed articles highlight different approaches to pump maintenance, emphasizing the role of automation, data-driven decision-making, and advanced control systems. The key insights from these case studies provide a comprehensive perspective on optimizing maintenance strategies to enhance pump performance and reduce operational risks. The six research works collectively demonstrate that effective pump maintenance management in mining requires a combination of automation, data-driven decision-making, and asset lifecycle optimization. Future advancements in AI, IoT, and sustainability-focused technologies will further revolutionize maintenance strategies, ensuring more efficient, cost-effective, and environmentally friendly mining operations (Figure 22).

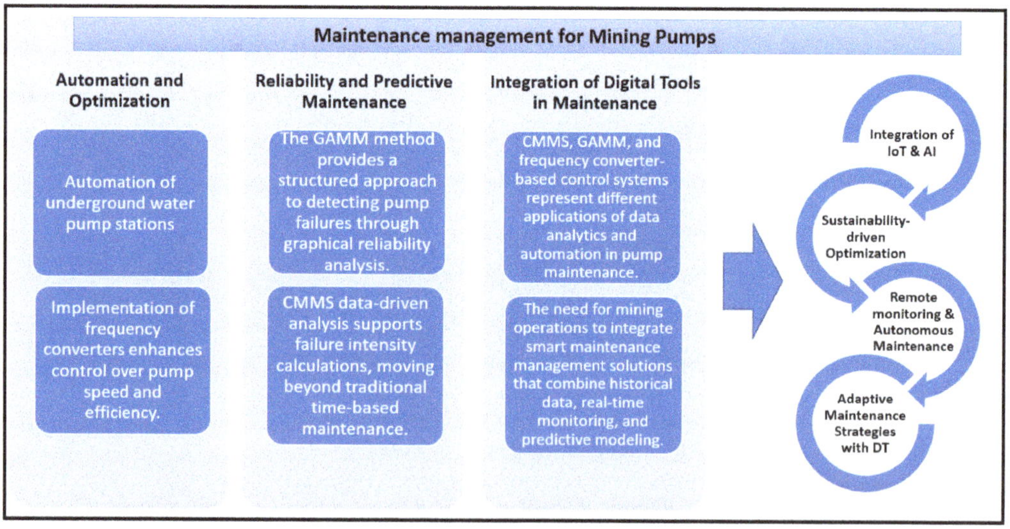

Figure 22. Main aspects of maintenance management for mining pump maintenance. Source: own contribution.

4.2.6. Intelligent Mining

In this research area, the authors focus on the aspects of intelligent mining. Intelligent mining refers to integrating advanced digital technologies to optimize and automate mining operations, enhancing efficiency, safety, and sustainability. It combines big data analytics, artificial intelligence (AI), the Internet of Things (IoT), cloud computing, automation, robotics, and digital twin technology to enable real-time monitoring, predictive maintenance, and autonomous decision-making [181,182].

A key aspect of intelligent mining is data-driven decision-making, where IoT sensors and AI algorithms continuously collect and analyze operational data. This enables predictive maintenance strategies that minimize downtime and extend equipment lifespan. Automation plays a crucial role by deploying autonomous vehicles, drilling systems, and robotic machinery, reducing the need for human intervention in hazardous environments. Additionally, intelligent mining leverages remote monitoring and control, allowing centralized operation centers to oversee and optimize mining processes in real-time.

The transition from reactive maintenance to predictive and proactive strategies is a defining feature of intelligent mining. Instead of responding to equipment failures after they happen, mining companies can now anticipate and prevent issues, leading to more resilient and cost-effective operations. Moreover, by integrating digital ecosystems, different mining components—such as pumps, conveyors, drilling equipment, and ventilation systems—can communicate and adapt dynamically, improving overall mine productivity.

In the context of pump maintenance, intelligent mining transforms traditional approaches by introducing real-time condition monitoring, AI-based fault detection, and automated control systems. IoT-enabled sensors track pump performance, while AI-driven models predict potential failures before they occur, reducing unplanned downtime. Digital twin technology further enhances maintenance strategies by simulating pump behavior under different conditions, allowing for optimized operational planning. These advancements ensure higher reliability, reduced maintenance costs, and improved operational efficiency [183,184].

As a result, in the defined research area, the authors identified four main research works that focus on intelligent mining in the context of pump maintenance. These articles focus on different aspects of intelligent pump system management in the mining industry, emphasizing the use of advanced technologies such as big data analysis, automation, adaptive control, and digital ecosystems. The common goal across all papers is to improve the reliability and efficiency of pump systems through intelligent monitoring, fault diagnosis, and operational optimization.

In the first article [185], the authors discuss the application of big data analysis in pump fault diagnosis. By leveraging machine learning and data-driven models, the fault diagnosis process becomes more intelligent, autonomous, and accurate. The study highlights how big data enable self-learning fault identification, reducing downtime and improving predictive maintenance strategies.

In another work [66], the development of an intelligent automated water intake system for open-pit mines, designed using LabVIEW and Siemens 1200 PLCs, is presented. The system integrates electric gate valves, ball valves, and various sensors to enhance operational efficiency. The software enables real-time monitoring, automated pump rotation control, and fault detection, reducing manual labor and increasing the system's reliability.

A different approach is given in [186]. This article explores the implementation of frequency conversion constant pressure control systems for emulsion pump stations in mining operations. The research integrates IoT, 3D laser scanning, and 5G communication to enhance automation, fault self-diagnosis, and remote monitoring. By analyzing the friction coefficients of different drawing fluids, the study emphasizes how environmental factors influence the selection of pump control strategies.

The fourth article [187] discusses the concept of a digital ecosystem in mining, where networks of sensors, pumps, and control systems interact dynamically. The paper highlights the scalability and adaptability of digital ecosystems, from monitoring a localized pump system to coordinating multiple mining operations worldwide. Such integration enhances real-time decision-making, predictive maintenance, and operational efficiency.

All four papers contribute to the broader theme of intelligent mining and pump system maintenance, but they differ in their technological focus and approach (Figure 23).

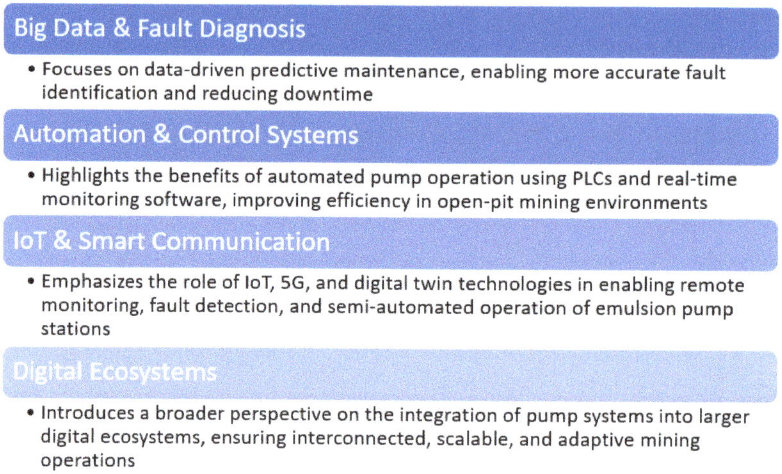

Figure 23. Main aspects of intelligent mining in the context of pump maintenance. Source: own contribution.

The evolution of pump maintenance in mining reflects a broader shift towards intelligent mining operations, where real-time data analysis, automation, and digital interconnectivity play a central role. These papers collectively illustrate how the industry is moving from reactive maintenance towards proactive and predictive approaches, ultimately leading to increased efficiency, reliability, and sustainability in mining operations. Based on their analysis, we may state that the evolution of intelligent mining reflects a clear shift from traditional, reactive maintenance approaches to fully automated, data-driven, and predictive maintenance strategies. This transformation is driven by the integration of big data analytics, IoT, AI, and digital twin technology, allowing mining operations to move from manual fault detection toward real-time monitoring, automated diagnostics, and remote control. Adopting these technologies ensures higher operational efficiency, improved equipment longevity, and reduced downtime.

A crucial element of this transition is the integration of IoT and smart communication technologies, such as 5G-enabled remote monitoring and cloud-based data management, which enable centralized control and seamless coordination between different mining components. This interconnectivity enhances operational scalability, allowing companies to optimize their mining infrastructure while reducing labor-intensive processes. Additionally, the scalability of digital ecosystems plays a vital role as mining companies increasingly connect localized pump systems to global networks, ensuring synchronized operations and predictive decision-making across multiple sites.

Ultimately, in the mining industry, the evolution of maintenance strategies has been increasingly influenced by intelligent mining technologies, which aim to optimize efficiency, reduce downtime, and enhance the sustainability of operations. Intelligent mining integrates advanced technologies like big data analytics, Internet of Things (IoT), artificial intelligence (AI), digital twins, and automation to support real-time monitoring, predictive maintenance, and autonomous decision-making. The research on intelligent pump system management in mining explores various technological approaches to enhance the reliability and efficiency of pump maintenance. Below, in Table 3, we summarize and critically analyze the main approaches presented in the literature, comparing their strengths, weaknesses, and practical implications.

Table 3. Comparison of technological approaches for intelligent pump maintenance in mining operations.

Aspect	Approach 1 (Big Data and Machine Learning)	Approach 2 (IoT and 5G for Remote Monitoring)	Approach 3 (Automation and Digital Twin)	Approach 4 (PLC and Automation)
Technologies	Big data, machine learning, fault prediction	IoT, 5G, remote monitoring, automation	IoT, digital twin, operational condition analysis	PLC, automation, real-time monitoring
Advantages	High accuracy in fault detection, autonomous diagnostics	Remote control, real-time monitoring, low operational costs	Optimized planning, predictive failure analysis	Reduces manual labor, easy to integrate with existing systems
Disadvantages	High data quality requirements, integration challenges	High initial cost, integration issues with legacy systems	High technology requirements, incompatibility with older equipment	Limited scalability, dependency on PLC infrastructure
Applications	Fault diagnosis, predictive maintenance using data models	Real-time monitoring, fault detection and control	Simulation and optimization of pump operation	Automation in water intake systems, pump rotation control
Infrastructure Compatibility	Challenging integration with legacy systems	Difficult adaptation for older systems	High technological demands, not always compatible with older setups	Easy integration with existing PLC systems
Scalability	Limited by data availability, dependent on large data sets	Suitable for remote operations, limited local control	High infrastructure demands, scalability challenges	Scalability depends on PLC size and budget
General Comparison	Effective in predictive fault diagnosis and autonomous decision-making	Focuses on remote monitoring and real-time system adjustments via IoT and 5G	Emphasizes automation and optimization using simulations (digital twins) for maintenance and control	Focuses on basic automation using PLC systems and sensors for real-time fault detection and control
Example References	[185]	[186]	[183,184]	[66]

The presented table summarizes various technological approaches to intelligent mining in the context of pump maintenance, illustrating the benefits and limitations of each. Approach 1, based on big data and machine learning, provides accurate fault detection and autonomous diagnostics, but it requires high-quality data and faces integration challenges with older systems. Approach 2 utilizes IoT and 5G technology for remote monitoring, offering the advantage of real-time control and reduced operational costs. However, the initial costs and integration with legacy systems may pose challenges. Approach 3, focusing on automation and digital twin technology, allows for operational optimization and predictive maintenance. While it offers high efficiency, it demands significant investments in infrastructure and may not be easily implemented in older systems. Approach 4, relying on PLC and automation, reduces manual labor and is easy to integrate with existing systems. However, it has limited scalability in large installations and depends on the specific PLC infrastructure in place.

The comparisons show that while all approaches aim to improve efficiency and reduce downtime, the choice of technology must align with the specific operational context of the mining company. Larger and more technologically advanced operations may benefit from integrating multiple technologies, such as combining IoT, AI, and digital twin technology. However, these approaches require substantial investments and may not be feasible for all mining operations.

In conclusion, the evolution of intelligent mining reflects a shift from reactive to proactive and predictive maintenance strategies. By integrating automation, real-time data analysis, and digital connectivity, mining operations can move toward more resilient, cost-effective, and sustainable practices. Future advancements will likely focus on further improving the scalability and interoperability of these technologies, ensuring that even smaller mining operations can reap the benefits of intelligent maintenance practices.

5. Discussion

The main aim of this paper is to conduct a comprehensive review of the existing literature to provide a substantive analysis of the key areas of pump maintenance in relation to the mining industry. A total of 88 articles meeting the established selection criteria were reviewed, allowing for an in-depth examination of the analyzed issue. Such deep analysis makes it possible to answer the stated research questions:

RQ1 intended to discover the state of the literature between 2005 and 2024 on mining equipment maintenance, with a particular focus on pump systems. The main research outputs here are discussed broadly in Sections 4.1 and 4.2.

The mining equipment maintenance literature, specifically on pump systems, has evolved significantly between 2005 and 2024. Research has primarily concentrated on improving efficiency, reducing downtime, and enhancing predictive maintenance strategies through advanced sensor-based monitoring, AI-driven analytics, and automation technologies.

The reviewed literature has been categorized into six major research areas, each addressing different aspects of pump reliability, performance optimization, and failure prevention.

Dewatering systems are essential in both open-pit and underground mining to control water levels, prevent flooding, and ensure smooth operations. The literature extensively discusses the following:

- Optimization of pumping systems—research explores the selection of pump configurations, balancing flow rates, and designing multi-stage pumping stations to enhance water removal efficiency.
- Impact of harsh mining environments—studies highlight the abrasive nature of mine water, which accelerates wear in pump components such as impellers and seals.
- Remote monitoring of dewatering pumps—the use of IoT-enabled sensors and automated diagnostics has been explored to improve operational oversight and reduce manual inspections.

In this area, across multiple studies, the challenge of pump reliability in extreme environments (e.g., high-pressure conditions, variable water loads) emerges as a critical issue. Solutions often involve integrating real-time monitoring with predictive maintenance techniques.

In addition, mining operations depend heavily on the continuous and efficient operation of pumps, making performance optimization a key research area. Indeed, the key contributions in the area of **operational efficiency and reliability optimization** mostly include the following:

- Energy consumption analysis—studies have quantified the energy footprint of pumps in mining operations, emphasizing the potential for efficiency improvements through variable-speed drives (VSDs) and frequency conversion technologies.
- Computational fluid dynamics (CFD) in pump design—CFD has been widely applied to simulate fluid flow, identify turbulence effects, and optimize pump impeller designs.
- Impact of wear on pump performance—research frequently examines degradation mechanisms (e.g., cavitation, erosion, and corrosion), providing insights into material selection and protective coatings.

Condition/health status monitoring was the third research area under investigation. The shift towards real-time condition monitoring and predictive maintenance is one of the most notable trends in pump maintenance research. The key insights from the articles in this research area are as follows:

- Early detection of faults: Several studies emphasize the importance of monitoring pump performance to identify potential issues before they result in significant breakdowns or inefficiencies. Predictive techniques help optimize maintenance schedules and minimize unplanned downtime.
- Wear and tear monitoring: Many studies focus on slurry pump wear, particularly the wear of impellers. Wear patterns are influenced by factors like flow rate, particle concentration, and particle size, which help inform better pump design, maintenance, and operational strategies.
- Advanced monitoring methods: The use of diverse methodologies, including unsupervised clustering, multiphase flow models, nodal analysis, and deep learning-based fault detection, is central to these studies. These methods allow for the prediction of pump health and early intervention to prevent failures.
- Real-time and preventive maintenance: t = Technologies such as intelligent electronic devices (IEDs), simulation-based monitoring, and real-time data analysis (e.g., power consumption and pressure measurements) are highlighted as essential for ensuring optimal performance and preventing breakdowns.
- Multisensor and intelligent systems: The integration of multisensor systems in monitoring and the use of intelligent systems, such as IEDs and simulation models, play a crucial role in enhancing preventive maintenance capabilities and ensuring the safety of mining operations.
- Focus on safety: Real-time monitoring of various equipment, including pumps, is critical to ensuring safety in mining operations, particularly in relation to issues like gas leaks, temperature, and pressure monitoring.

In summary, the research underscores the significance of advanced monitoring techniques and real-time data analysis in improving pump reliability, extending lifespan, and ensuring the safety of mining operations.

The fourth research area is connected with **health diagnosis and prognosis**. Health diagnosis involves detecting faults and wear in pumps through real-time monitoring using sensors and diagnostic algorithms. Prognosis, on the other hand, predicts the future health of the pump, estimating its remaining useful life (RUL) and potential failures using data from diagnostic tools and environmental conditions.

The research highlights the growing importance of data-driven approaches such as machine learning, predictive analytics, and condition-based monitoring. These tools enhance diagnostic accuracy and predictive maintenance, helping shift from reactive to proactive maintenance strategies in mining operations. This transition is vital for improving the sustainability and efficiency of the mining industry. Key research areas include the following:

- Prediction of remaining useful life: Advanced models like hidden semi-Markov models, relevance vector machines, and support vector machines have been used to predict the RUL of hydraulic pumps, slurry pump impellers, and other components. These models improve the accuracy of predictions, reducing unplanned breakdowns and optimizing maintenance schedules.
- Condition monitoring and diagnostics: Research has focused on real-time monitoring of pump health, with applications like vibration-based monitoring, thermal imaging, and automated diagnostics using PLC control. These systems aim to identify faults early, preventing failures and extending the operational life of pumps.

- Technological integration and innovation: The incorporation of advanced technologies, including AR, dynamic modeling, and machine learning, has been explored to enhance diagnostic capabilities and predictive maintenance. These innovations allow for more efficient maintenance strategies, particularly in remote and extreme environments like oil rigs and mining operations.
- Health monitoring of specific pump types: Studies have also focused on the health monitoring of specific pump types, such as hydraulic plunger pumps and piston diaphragm pumps, exploring failure mechanisms and proposing solutions to improve reliability and reduce operational costs.

In summary, the research underscores the importance of advanced simulation, predictive analytics, and data-driven maintenance in improving the performance and reliability of pumps in challenging industrial environments. These methods prevent costly downtimes and contribute to cost-effective and sustainable mining operations.

A well-structured maintenance strategy is crucial for cost reduction and equipment longevity. Thus, **maintenance management** in the mining industry, particularly for pump systems, is a critical practice aimed at ensuring the reliable, efficient, and cost-effective operation of pumps used in water management, slurry transport, and hydraulic systems. It involves the integration of preventive, predictive, and condition-based maintenance strategies, which minimize unplanned failures, optimize energy consumption, and extend equipment lifespan.

Research in this area identifies six key studies that explore various aspects of maintenance management for mining pumps, with a focus on automation, data-driven decision-making, and advanced maintenance strategies. These articles collectively demonstrate that effective maintenance management for mining pumps requires a combination of automation, data-driven decision-making, and lifecycle optimization of assets. These studies provide valuable insights into improving maintenance strategies to enhance pump performance, reduce operational risks, and lower costs. Looking ahead, advancements in AI, IoT, and sustainability-focused technologies will further refine maintenance strategies, driving more efficient, cost-effective, and environmentally friendly mining operations.

The last research area refers to **intelligent mining** recognized as the integration of advanced digital technologies to optimize and automate mining operations, enhancing efficiency, safety, and sustainability. This field combines big data analytics, artificial intelligence (AI), the Internet of Things (IoT), cloud computing, automation, robotics, and digital twin technology to enable real-time monitoring, predictive maintenance, and autonomous decision-making in mining operations.

A central aspect of intelligent mining is data-driven decision-making, where IoT sensors and AI algorithms continuously collect and analyze operational data. This approach allows for predictive maintenance strategies that minimize downtime and extend equipment lifespan. Automation plays a crucial role by deploying autonomous vehicles, drilling systems, and robotic machinery, reducing human intervention, particularly in hazardous environments. Additionally, remote monitoring and control enable centralized operation centers to optimize mining processes in real-time.

The transition from reactive maintenance to predictive and proactive maintenance strategies is a defining feature of intelligent mining. By anticipating and preventing issues before they occur, mining companies can achieve more resilient and cost-effective operations. The integration of digital ecosystems enables different mining components—such as pumps, conveyors, drilling equipment, and ventilation systems—to communicate and adapt dynamically, improving overall mine productivity.

In the context of pump maintenance, intelligent mining transforms traditional approaches by introducing real-time condition monitoring, AI-based fault detection, and automated control systems. IoT-enabled sensors track pump performance, while AI-driven models predict potential failures before they occur, reducing unplanned downtime. Digital twin technology further enhances maintenance strategies by simulating pump behavior under various conditions, enabling optimized operational planning. These advancements ensure higher reliability, reduced maintenance costs, and improved operational efficiency.

The research identified four key papers in intelligent mining, specifically focusing on pump maintenance. These papers cover various aspects of intelligent pump system management in the mining industry, highlighting the use of technologies such as big data analysis, automation, adaptive control, and digital ecosystems. All the studies aim to improve the reliability and efficiency of pump systems through intelligent monitoring, fault diagnosis, and operational optimization.

Ultimately, intelligent mining reshapes pump maintenance by combining automation, predictive analytics, and real-time data processing. The ability to anticipate failures before they occur, automate routine operations, and interconnect mining infrastructure leads to a more resilient, cost-effective, and sustainable industry.

The conducted systematic analysis of the selected literature makes it possible to answer the second research question. RQ2 intended to define the main research and knowledge gaps in pump systems maintenance, especially in the mining industry sector. Several key gaps can be highlighted based on the literature reviewed in the context of pump maintenance (presented in Section 4).

The first gap is the aspect of integration of advanced predictive models. While recent advancements in data-driven models, artificial intelligence (AI), and the Internet of Things (IoT) for predictive maintenance have made significant progress, there is still a gap in integrating more advanced predictive models applied in the mining industry. Current approaches often focus on monitoring basic pump parameters and do not consider the full range of influencing factors, such as environmental variables (e.g., temperature, humidity) or detailed data from other systems within the mine. There is a need for the development of more advanced AI algorithms that take into account a broader operational context and can predict failures more precisely. This would help minimize downtime and repair costs by identifying potential issues before they arise and improving the accuracy of maintenance strategies.

Next, comprehensive automation solutions development. Automation is a key element in improving the efficiency of pump systems in mining; however, there are still gaps in fully automating pump operations in the most challenging, often remote or hazardous locations within mines. While some technologies enable remote control, many systems still require human intervention in critical stages. There is a need for further development in autonomous systems that can manage the entire lifecycle of pump operations, including automated fault detection, performance optimization, and self-adjustment based on varying environmental conditions. Expanding automation capabilities would help reduce the dependence on human labor in high-risk areas and improve overall operational efficiency.

Data-driven maintenance strategies for complex systems are another area of research interest worth further development. While condition-based and predictive maintenance strategies have been widely applied to pumps, many mining operations still lack comprehensive systems for data-driven maintenance across complex systems. Pump systems in mining are often interconnected with other critical systems, such as conveyors, hydraulic equipment, and ventilation systems, and maintaining them effectively requires considering the interactions and dependencies between these systems. Future research should focus on developing integrated data ecosystems that allow for real-time monitoring, fault diagnosis, and predictive maintenance of pumps in conjunction with other equipment. This would lead to a more holistic and efficient approach to maintenance, improving overall mine productivity and reducing unplanned downtime.

According to recent developments, one should also focus on the problem of digital ecosystems scalability. The concept of digital ecosystems, where networks of sensors, pumps, and control systems interact dynamically, is still in its infancy in mining operations. While localized pump systems can be monitored and controlled remotely, scaling these systems to operate across multiple mine sites remains a challenge. More research is needed into scalable digital ecosystem architectures that enable seamless communication and integration between individual systems at a global level. This would allow for centralized control, predictive maintenance, and real-time decision-making across a mining company's entire infrastructure, improving coordination and operational efficiency across various sites.

Additionally, one of the research gaps is environmental and external influences on pump performance investigation. Environmental factors such as temperature, humidity, and even the chemical composition of the fluid being pumped can have significant impacts on pump performance and lifespan. However, many current maintenance strategies do not fully account for these variables. Future research could explore how to integrate environmental data into predictive maintenance models, taking into consideration how these factors affect pump performance. This would lead to more tailored and effective maintenance strategies that extend the operational lifespan of pumps and reduce failures caused by overlooked external influences.

The last research gap that could be identified in the investigated research area is the focus on human factors and user-centered maintenance. Despite the rapid advancement of automation, AI-driven diagnostics, and predictive maintenance strategies, human factors remain a crucial yet often underexplored element in the maintenance of pump systems [188,189]. While modern technologies enable real-time monitoring and data-driven decision-making, human operators continue to play a key role in interpreting diagnostics, executing maintenance actions, and responding to unforeseen system behaviors. Integrating proactive maintenance approaches introduces several challenges related to human factors, which must be addressed to ensure their effective implementation [190,191]. The problem is worth investigating, as we can see from the mining accident causality, where human error is one of the most significant causes (see, e.g., [7,192]). In addition, a comprehensive analysis of human errors in pump maintenance is given in [193].

One of the key challenges is the growing skills gap among maintenance personnel. The increasing complexity of predictive maintenance tools, such as IoT-enabled monitoring systems and AI-based diagnostics, requires specialized knowledge that is not always covered in traditional training programs. Without adequate preparation, maintenance staff may struggle to interpret diagnostic outputs correctly, leading to inefficiencies in maintenance execution and potential equipment failures. Additionally, adopting advanced maintenance technologies often encounters resistance from employees who are unfamiliar with digital tools or uncertain about their reliability. A lack of trust in automated decision-making systems can hinder their widespread use, emphasizing the need for strategies that enhance confidence in AI-assisted diagnostics.

Another critical issue is the cognitive load imposed on operators who must process vast amounts of diagnostic data from multiple sources. Poorly designed human-machine interfaces (HMI) can lead to information overload, increasing the likelihood of errors and reducing the effectiveness of predictive maintenance strategies. Ensuring that diagnostic tools present data clearly and intuitively is essential for enabling quick and informed decision-making. Furthermore, while AI-driven systems can generate predictive insights, human expertise remains indispensable in validating recommendations and making final maintenance decisions. The absence of well-defined interaction models between human operators and automated maintenance systems can limit the effectiveness of proactive maintenance approaches.

The concept of Maintenance 5.0 introduces a more human-centric approach in which automation and AI do not replace human expertise but rather enhance it. However, research on how to effectively integrate human decision-making with automated diagnostic tools in the context of mining pump maintenance remains limited. Addressing this gap requires a stronger focus on workforce training, user-friendly interface design, and hybrid decision-making frameworks that combine human intuition with AI-driven analytics.

To fully realize the potential of proactive maintenance, future research should explore strategies for improving the skills of maintenance personnel, developing intuitive interfaces that minimize cognitive overload, and fostering trust in AI-assisted maintenance tools. Implementing Maintenance 5.0 principles in pump maintenance should prioritize human-system collaboration, ensuring that operators remain central to the decision-making process while benefiting from the efficiency and precision offered by automation. By incorporating human factors into predictive maintenance strategies, organizations can enhance pump systems' reliability and operational efficiency, ultimately contributing to safer and more resilient mining operations.

In summary, these knowledge gaps represent opportunities for innovation in the field of pump system maintenance in the mining industry. Advancing predictive maintenance models, automating more complex systems, integrating environmental factors, and expanding digital ecosystem scalability will drive the next generation of maintenance strategies, leading to more efficient, cost-effective, and sustainable mining operations.

RQ3 is intended to discover the main trends that can be identified in proactive maintenance approaches and how they have evolved over recent years in the mining sector. According to the conducted literature review, the main trends in proactive maintenance approaches in the mining sector (pump maintenance) and their evolution in recent years reflect the growing integration of advanced technologies, data-driven decision-making, and automation. Table 4 provides a structured comparison of key research areas' current trends and challenges, offering a clear roadmap for future research in mining pump maintenance.

Over the past few years, proactive maintenance approaches in the mining sector have undergone significant transformation, driven by advancements in technology, data-driven decision-making, and automation. Traditionally, mining maintenance was largely reactive, relying on repairs and interventions after equipment failures. However, there has been a marked shift towards predictive and proactive maintenance strategies, which focus on anticipating problems before they occur, reducing downtime, and extending equipment lifespan.

Following Table 2, one of the key trends is the movement from reactive maintenance to predictive and proactive approaches. In the past, maintenance was often based on scheduled inspections or took place after failures occurred, leading to inefficiencies. With the rise of predictive maintenance powered by data analytics and machine learning, maintenance is now driven by data collected from equipment sensors and historical performance. This allows operators to predict when maintenance is needed, enabling them to act before problems arise. Using AI and machine learning models enhances the accuracy of these predictions, providing more reliable and effective maintenance strategies.

Another major trend is the increasing use of the Internet of Things (IoT) and real-time monitoring systems. IoT-enabled sensors are now widely used to track critical parameters such as real-time temperature, pressure, and vibration of mining equipment. These data are continuously analyzed, allowing operators to identify potential failures early, optimize maintenance schedules, and improve operational efficiency. Real-time monitoring has become essential for maintaining the health of mining equipment, reducing unplanned downtime, and ensuring a more resilient operation.

Integrating AI and machine learning in maintenance decision-making is also a key development. Traditional maintenance approaches were often based on simple inspections or time-based schedules, which sometimes missed early signs of potential failures. With AI, mining companies can now rely on data-driven models to anticipate equipment failures, optimize schedules, and ensure that maintenance is performed only when needed rather than following a fixed schedule. This shift significantly reduces unnecessary downtime and maintenance costs.

Table 4. A summary of the main trends and challenges identified in mining pump maintenance (2005–2024).

Research Area	Main Trends	Challenges
Dewatering System Operation and Maintenance	- Increasing use of automated water management - Integration of IoT-enabled sensors for real-time monitoring - AI-driven predictive dewatering models	- Abrasive and corrosive conditions reducing pump lifespan - Need for real-time system adjustments under changing mine conditions - High cost of automated dewatering solutions
Operational Efficiency and Reliability Optimization	- Adoption of variable-speed drives (VSDs) to optimize energy use - AI-based pump performance optimization - Material science advancements to reduce wear	- High energy costs of pumping operations - Need for customized CFD models for different pump types - Complex relationship between fluid dynamics and wear mechanisms
Condition/Health Status Monitoring	- Widespread use of sensor-based monitoring - Integration of cloud-based diagnostics - AI-driven failure prediction models improving accuracy	- Managing large-scale sensor data efficiently - False positives in anomaly detection - Standardization issues in monitoring technologies
Health Diagnosis and Prognosis	- Use of relevance vector machines (RVMs), hidden semi-Markov models (HSMMs), and support vector machines (SVMs) for RUL estimation - Combining historical failure data with real-time analytics - AI-powered predictive maintenance platforms	- Data quality and availability issues for AI training - Variability in failure patterns under different operating conditions - High computational costs for real-time RUL modeling
Maintenance Management	- Shift from preventive to predictive maintenance - Widespread adoption of computerized maintenance management systems (CMMS) - AI-assisted fault classification for automated decision-making	- Resistance to the adoption of AI-driven maintenance in traditional mining setups - Lack of standard frameworks for maintenance cost analysis - High initial investment costs
Intelligent Mining and Digital Transformation	- Increasing adoption of digital twins for failure prediction - AI-driven predictive analytics becoming standard - Remote and autonomous mining operations reducing human intervention	- Data security concerns with cloud-based diagnostics - Integration challenges with legacy pump monitoring systems - High costs of AI-based automation and digital twin implementation

Automation and autonomous systems have also played a crucial role in evolving maintenance strategies. In the past, maintenance tasks were largely manual, requiring human intervention for inspections and repairs. Today, autonomous systems, such as robotic maintenance platforms and drones, can conduct inspections, perform routine maintenance tasks, and even diagnose faults in hazardous environments. This improves operational efficiency and enhances safety by reducing the need for human workers in dangerous areas.

Digital twin technology is another important trend that has emerged in proactive maintenance. A digital twin is a virtual representation of physical assets, such as pumps or machinery, enabling real-time monitoring and simulation of equipment performance. By creating a digital replica of the equipment, mining companies can simulate various operating conditions and predict potential failures before they occur. This allows for better planning, optimized maintenance schedules, and improved equipment performance.

Big data analytics and data-driven decision-making have become integral to proactive maintenance approaches. In the past, maintenance decisions were often based on limited data or subjective assessments. Mining companies can now make more informed decisions by integrating sensor data, historical performance records, and environmental conditions. Big data analytics help identify subtle patterns that may indicate potential issues, allowing for timely interventions that prevent breakdowns and optimize equipment performance.

Sustainability and energy efficiency are also increasingly influencing maintenance strategies. As environmental concerns grow, mining companies focus on reducing energy consumption and extending the lifespan of equipment. Proactive maintenance now includes monitoring energy usage, identifying inefficiencies, and taking steps to reduce energy waste. This aligns with broader sustainability goals and helps minimize mining operations' environmental impact.

In summary, proactive maintenance strategies in the mining sector have evolved significantly, shifting from reactive approaches to more predictive and data-driven models. Integrating advanced technologies such as IoT, AI, machine learning, automation, and digital twins has improved maintenance efficiency, reliability, and cost-effectiveness. These trends enhance the overall performance of mining operations and contribute to more sustainable and energy-efficient practices.

The conducted systematic analysis of the selected literature makes it possible to answer the last research question. RQ4 is intended to define the framework's scope for proactive maintenance of pump systems in the mining industry.

The proactive maintenance framework for pump systems in the mining industry aims to enhance mining operations' reliability, efficiency, and sustainability by utilizing advanced digital technologies for maintenance management. The framework's scope should be built on integrating real-time data monitoring, predictive analytics, and performance optimization strategies to ensure continuous, efficient, and cost-effective operations.

Key elements that should be included in the framework are as follows:

1. Real-Time Data Acquisition and Integration:
    - Teal-time data collection through IoT sensors embedded in the pump systems is a critical component. Data from these sensors will provide a continuous flow of information on pump health, operational performance, and environmental factors.
    - Data must be integrated from various sources, including maintenance history, sensor data, operational logs, and environmental conditions, to comprehensively view the pump system's performance.
2. Predictive Maintenance Algorithms:
    - The core of the framework is the implementation of predictive maintenance strategies. Using machine learning models and statistical analysis, these algorithms will predict potential failures before they occur, allowing the operators to schedule maintenance tasks proactively.
    - Predictive maintenance will reduce unplanned downtimes, enhance the lifespan of pumps, and optimize resource allocation.
3. Data Analytics and Visualization:
    - The framework should include advanced data analytics tools to identify patterns and anomalies in the operational data, providing actionable insights.
    - Visualization dashboards will allow operators and maintenance personnel to easily interpret data, spot emerging issues, and make informed maintenance and operational adjustments decisions.

4. Simulation and Scenario Testing:
    - Simulation models of the pump systems will enable the testing of various maintenance strategies and failure scenarios. This allows the mining operators to evaluate the effectiveness of different approaches without disrupting actual operations.
    - This component also aids in optimizing maintenance schedules and identifying the best timing for interventions to minimize the impact on overall productivity.
5. Condition Monitoring and Performance Tracking:
    - Continuous monitoring of the pump systems, including key parameters such as pressure, temperature, flow rate, and vibration levels, will provide real-time insight into their operational health.
    - Automated condition monitoring tools will send alerts for any deviations from normal operating conditions, allowing quick actions to be taken before a potential failure occurs.
6. Feedback Mechanisms for Continuous Improvement:
    - The framework should incorporate feedback loops that capture the outcomes of maintenance activities and compare them with the predictions made by the system. By continually analyzing maintenance results, organizations can refine their predictive models and improve maintenance strategies.
    - A continuous improvement process helps the organization enhance the accuracy and effectiveness of its predictive maintenance capabilities.
7. Collaboration and Communication Tools:
    - Efficient communication tools are essential for coordinating maintenance activities, especially in complex mining operations. The framework should include collaboration platforms that allow seamless information sharing between maintenance teams, management, and operators to ensure that all stakeholders are informed about the pump systems' status and any required actions.
8. Compliance and Reporting:
    - The framework must ensure compliance with industry standards and regulatory requirements for maintenance practices. It should include reporting functionalities to generate maintenance logs, compliance reports, and performance reviews.
    - The data-driven approach will help demonstrate environmental, safety, and operational standards adherence.
9. Scalability and Adaptability:
    - The framework should be designed to be scalable, supporting various pump systems used in mining operations of different sizes and complexities.
    - It should also be adaptable to future technological advancements, ensuring that the system can evolve with emerging technologies such as 5G communications, AI, and enhanced data analytics tools.

As the analysis outlines, the proactive maintenance framework for pump systems in the mining industry should focus on integrating advanced digital technologies, such as IoT, data analytics, predictive maintenance, and real-time monitoring. Additionally, this framework should minimize downtime, reduce maintenance costs, and enhance operational efficiency, ultimately improving asset longevity and sustainability in mining operations.

By incorporating these core elements, the framework will maintain pump systems proactively, mitigating risks and avoiding unplanned failures. It also aligns with the broader industry trend of moving from reactive maintenance strategies to more sophisticated, data-driven, and predictive approaches.

6. Framework for Proactive Maintenance in Pump Systems from the Mining Industry

After ventilation systems, pumping systems are the most important elements of mine operation. They ensure the safety of workers and allow for the proper conduct of work in the mine. For example, pumping stations for the main drainage system must allow water to be continuously removed from mine workings regardless of emergency conditions or the suspension of mining operations. Two independent sources of electricity must power systems of this type. The control of such pumps should allow switching between pumps automatically or manually. Depending on the requirements and design of these systems, various measurements of operating parameters are used, i.e., capacity, water temperature, bearing temperature, supply voltage, motor supply current, head, and operating pressure. Depending on these indications, pumps are adjusted, or diagnostic and maintenance activities are undertaken.

Determination of the unreliability of pumping systems largely depends on the technical evaluation of electrical, mechanical, or electromechanical parameters. At the same time, one of the key problems in the operation of mining pumps is correctly identifying the first symptoms of impending failure and making economically rational operating decisions. Due to the issue's complexity and the difficult operating conditions of the pumps, proper management of mining pump maintenance should be based on a proactive approach using a predictive strategy.

The general concept of the proactive approach in mining machinery maintenance, proposed by the authors, is shown in Figure 24. It includes two basic elements—the development of a diagnostic-prognostic model and the definition of stages in making operational decisions based on the determined level of operational risk and the incurred operational costs. It was developed for the pumps performing in the selected Polish mine. Preliminary studies also made it possible to determine the conditions of applicability of the various detection methods and refine the plan for in-service testing to ensure the efficiency of the diagnostic work carried out (see [129]). In practice, this means that mining operators in the selected mine now use a diagnostic model developed through the insights from the literature review to assess pump conditions. The implementation of predictive maintenance strategies allows for better resource allocation, timely detection of anomalies, and more informed decision-making regarding equipment repairs and replacements.

The first step in the implementation of the proactive maintenance framework should focus on condition forecasting and rational decision-making processes for operations. This entails refining assumptions and requirements based on detailed diagnostic testing, such as vibroacoustic diagnostics, acoustic tests, and thermal measurements. The designed solution assumes the following operational conditions:

Cascade shaft drainage system:

- Pumps operating in cascade should be responsible for transporting water from different depth levels to the surface.
- The system should allow automatic start-up of subsequent pumps in the event of failure of one of the lower level pumps to avoid interruption of dewatering.
- Specific Recommendation: Implement an IoT-based monitoring system to continuously assess pump health (vibration, temperature, etc.) and trigger automatic start-up of backup pumps when necessary.

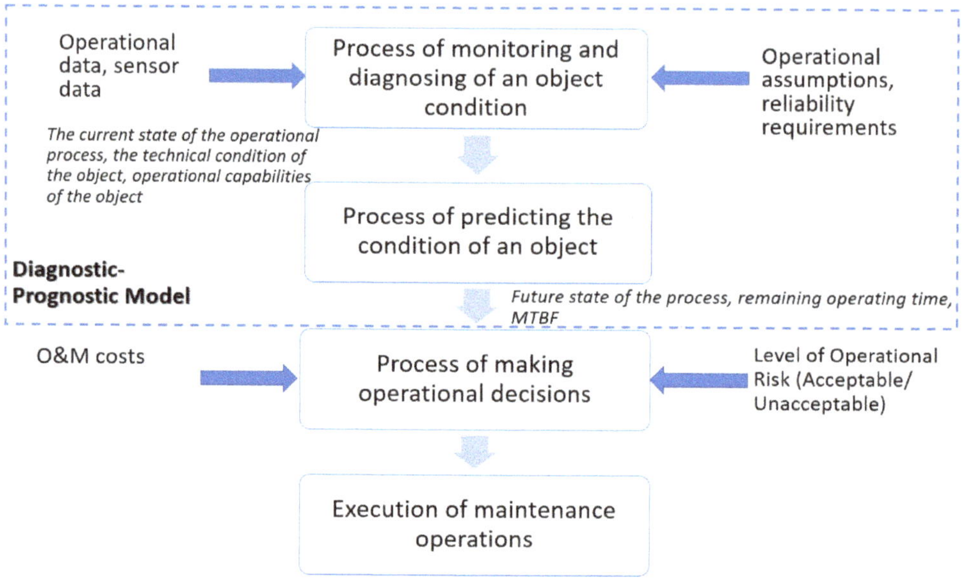

Figure 24. Proposed proactive approach in mining pump maintenance.

Surface pumps:

- Pumps on the surface responsible for removing process water to the main pipeline must work reliably, as their downtime could lead to flooding of the mine.
- In case of pump overload, the system should automatically switch to another pump to ensure continuous operation.
- Specific Recommendation: Use vibration and thermal sensors to monitor real-time operational conditions and integrate these with an automated switching mechanism that activates alternative pumps when needed.

Proactive diagnostics:

- The maintenance system should use various diagnostic methods to monitor the condition of pumps, including vibroacoustic diagnostics, acoustic tests, and thermal measurements that have been pre-tested.
- Monitoring of the condition of rolling elements should be regular and allow detection of anomalies at an early stage.
- Specific Recommendation: Develop a predictive maintenance dashboard that integrates all sensor data to provide early warnings for potential failures, allowing for a data-driven, proactive response.

Requirements for the proactive maintenance method are presented in Table 5.

Additionally, it is necessary to define the predictive analysis assumptions to accurately predict future failures, enabling maintenance planning in a way that minimizes production disruption and extends equipment life.

At the same time, correct inference requires summarizing the process of evaluating the technical condition of pumps and their components based on accumulated diagnostic test results. This includes analysis of historical failures and diagnostic results, such as vibroacoustic or thermal measurements. To this end, a set of diagnostic indicators (e.g., RMS vibration level and bearing temperature) was proposed to assess the technical condition of pumps. In the next step, permissible values and alarm thresholds (for normal, warning, and critical conditions) should be proposed as a reference for evaluating the measurement results. The cooperating entity is currently verifying the alarm thresholds.

Table 5. Requirements for the proactive maintenance method adopted for the case mining company.

Defined Area	Main Requirements
Monitoring and diagnostics module	• The system must be equipped with sensors to measure vibration, noise, and temperature, allowing real-time data collection; these sensors should be installed on critical components (e.g., motor bearings, pump casing) to ensure accurate, real-time condition monitoring. • The system must analyze trends in diagnostic data to predict failures, such as from increasing vibration and temperature levels. • Actionable Step: Calibrate sensors regularly to ensure data reliability, and integrate them into a centralized system for continuous analysis.
Predictive analytics	• The predictive analysis should consider the characteristics of each pump (cascade and surface) and their operating loads, which will allow a more accurate prediction of the operating time remaining before failure. • Actionable Step: Develop a machine learning-based predictive model that learns from past pump failures and operational data to more accurately forecast the remaining life of each component.
Automatic response to alarms	• The system should generate automatic alarms when the operating parameters of the pumps exceed the established safety thresholds (e.g., temperature of rolling elements or vibration level). • In case of critical anomalies, it should be possible to remotely turn the pump off and on or switch it to standby operation. • Actionable Step: Implement an automated shutdown mechanism to safely stop pumps if critical thresholds are exceeded and to prevent further damage to the equipment.
Service Scheduling	• The system should automatically generate maintenance schedules based on real-time pump data. These schedules should reflect the true condition of the pumps rather than fixed intervals. • Service should include regular inspection of rolling elements and lubrication and replacement of components as needed. • Actionable Step: Create a dynamic scheduling system that adjusts the maintenance frequency based on wear patterns detected through sensor data (e.g., more frequent inspections for pumps with higher vibration levels).
Reporting and historical analysis	• The system should generate detailed reports on pump operation and failures occurring to enable analysis of the causes of failures and optimization of maintenance activities. • Reports should also include recommendations for maintenance strategy changes based on historical data analysis. • Actionable Step: Use a dashboard for real-time data visualization, allowing engineers to access historical trends and failure analysis quickly, thus enabling more effective decision-making.
Integration with the mine management system	• The system should be integrated with the mine's central management system, allowing real-time monitoring of pumps. • Visualizing pump operation and alarms at the management level should be possible so mine operators have full control over the dewatering system. • Actionable Step: Set up cloud-based connectivity that allows the central control room to monitor pump health and operation remotely, enabling quick response in case of system failure.

A classification of damage types based on diagnostic tests must be developed to prepare a proactive maintenance method.

All these aspects should be prepared according to the operational and technical conditions of the given mining company.

The final component of the method is the operational decision-making process. In the developed decision-making process based on the diagnostic-predictive model, recommendations include different types of recommendations adapted to the characteristics of detected anomalies, failure predictions, and the overall condition of the machine. The recommendations are initially divided into eight categories: recommendations for preventive services, recommendations for corrective operational measures, recommendations for emergency repairs, recommendations for automatic actions, recommendations for resources and materials, recommendations for optimizing the operation process, long-term recommendations, and recommendations for security. These recommendations should be tailored to the specific operation of the machine or equipment in question, considering current operating conditions and the long-term goals of optimizing the operation process.

In addition, based on the findings of the literature review and referring to the defined framework, mining companies can implement the following specific steps to enhance pump maintenance practices:

- Use of vibroacoustic diagnostics, acoustic tests, and thermal measurements to continuously monitor the condition of pumps;
- Development of a predictive analytics model that considers pump characteristics and operational loads to forecast remaining operating time before failure;
- Implementation of an automatic response system that triggers alarms and initiates corrective actions (e.g., switching pumps or shutting them down remotely in case of critical anomalies);
- Establishment of a service scheduling system that adapts to the real-time condition of the pumps (i.e., "maintenance on condition");
- Integration of these diagnostic and predictive techniques into the mine's central management system, allowing operators to make real-time decisions regarding maintenance and operational strategies.

7. Conclusions

This article presents a systematic literature review focused on the proactive maintenance of pump systems in the mining industry. By analyzing key publications and case studies from 2005 to 2024, the review provides a comprehensive overview of the evolving approaches and best practices for maintaining pump systems in the mining sector. This study aimed to identify the main trends, research gaps, and emerging methodologies in proactively maintaining these critical assets.

The findings highlight the increasing importance of predictive maintenance strategies, the integration of digital technologies such as IoT and data analytics, and the role of real-time monitoring in improving operational efficiency and reducing downtime. Additionally, the research identifies the growing trend of leveraging simulation models and AI-driven algorithms for forecasting equipment failures and optimizing maintenance schedules.

However, the study has several limitations. The scope of the review is primarily focused on pump systems within the mining sector, which may exclude valuable insights from other industries or related sectors that employ similar maintenance strategies. The literature search and analysis were mostly restricted to peer-reviewed articles and conference papers, which might have overlooked significant contributions from industry reports, technical papers, or other non-academic sources. Furthermore, the research methodology did not include a quality assessment of the reviewed publications based on citation counts or relevance, which could affect the generalizability of the findings.

The review also highlights that although significant progress has been made in applying proactive maintenance strategies in mining, there are still gaps in understanding the full potential of emerging technologies, such as digital twins, in this specific context. The integration of advanced predictive maintenance systems, the development of standardized frameworks, and the establishment of industry-wide best practices are key areas requiring further research. Additionally, while this study provides a technical perspective on proactive maintenance, it does not fully address the human factors involved in its implementation, such as staff training, the adoption of new technologies, and organizational readiness. These factors play a crucial role in the success of proactive maintenance programs, and future research should consider these aspects to provide a more comprehensive approach to maintenance system implementation.

In conclusion, the study emphasizes the need for continued advancements in maintenance management practices, particularly in terms of embracing digital transformation and incorporating innovative technologies to improve asset reliability and performance. Future research should focus on exploring the challenges and opportunities of implementing these frameworks in real-world mining operations, addressing issues such as system integration, data security, and scalability. Moreover, further studies are needed to investigate the environmental and economic impacts of adopting proactive maintenance strategies and the potential for new business models that leverage emerging technologies in the mining sector.

Supplementary Materials: The following supporting information can be downloaded at https://www.mdpi.com/article/10.3390/s25082365/s1: Table S1: The PRISMA Checklist.

Author Contributions: Conceptualization, S.W.-W.; methodology, S.W.-W.; formal analysis, S.W.-W. and R.R.; resources, S.W.-W. and R.R.; data curation, S.W.-W. and R.R.; writing—original draft preparation, S.W.-W. and R.R.; writing—review and editing, S.W.-W. and R.R.; visualization, S.W.-W. and R.R.; supervision, S.W.-W. All authors have read and agreed to the published version of the manuscript.

Funding: This research received partial funding from the framework of the program "Implementation Doctorate," 5th edition, funded by the Ministry of Education and Science (Grant no. DWD/5/0030/2021).

Institutional Review Board Statement: Not applicable.

Informed Consent Statement: Not applicable.

Data Availability Statement: Not applicable.

Acknowledgments: The authors would like to thank the reviewers for their insightful comments.

Conflicts of Interest: The authors declare no conflicts of interest.

Abbreviations

The following abbreviations are used in this manuscript:

AI	artificial intelligence
AR	augmented reality
ARIMA	auto-regressive integrated moving average
BNN	backpropagation neural network
CBM	condition-based maintenance
CFD	computational fluid dynamics
CM	condition monitoring
CMMS	computerized maintenance management systems
CR	corrective maintenance
DeT	decision tree
DT	digital twin
ESP	electrical submersible pump
FSI	fluid-structure interaction
GAMM	graphical analysis for maintenance management
HSMM	hidden semi-Markov model
HSM	health status monitoring
HVAC	heating, ventilation, air conditioning
IEDs	intelligent electronic devices
IMs	induction motors
IoT	Internet of Things
LiDAR	light detection and ranging
MED	minimum entropy deconvolution
ML	machine learning
MoASoID	method of areas selection of image differences
MTBF	mean time between failures
MTTF	mean time to failure

MTTR	mean time to repair
NN	nearest neighbor
O&M	operation and maintenance
PAT	pump as turbine
Pdis	pump discharge pressure
PdM	predictive maintenance
PIP	pump intake pressure
PM	preventive maintenance
PRISMA	Preferred Reporting Items for Systematic Reviews and Meta-Analyzes
RBFNN	radial basis function neural networks
RVM	relevance vector machine
RUL	residual lifetime
SLR	systematic literature review
SVM	support vector machines
WPD	wavelet packet decomposition

References

1. Dhillon, B.S. *Mining Equipment Reliability, Maintainability, and Safety*; Springer-Verlag London Limited: London, UK, 2008; ISBN 9781848002876.
2. Christiansen, B. How Maintenance Tech Assists in Reducing Mining Project Costs. Available online: https://empoweringpumps.com/how-maintenance-tech-assists-in-reducing-mining-project-costs/?utm_source=chatgpt.com (accessed on 21 December 2024).
3. Dhillon, B.S. Mining Equipment Safety: A Review, Analysis Methods and Improvement Strategies. *Int. J. Min. Reclam. Environ.* **2009**, *23*, 168–179. [CrossRef]
4. Shi, X. Operational State Monitoring and Fuzzy Fault Diagnostic System of Mine Drainage. *Min. Sci. Technol.* **2010**, *20*, 581–584. [CrossRef]
5. Dhillon, B.S. Mining Equipment Maintenance. In *Mining Equipment Reliability, Maintainability, and Safety*; Springer Series in Reliability Engineering; Springer: London, UK, 2008; Volume 26, pp. 115–133. [CrossRef]
6. Rogers, W.P.; Kahraman, M.M.; Drews, F.A.; Powell, K.; Haight, J.M.; Wang, Y.; Baxla, K.; Sobalkar, M. Automation in the Mining Industry: Review of Technology, Systems, Human Factors, and Political Risk. *Min. Metall. Explor.* **2019**, *36*, 607–631. [CrossRef]
7. Kovacevic, S.; Papic, L.; Janackovic, G.L.; Savic, S. The Analysis of Human Error as Causes in the Maintenance of Machines: A Case Study in Mining Companies. *S. Afr. J. Ind. Eng.* **2016**, *27*, 193–202. [CrossRef]
8. Odeyar, P.; Apel, D.B.; Hall, R.; Zon, B.; Skrzypkowski, K. A Review of Reliability and Fault Analysis Methods for Heavy Equipment and Their Components Used in Mining. *Energies* **2022**, *15*, 6263. [CrossRef]
9. Sottile, J.; Holloway, L.E. An Overview of Fault Monitoring and Diagnosis in Mining Equipment. *IEEE Trans. Ind. Appl.* **1994**, *30*, 1326–1332. [CrossRef]
10. Dayo-Olupona, O.; Genc, B.; Celik, T.; Bada, S. Adoptable Approaches to Predictive Maintenance in Mining Industry: An Overview. *Resour. Policy* **2023**, *86*, 104291. [CrossRef]
11. Dayo-Olupona, O.; Genc, B.; Bada, S.; Celik, T. Predictive Maintenance: A Viable Maintenance Option for Machines/Equipment/Plants in Mining and Mineral Processing. In Proceedings of the 27th International Mining Congress and Exhibition of Turkey IMCET2022, Antalya, Turkey, 22–25 March 2022; pp. 759–765.
12. Kokkinis, A.; Frantzis, T.; Skordis, K.; Nikolakopoulos, G.; Koustoumpardis, P. Review of Automated Operations in Drilling and Mining. *Machines* **2024**, *12*, 845. [CrossRef]
13. Pourjavad, E.; Shirouyehzad, H.; Shahin, A. Selecting Maintenance Strategy in Mining Industry by Analytic Network Process and TOPSIS. *Int. J. Ind. Syst. Eng.* **2013**, *15*, 171–192. [CrossRef]
14. Tripathy, D.P. Effective Maintenance Practices for Mining Equipments. 2011, pp. 4–7. Available online: http://dspace.nitrkl.ac.in/dspace/bitstream/2080/1947/1/geominetech.pdf (accessed on 21 December 2024).
15. Angeles, E.; Kumral, M. Optimal Inspection and Preventive Maintenance Scheduling of Mining Equipment. *J. Fail. Anal. Prev.* **2020**, *20*, 1408–1416. [CrossRef]
16. Chaowasakoo, P.; Seppala, H.; Koivo, H. Age-Based Maintenance for a Fleet of Haul Trucks. *J. Qual. Maint. Eng.* **2018**, *24*, 511–528. [CrossRef]
17. Dreyer, E. Cost-Effective Prevention of Equipment Failure in the Mining Industry. *Int. J. Press. Vessel. Pip.* **1995**, *61*, 329–347. [CrossRef]

18. Online Condition Monitoring Offers Early Warning of Impending Equipment Problems. *Eng. Min. J.* **2011**, *212*, 106–108. Available online: https://www.e-mj.com/departments/operating-strategies/online-condition-monitoring-offers-early-warning-of-impending-equipment-problems/ (accessed on 21 December 2024).
19. Bauer, B.; Geropp, B.; Seeliger, A. Condition Monitoring and Predictive Maintenance in Mining Industry Using Vibration Analysis for Diagnosis of Gear Boxes. *IFAC Proc. Vol.* **1997**, *30*, 989–992. [CrossRef]
20. Bartelmus, W. Object and Operation Supported Maintenance for Mining Equipment. *Min. Sci.* **2014**, *21*, 7–21. [CrossRef]
21. Hu, H. Use of Data Mining Techniques for Mine Machine Condition Monitoring. Ph.D. Thesis, University of Missouri-Rolla, Rolla, MO, USA, 2003.
22. Beckman, M. Predictive Maintenance Polishes up Mining. *Tribol. Lubr. Technol.* **2021**, *77*, 38–40.
23. Robatto Simard, S.; Gamache, M.; Doyon-Poulin, P. Current Practices for Preventive Maintenance and Expectations for Predictive Maintenance in East-Canadian Mines. *Mining* **2023**, *3*, 26–53. [CrossRef]
24. Yu, S.; Rong, X.; Wei, L.; Shi, X. Review of Fault Diagnosis and Early Warning of Coal Mine Ventilator. In Proceedings of the 2019 Chinese Automation Congress, CAC 2019, Hangzhou, China, 22–24 November 2019; pp. 5226–5230.
25. Gerike, B.L.; Mokrushev, A.A. Diagnostics of the Technical State of Bearings of Mining Machines Base Assemblies. In Proceedings of the IOP Conference Series: Materials Science and Engineering Volume 253, Kemerovo, Russia, 18–21 April 2017; pp. 1–8.
26. Burrows, J.H. Predictive and Preventive Maintenance of Mobile Mining Equipment Using Vibration. Master's Thesis, McGill University, Montreal, QC, Canada, 1996.
27. Dos Santos Silva, F.J.; Garcia Viana, H.R.; Aquino Queiroz, A.N. Availability Forecast of Mining Equipment. *J. Qual. Maint. Eng.* **2016**, *22*, 418–432. [CrossRef]
28. Curilem, M.; Huanquilef, C.; Acuña, G.; Cubillos, F.; Araya, B.; Segovia, G.; Pérez, C. Prediction of the Criticality of a Heavy Duty Mining Equipment. In Proceedings of the 2015 Latin America Congress on Computational Intelligence (LA-CCI), Curitiba, Brazil, 13–16 October 2015; pp. 1–5.
29. Razor Labs. AI Enabled Predictive Maintenance in Mining. 2022. Available online: https://www.razor-labs.com/wp-content/uploads/2022/03/Razor-Labs-AI-enabled-predictive-maintenance-in-mining-Report-2022.pdf (accessed on 21 December 2024).
30. Dong, L.; Mingyue, R.; Guoying, M. Application of Internet of Things Technology on Predictive Maintenance System of Coal Equipment. *Procedia Eng.* **2017**, *174*, 885–889. [CrossRef]
31. Wu, T.; Chen, S.; Wu, P. Intelligent Fault Diagnosis System Based on Big Data. *J. Eng.* **2019**, *2019*, 8980–8985. [CrossRef]
32. Isermann, R. Fault-Diagnosis of Pumps. In *Fault-Diagnosis Applications, Model-Based Condition Monitoring: Actuators, Drives, Machinery, Plants, Sensors, and Fault-tolerant Systems*; Isermann, R., Ed.; Springer: Berlin/Heidelberg, Germany, 2011; pp. 143–179. ISBN 9783642127670.
33. Jain, S.V.; Patel, R.N. Investigations on Pump Running in Turbine Mode: A Review of the State-of-the-Art. *Renew. Sustain. Energy Rev.* **2014**, *30*, 841–868. [CrossRef]
34. Olesen, J.F.; Shaker, H.R. Predictive Maintenance for Pump Systems and Thermal Power Plants: State-of-the-Art Review, Trends and Challenges. *Sensors* **2020**, *20*, 2425. [CrossRef] [PubMed]
35. Aguilera, J.J.; Meesenburg, W.; Ommen, T.; Markussen, W.B.; Poulsen, J.L.; Zühlsdorf, B.; Elmegaard, B. A Review of Common Faults in Large-Scale Heat Pumps. *Renew. Sustain. Energy Rev.* **2022**, *168*, 112826. [CrossRef]
36. Mallioris, P.; Aivazidou, E.; Bechtsis, D. Predictive Maintenance in Industry 4.0: A Systematic Multi-Sector Mapping. *CIRP J. Manuf. Sci. Technol.* **2024**, *50*, 80–103. [CrossRef]
37. Qu, J.; Kizil, M.S.; Yahyaei, M.; Knights, P.F. Digital Twins in the Minerals Industry–a Comprehensive Review. *Min. Technol. Trans. Inst. Min. Metall.* **2023**, *132*, 267–289. [CrossRef]
38. Choudhary, A.; Goyal, D.; Shimi, S.L.; Akula, A. Condition Monitoring and Fault Diagnosis of Induction Motors: A Review. *Arch. Comput. Methods Eng.* **2019**, *26*, 1221–1238. [CrossRef]
39. Bagri, I.; Tahiry, K.; Hraiba, A.; Touil, A.; Mousrij, A. Vibration Signal Analysis for Intelligent Rotating Machinery Diagnosis and Prognosis: A Comprehensive Systematic Literature Review. *Vibration* **2024**, *7*, 1013–1062. [CrossRef]
40. Sharma, A.K.; Kumar, N.; Das, A.K. A Review on Wear Failure of Hydraulic Components: Existing Problems and Possible Solutions. *Eng. Res. Express* **2024**, *6*, 012502. [CrossRef]
41. Rajak, P.K.; Sarkar, S.; Paul, B. Comparison of the Dewatering of Underground and Open Pit Coal Mine Pumping Systems in (BCCL), Dhanbad, Jharkhand, India. In Proceedings of the IOP Conference Series: Materials Science and Engineering Volume 377, Sikkim, India, 8–10 December 2017.
42. Page, M.J.; McKenzie, J.E.; Bossuyt, P.M.; Boutron, I.; Hoffmann, T.C.; Mulrow, C.D.; Shamseer, L.; Tetzlaff, J.M.; Akl, E.A.; Brennan, S.E.; et al. The PRISMA 2020 Statement: An Updated Guideline for Reporting Systematic Reviews. *BMJ* **2021**, *372*, n71. [CrossRef]
43. Agboola, O.; Babatunde, D.E.; Isaac Fayomi, O.S.; Sadiku, E.R.; Popoola, P.; Moropeng, L.; Yahaya, A.; Mamudu, O.A. A Review on the Impact of Mining Operation: Monitoring, Assessment and Management. *Results Eng.* **2020**, *8*, 100181. [CrossRef]

44. Asr, E.T.; Kakaie, R.; Ataei, M.; Tavakoli Mohammadi, M.R. A Review of Studies on Sustainable Development in Mining Life Cycle. *J. Clean. Prod.* **2019**, *229*, 213–231. [CrossRef]
45. Zhironkin, S.; Szurgacz, D. *Mining Technologies Innovative Development*; MDPI: Basel, Switzerland, 2022; ISBN 9783036532257.
46. Afum, B.O.; Ben-Awuah, E. A Review of Models and Algorithms for Surface-Underground Mining Options and Transitions Optimization: Some Lessons Learnt and the Way Forward. *Mining* **2021**, *1*, 112–134. [CrossRef]
47. Rogowski, R.; Werbińska-Wojciechowska, S. Problems of Maintenance of Mining Equipment on the Example of a Shaft Dewatering System. In *Issues of Operation and Reliability of Selected Anthropotechnical Systems: Monograph: Collective Work*; e Rosiński, A., Siergiejczyk, M., Eds.; Faculty of Transport, Warsaw University of Technology: Warsaw, Poland, 2023; pp. 101–120.
48. Bridgwood, E.W.; Singh, R.N.; Atkins, A.S. Selection and Optimization of Mine Pumping Systems. *Int. J. Mine Water* **1983**, *2*, 1–19. [CrossRef]
49. Korbiel, T.; Wojciechowski, J. Analiza Kosztów Eksploatacji Systemu Głównego Odwadniania Kopalni Węgla Kamiennego. *Syst. Wspomagania W Inżynierii Prod.* **2019**, *8*, 413–419.
50. Andrzej, K.; Mikuła, S. Badania Eksploatacyjne Górniczych Pomp Odwadniających. *Zesz. Nauk. Politech Śląskiej Ser. Górnictwo Z. 269* **2005**, *1697*, 465–474.
51. Gogolewska, A.B. *Surface and Underground Mining Technology*; Wrocław University of Technology: Wroclaw, Poland, 2010.
52. Tse, P.W.; Shen, C. Remaining Useful Life Estimation of Slurry Pumps Using the Health Status Probability Estimation Provided by Support Vector Machine. In *Engineering Asset Management-Systems, Professional Practices and Certification, Lecture Notes in Mechanical Engineering*; Tse, P.W., Mathew, J., Wong, K., Lam, R., Ko, C.N., Eds.; Springer International Publishing: Cham, Switzerland, 2015; pp. 87–98, ISBN 978-3-319-09506-6.
53. Di Maio, F.; Hu, J.; Tse, P.; Pecht, M.; Tsui, K.; Zio, E. Ensemble-Approaches for Clustering Health Status of Oil Sand Pumps. *Expert Syst. Appl.* **2012**, *39*, 4847–4859. [CrossRef]
54. Tse, P.W.; Hu, J. A Fusion Approach with Application to Oil Sand Pump Prognostics. In *Engineering Asset Management-Systems, Professional Practices and Certification. Lecture Notes in Mechanical Engineering*; Tse, P.W., Mathew, J., Wong, K., Lam, C.N., Eds.; Springer International Publishing: Cham, Switzerland, 2015; pp. 31–41.
55. Beck, D.; Wulff, S.; Thamsen, P.U.; Reich, M.; Worringer, J.P. Recommendations for Long-Term Operation of High-Performance Submersible Motor Pumps in Mine Dewatering. In Proceedings of the American Society of Mechanical Engineers, Fluids Engineering Division (Publication) FEDSM, Virtual, Online, 13–15 July 2020; Volume 1.
56. Morgan, S.E. The ABCs of Pump Selection for Mine Dewatering. *Coal Age* **2008**, *113*, 46–47.
57. Boyce, K. Selecting a Reliable Dewatering Pump for Mining Applications. *Eng. Min. J.* **2013**, *214*, 36–37.
58. Vujic, S.B. Adaptive Computer Supported Surveilance-Management Model of Dewatering System at Coal Open Pit Mine. *Therm. Sci.* **2006**, *10*, 33–42. [CrossRef]
59. Majumdar, S.R. *Oil Hydraulic Systems: Principles and Maintenance*; McGraw-Hill Education: New York, NY, USA, 2013.
60. Minav, T. Electric-Drive-Based Control and Electric Energy Regeneration in a Hydraulic System. Ph.D. Thesis, Lappeenranta University of Technology, Lappeenranta, Finland, 2011. ISBN 9789522651037.
61. Dutta, N.; Kaliannan, P.; Paramasivam, S. A Comprehensive Review on Fault Detection and Analysis in the Pumping System. *Int. J. Ambient. Energy* **2022**, *43*, 6878–6898. [CrossRef]
62. Premium Pumping Technology for the Mining Industry. IIoT Enabled Pump Solutions. Available online: https://www.andritz.com/resource/blob/34958/4fcd8d1f5334d6a8255da3aa28b5f9f9/sammlung-pumps-mining-brochure-en-data.pdf (accessed on 21 December 2024).
63. Hermawan, Y.; Danang Wijaya, F.; Setiawan, N.A.; Dharmastiti, R. Prediction of Lubricant Service Life Using Data Mining to Improve Reliability of Water Injection Pumps in Crude Oil Production Facility. In Proceedings of the 7th International Conference on Electrical, Electronics and Information Engineering: Technological Breakthrough for Greater New Life, ICEEIE 2021, Malang, Indonesia, 2 October 2021; pp. 2–6.
64. ALTobi, M.A.S.; Bevan, G.; Wallace, P.; Harrison, D.; Ramachandran, K.P. *Centrifugal Pump Condition Monitoring and Diagnosis Using Frequency Domain Analysis*; Springer International Publishing: Berlin/Heidelberg, Germany, 2019; Volume 15, ISBN 9783030112202.
65. Rogowski, R.; Werbińska-Wojciechowska, S. *Maintenance Problems of Mining Equipment on the Example of a Mine Dewatering System (in Polish)*; accepted for, publication; Nowakowski, T., Rosiński, A., Sieriejczyk, M., Eds.; Warsaw University of Technology Publ. House: Warsaw, Poland, 2023.
66. Wang, Q.; Geng, P.; Qiu, H.; Zhang, J.; Zhao, R. Development of Intelligent Automatic Water Intake System for Open-Pit Mine. *J. Phys. Conf. Ser.* **2023**, *2591*, 012057. [CrossRef]
67. Werbińska-Wojciechowska, S. *Technical System Maintenance. Delay-Time-Based Modelling*; Springer Nature Switzerland AG: Cham, Switzerland, 2019; ISBN 9783030107871.
68. *BS EN 13306:2017*; Maintenance. Maintenance Terminology. BSI: London, UK, 2018.

69. Ran, Y.; Zhou, X.; Lin, P.; Wen, Y.; Deng, R. A Survey of Predictive Maintenance: Systems, Purposes and Approaches. *arXiv* **2019**, arXiv:1912.07383.
70. Ahmad, R.; Kamaruddin, S. An Overview of Time-Based and Condition-Based Maintenance in Industrial Application. *Comput. Ind. Eng.* **2012**, *63*, 135–149. [CrossRef]
71. Prajapati, A.; Bechtel, J.; Ganesan, S. Condition Based Maintenance: A Survey. *J. Qual. Maint. Eng.* **2012**, *18*, 384–400. [CrossRef]
72. VOSviewer. Available online: https://www.vosviewer.com/ (accessed on 21 December 2024).
73. McKee, K.K.; Forbes, G.; Mazhar, I.; Entwistle, R.; Howard, I. A Review of Major Centrifugal Pump Failure Modes with Application to the Water Supply and Sewerage Industries. In Proceedings of the ICOMS Asset Management Conference, Gold Coast, QLD, Australia, 16–20 May 2011; pp. 1–12.
74. McKee, K.K.; Forbes, G.L.; Mazhar, I.; Entwistle, R.; Howard, I. A Review of Machinery Diagnostics and Prognostics Implemented on a Centrifugal Pump. In *Engineering Asset Management 2011. Lecture Notes in Mechanical Engineering*; Lee, J., Ni, J., Sarangapani, J., Mathew, J., Eds.; Springer: London, UK, 2014; pp. 593–614, ISBN 978-1-4471-4992-7.
75. Aliyu, R.; Mokhtar, A.A.; Hussin, H. Prognostic Health Management of Pumps Using Artificial Intelligence in the Oil and Gas Sector: A Review. *Appl. Sci.* **2022**, *12*, 11691. [CrossRef]
76. Hallaji, S.M.; Fang, Y.; Winfrey, B.K. Predictive Maintenance of Pumps in Civil Infrastructure: State-of-the-Art, Challenges and Future Directions. *Autom. Constr.* **2022**, *134*, 104049. [CrossRef]
77. Lakal, N.; Shehri, A.H.; Brashler, K.W.; Wankhede, S.P.; Morse, J.; Du, X. Sensing Technologies for Condition Monitoring of Oil Pump in Harsh Environment. *Sens. Actuators A Phys.* **2022**, *346*, 113864. [CrossRef]
78. Sunal, C.E.; Dyo, V.; Velisavljevic, V. Review of Machine Learning Based Fault Detection for Centrifugal Pump Induction Motors. *IEEE Access* **2022**, *10*, 71344–71355. [CrossRef]
79. Yang, Y.; Ding, L.; Xiao, J.; Fang, G.; Li, J. Current Status and Applications for Hydraulic Pump Fault Diagnosis: A Review. *Sensors* **2022**, *22*, 9714. [CrossRef]
80. Verhulst, T.; Judt, D.; Lawson, C.; Chung, Y.; Al-Tayawe, O.; Ward, G. Review for State-of-the-Art Health Monitoring Technologies on Airframe Fuel Pumps. *Int. J. Progn. Heal. Manag.* **2022**, *13*, 1–20. [CrossRef]
81. Mohamed Almazrouei, S.; Dweiri, F.; Aydin, R.; Alnaqbi, A. A Review on the Advancements and Challenges of Artificial Intelligence Based Models for Predictive Maintenance of Water Injection Pumps in the Oil and Gas Industry. *SN Appl. Sci.* **2023**, *5*, 391. [CrossRef]
82. Klis, P.; Xing, Y.; Abd, A.H.; Ren, C. A Survey of Control Systems for Offshore Well Service Pumps: State-of-the-Art, Challenges and Future Perspectives. In *Volume 8: Offshore Geotechnics; Petroleum Technology, Proceedings of the ASME 2024 43rd International Conference on Ocean, Offshore and Arctic Engineering, Singapore, 9–14 June 2024*; V008T11A018; ASME: New York, NY, USA, 2024.
83. Zhu, Y.; Zhou, L.; Lv, S.; Shi, W.; Ni, H.; Li, X.; Tao, C.; Hou, Z. Research Progress on Identification and Suppression Methods for Monitoring the Cavitation State of Centrifugal Pumps. *Water* **2024**, *16*, 52, ISBN 2210310024.
84. Blischke, W.R.; Murthy, P.D.N. *Reliability: Modeling, Prediction, and Optimization*; John Wiley & Sons, Inc.: Hoboken, NJ, USA, 2000.
85. Ali, A.; Abdelhadi, A. Condition-Based Monitoring and Maintenance: State of the Art Review. *Appl. Sci.* **2022**, *12*, 688. [CrossRef]
86. Discenzo, F.M.; Chung, D.; Loparo, K.A. Pump Condition Monitoring Using Self-Powered Wireless Sensor Nodes. *Shock. Vib. Dig.* **2006**, *38*, 500–502.
87. Ranjbar, A.; Suratgar, A.; Ghidary, S.S.; Milimonfared, J. Condition monitoring of an industrial oil pump using a learning based technique. *Sound Vib.* **2020**, *54*, 257–267. [CrossRef]
88. Kalmár, C.; Hegedus, F. Condition Monitoring of Centrifugal Pumps Based on Pressure Measurements. *Period. Polytech. Mech. Eng.* **2019**, *63*, 80–90. [CrossRef]
89. Pech, M.; Vrchota, J.; Bednář, J. Predictive Maintenance and Intelligent Sensors in Smart Factory: Review. *Sensors* **2021**, *21*, 1470. [CrossRef]
90. Divya, D.; Marath, B.; Santosh Kumar, M.B. Review of Fault Detection Techniques for Predictive Maintenance. *J. Qual. Maint. Eng.* **2022**, *29*, 420–441. [CrossRef]
91. Harihara, P.P.; Parlos, A.G. Fault Diagnosis of Centrifugal Pumps Using Motor Electrical Signals. *Centrif. Pumps* **2012**, *2012*, 15–32. [CrossRef]
92. Chen, L.; Wei, L.; Wang, Y.; Wang, J.; Li, W. Monitoring and Predictive Maintenance of Centrifugal Pumps Based on Smart Sensors. *Sensors* **2022**, *22*, 2106. [CrossRef]
93. Budgen, D.; Brereton, P. Performing Systematic Literature Reviews in Software Engineering. *Proc.-Int. Conf. Softw. Eng.* **2006**, *2006*, 1051–1052. [CrossRef]
94. Madhukar, P.; McCulloch, M.; Gorman, J.; Pai, N.P.; Enanoria, W.T.A.; Kennedy, G.E.; Tharyan, P.; Colford, J.M. Systematic Reviews and Meta-Analyses: An Illustrated, Step-by-Step Guide. *Natl. Med. J. India* **2004**, *17*, 86–95.
95. Zamani, E.D.; Smyth, C.; Gupta, S.; Dennehy, D. Artificial Intelligence and Big Data Analytics for Supply Chain Resilience: A Systematic Literature Review. *Ann. Oper. Res.* **2022**, *327*, 605–632. [CrossRef] [PubMed]
96. Aromataris, E.; Pearson, A. The Systematic Review: An Overview. *Am. J. Nurs.* **2014**, *114*, 53–58. [CrossRef]

97. Tranfield, D.; Denyer, D.; Smart, P. Towards a Methodology for Developing Evidence-Informed Management Knowledge by Means of Systematic Review. *Br. J. Manag.* **2003**, *14*, 207–222. [CrossRef]
98. Moher, D.; Liberati, A.; Tetzlaff, J.; Altman, D.G. Preferred Reporting Items for Systematic Reviews and Meta-Analyses: The PRISMA Statement. *Int. J. Surg.* **2010**, *8*, 336–341. [CrossRef]
99. Moher, D.; Liberati, A.; Tetzlaff, J.; Altman, D.G. Preferred Reporting Items for Systematic Reviews and Meta-Analyses: The PRISMA Statement. *BMJ* **2009**, *339*, 332–336. [CrossRef]
100. Available online: https://omnis-pwr.primo.exlibrisgroup.com/discovery/search?Vid=48OMNIS_TUR:48TUR&lang=pl&mode=advanced (accessed on 21 December 2024).
101. Flores-Castañeda, R.O.; Olaya-Cotera, S.; López-Porras, M.; Tarmeño-Juscamaita, E.; Iparraguirre-Villanueva, O. Technological Advances and Trends in the Mining Industry: A Systematic Review. *Miner. Econ.* **2024**. [CrossRef]
102. Watson, R.T.; Webster, J. Analysing the Past to Prepare for the Future: Writing a Literature Review a Roadmap for Release 2.0. *J. Decis. Syst.* **2020**, *29*, 129–147. [CrossRef]
103. Werbińska-Wojciechowska, S.; Winiarska, K. Maintenance Performance in the Age of Industry 4.0: A Bibliometric Performance Analysis and a Systematic Literature Review. *Sensors* **2023**, *23*, 1409. [CrossRef]
104. Amin, M.T.; Khan, F.; Zuo, M.J. A Bibliometric Analysis of Process System Failure and Reliability Literature. *Eng. Fail. Anal.* **2019**, *106*, 104152. [CrossRef]
105. Midha, A.; Maheshwari, I.; Ojha, K.; Gupta, K.; Deshpande, S.V. Bibliometric Review of Predictive Maintenance Using Vibration Analysis. *Libr. Philos. Pract.* **2021**, *5790*. Available online: https://digitalcommons.unl.edu/libphilprac/5790 (accessed on 21 December 2024).
106. Grubisic, V.V.F.; Aguiar, J.P.F.; Simeu-Abazi, Z. A Review on Intelligent Predictive Maintenance: Bibliometric Analysis and New Research Directions. In Proceedings of the 2020 International Conference on Control, Automation and Diagnosis (ICCAD), Paris, France, 7–9 October 2020; pp. 1–6. [CrossRef]
107. Vithi, N.L.; Chibaya, C. Advancements in Predictive Maintenance: A Bibliometric Review of Diagnostic Models Using Machine Learning Techniques. *Analytics* **2024**, *3*, 493–507. [CrossRef]
108. Lorenzetti, D.L.; Ghali, W.A. Reference Management Software for Systematic Reviews and Meta-Analyses: An Exploration of Usage and Usability. *BMC Med. Res. Methodol.* **2013**, *13*, 141. [CrossRef]
109. da Silva, A.A.; Pedrosa, M.R. *Using Reference Manager (Mendeley) in Systematic Reviews*; Cochrane Brasil: São Paulo, Brazil, 2018.
110. van Eck, N.J.; Waltman, L. Software Survey: VOSviewer, a Computer Program for Bibliometric Mapping. *Scientometrics* **2010**, *84*, 523–538. [CrossRef] [PubMed]
111. Peng, G.; Chen, Q.; Bai, L.; Hu, Z.; Zhou, L.; Huang, X. Wear Mechanism Investigation in a Centrifugal Slurry Pump Impeller by Numerical Simulation and Experiments. *Eng. Fail. Anal.* **2021**, *128*, 105637. [CrossRef]
112. Peng, G.; Du, J.; Chang, H.; Chen, Q.; Li, J.; Pan, C.; Yang, Y. Numerical and Experimental Analysis of Influence of Impeller Structures on Slurry Pump Performance. *Front. Energy Res.* **2021**, *9*, 762159. [CrossRef]
113. Peng, G.; Tian, L.; Chang, H.; Hong, S.; Ye, D.; You, B. Numerical and Experimental Study of Hydraulic Performance and Wear Characteristics of a Slurry Pump. *Machines* **2021**, *9*, 373. [CrossRef]
114. Gierlotka, S. The Drive of Mining Electric Pumps. *Silesian Electr. J.* **2016**, *XXIII*, 28–31.
115. Chandhini, G.; Chithra, B.; Kiruthikadevi, P.; Bhagya, S.; Kumar, P.V.K. IoT Based Underground Drainage Monitoring System. *Int. J. Recent Technol. Eng.* **2020**, *9*, 247–249. [CrossRef]
116. Myszkowski, P. Money-Saving and Innovative Products and Systems Used in Minerals Processing—Selection. *J. Pol. Miner. Eng. Soc.* **2017**, *2017*, 73–80.
117. Carter, R.A. Latest Dewatering Pumps Put a New Spin on Performance. *Eng. Min. J.* **2012**, *213*, 75–76.
118. Walker, S. Pumping Technology: For Water, Slurries, Solids and More. *Eng. Min. J.* **2012**, 34.
119. Treutel, C. Hose Pumps Becoming an Option for Abrasive Mining Applications. *Eng. Min. J.* **2007**, *208*, 82–85.
120. Emal Qazizada, M.; Pivarčiová, E. Reliability of Parallel and Serial Centrifugal Pumps for Dewatering in Mining Process. *Acta Montan. Slovaca* **2018**, *23*, 141–152.
121. Wright, M. Water Management in Mining Applications: Selecting the Correct Pumps to Meet a Mine Site's Needs. *Min. Eng.* **2021**, *73*, 18–23.
122. De Jager, J.P.; Groenewald, H.J.; Cilliers, C. Investigating the Effect of Pump Availability on Load Shift Performance. In Proceedings of the 2016 International Conference on the Industrial and Commercial Use of Energy (ICUE), Cape Town, South Africa, 16–17 August 2016; pp. 98–103.
123. Li, S.G.; Mei, S.B.; Gao, Z.Z. Design of Mining Automatic Drainage Monitoring and Control System Based on PLC and Kingview. In Proceedings of the 2013 Seventh International Conference on Image and Graphics, Qingdao, China, 26–28 July 2013; pp. 779–783. [CrossRef]
124. Michlowicz, E.; Wojciechowski, J. Analysis of Energy Consumption of Main Dewatering Pumps in Underground Mines. *Min.-Inform. Autom. Electr. Eng.* **2021**, *2*, 55–62.

125. Walker, S. Pumps and Pielines Keep Mines on the Move. *Eng. Min. J.* **2013**, *214*, 68–79.
126. Hasan, A.N. Evaluating the Performance of Single Classifiers Against Multiclassifiers in Monitoring Underground Dam Levels and Energy Consumption for a Deep Gold Mine Pump Station. In Proceedings of the IECON 2016—42nd Annual Conference of the IEEE Industrial Electronics Society, Florence, Italy, 23–26 October 2016; pp. 907–911.
127. Rea, I.; Monaghan, D. Dewatering Bore Pumps—Reducing Costs and Emissions by Maximising Pumping Efficiency over Time. *Trans. Inst. Min. Metall. Sect. A Min. Technol.* **2009**, *118*, 220–224. [CrossRef]
128. Huang, X.; Wu, X.; Tian, Z.; Lin, S.; Ji, J.; Guo, Y.; Xie, F. Fault Diagnosis Study of Mine Drainage Pump Based on MED–WPD and RBFNN. *J. Braz. Soc. Mech. Sci. Eng.* **2023**, *45*, 347. [CrossRef]
129. Rogowski, R.; Werbińska-Wojciechowska, S. Mining Equipment Diagnostics in a Mine Shaft Dewatering System—Case Study. *J. Konbin* **2023**, *53*, 69–86. [CrossRef]
130. Szymański, Z. Modern Control Method and Diagnostics Research of the Mine Main Pumps. *Napędy I Sterow* **2013**, *2*, 54–61.
131. Saha, D.; Prasad, S.V.A.V.; Saluja, N. Design of Coal Mine Underground Drainage Pump Monitoring and Controlling System Using ARM7 and Touch Screen. *Int. J. Comput. Appl.* **2013**, *71*, 25–28. [CrossRef]
132. Wu, J.; Chen, G.J. Design of Coal Mine Underground Drainage Pump Monitoring and Controlling System Based on PLC and Touch Screen. In Proceedings of the 2011 International Conference on Mechatronic Science, Electric Engineering and Computer (MEC), Jilin, China, 19–22 August 2011; pp. 1245–1247. [CrossRef]
133. Wu, Y.Z.; Wang, Z.J.; Lin, Y.; Pan, C.F.; Pan, G. ying Three Step-Drawdown Dewatering Test in Unsteady Flow Condition: A Case Study of the Siwan Coal Mine in North China Coalfield. *Environ. Earth Sci.* **2018**, *77*, 695. [CrossRef]
134. Mahamadou, O.; Abdoulkader, H.A.; Seibou, B.; Abdourazakou, M.H.; Harouna, S.; Souleymane, I.I.; Marou, I.H. Study of the Operating Cost of the Electric Pump for Mine Dewatering: Case of the SOMAÏR Uranium Mine. *Phys. Sci. Int. J.* **2024**, *28*, 1–13. [CrossRef]
135. Marchi, A.; Simpson, A.R.; Lambert, M.F. Pump Operation Optimization Using Rule-Based Controls. *Procedia Eng.* **2017**, *186*, 210–217. [CrossRef]
136. Lin, T.; Zhang, J.; Li, J.; Li, X.; Zhu, Z. Pressure Fluctuation–Vorticity Interaction in the Volute of Centrifugal Pump as Hydraulic Turbines (PATs). *Processes* **2022**, *10*, 2241. [CrossRef]
137. Zhang, Z.; He, X.; Kusiak, A. Data-Driven Minimization of Pump Operating and Maintenance Cost. *Eng. Appl. Artif. Intell.* **2015**, *40*, 37–46. [CrossRef]
138. Chu, T.; Nguyen, T.C.; Wang, J.; Vuong, D. New Correlations for Predicting Two-Phase Electrical Submersible Pump Performance under Downhole Conditions Using Field Data. *J. Pet. Explor. Prod. Technol.* **2022**, *12*, 1225–1235. [CrossRef]
139. Hu, J.; Li, K.; Su, W.; Zhao, X. Numerical Simulation of Drilling Fluid Flow in Centrifugal Pumps. *Water* **2023**, *15*, 992. [CrossRef]
140. Rutkowski, W. Possible Application of Low-Cost Laser Scanners and Photogrammetric Measurements for Inspections in Shaft Sinking Process. *J. Konbin* **2023**, *53*, 147–156. [CrossRef]
141. Zhang, D.; Liu, Y.; Kang, J.; Zhang, Y.; Meng, F. The Effect of Discharge Areas on the Operational Performance of a Liquid-Ring Vacuum Pump: Numerical Simulation and Experimental Verification. *Vacuum* **2022**, *206*, 111425. [CrossRef]
142. Pujades, E.; Orban, P.; Bodeux, S.; Archambeau, P.; Erpicum, S.; Dassargues, A. Underground Pumped Storage Hydropower Plants Using Open Pit Mines: How Do Groundwater Exchanges Influence the Efficiency? *Appl. Energy* **2017**, *190*, 135–146. [CrossRef]
143. Carter, R.A. Pumps and Valves: Ready for Prime Time. *Eng. Min. J.* **2021**, *222*, 50–56.
144. Pierce, W. An Overview of a Cost-Effective Method of Sealing Slurry Pumps: Centrifugal or Dynamic Expeller Seals. *Min. Eng.* **2021**, *73*, 21–25.
145. Casteel, K. Pumping Progress New Models, Materials and Methods Enhance Efficiency. *Eng. Min. J.* **2006**, *207*, 36–42.
146. High-Tech Materials Maximize Pump Safety and Performance. *Eng. Min. J.* **2014**, 136–138.
147. Carter, R.A. Spotlight on Slurry. *Eng. Min. J.* **2010**, *211*, 102–105.
148. Michalicka, P. Boosting Bearing and Pump Reliability in Fracking Operations. *Can. Min. J.* **2013**, *134*, 42–46.
149. Crane, J. How to Improve Reliability, Bottom Line, and Water Reduction. *Can. Min. J.* **2023**, 33–34.
150. Cai, J.; Wu, J.; Yuan, S.; Liu, Z.; Kong, D. Numerical Analysis of Multi-Factors Effects on the Leakage and Gas Diffusion of Gas Drainage Pipeline in Underground Coal Mines. *Process Saf. Environ. Prot.* **2021**, *151*, 166–181. [CrossRef]
151. Carter, R.A. What's Next for Slurry Pumps? *Eng. Min. J.* **2021**, 38–42.
152. Metso Updates Plant Control Monitoring Software. *Eng. Min. J.* **2018**, 98–99.
153. Iranzi, J.; Son, H.; Lee, Y.; Wang, J. A Nodal Analysis Based Monitoring of an Electric Submersible Pump Operation in Multiphase Flow. *Appl. Sci.* **2022**, *12*, 2825. [CrossRef]
154. Mazur, D.C.; Sottile, J.; Novak, T. An Electrical Mine Monitoring System Utilizing the IEC 61850 Standard. *IEEE Trans. Ind. Appl.* **2013**, *51*, 1317–1325. [CrossRef]
155. Kumar, N.; Kumar, R.; Sarkar, B.K.; Maity, S. Condition Monitoring of Hydraulic Transmission System with Variable Displacement Axial Piston Pump and Fixed Displacement Motor. *Mater. Today Proc.* **2019**, *46*, 9758–9765. [CrossRef]

156. Kumar, N.; Sarkar, B.K.; Maity, S. Leakage Based Condition Monitoring and Pressure Control of the Swashplate Axial Piston Pump. In Proceedings of the ASME 2019 Gas Turbine India Conference, GTINDIA 2019, Tamil Nadu, India, 5–6 December 2019; Volume 2, pp. 1–9.
157. Tiwari, R.; Bordoloi, D.J.; Dewangan, A. Blockage and Cavitation Detection in Centrifugal Pumps from Dynamic Pressure Signal Using Deep Learning Algorithm. *Meas. J. Int. Meas. Confed.* **2021**, *173*, 108676. [CrossRef]
158. Rosic, M.; Leiden, A.; Abraham, T.; Herrmann, C. Data Mining Approach for Device Detection Using Power Signatures and Manufacturing Execution System Data. *Procedia CIRP* **2022**, *107*, 1047–1052. [CrossRef]
159. Jiang, Y.; Chen, W.; Zhang, X.; Zhang, X.; Yang, G. Real-time monitoring of underground miners's status based on mine IoT system. *Sensors* **2024**, *24*, 739. [CrossRef]
160. Dong, M.; He, D. A Segmental Hidden Semi-Markov Model (HSMM)-Based Diagnostics and Prognostics Framework and Methodology. *Mech. Syst. Signal Process.* **2007**, *21*, 2248–2266. [CrossRef]
161. Dong, J.; Deng, Y.; Cao, W.; Wang, Z.; Ma, W.; Wu, D.; Ji, H.; Liu, Y. Wear Failure Analysis of Suction Valve for High Pressure and Large Flow Water Hydraulic Plunger Pump. *Eng. Fail. Anal.* **2022**, *134*, 106095. [CrossRef]
162. Liu, Q.; Dong, M.; Lv, W.; Geng, X.; Li, Y. A Novel Method Using Adaptive Hidden Semi-Markov Model for Multi-Sensor Monitoring Equipment Health Prognosis. *Mech. Syst. Signal Process.* **2015**, *64–65*, 217–232. [CrossRef]
163. Sharma, A.K.; Punj, P.; Kumar, N.; Das, A.K.; Kumar, A. Lifetime Prediction of a Hydraulic Pump Using ARIMA Model. *Arab. J. Sci. Eng.* **2023**, *49*, 1713–1725. [CrossRef]
164. Barszcz, T.; Urbanek, J.; Szumilas, Ł. Selection of Efficient Monitoring Methods for Machinery Generating High Vibration Signal Disturbance. *Diagnostics* **2010**, *4*, 55–58.
165. Glowacz, A.; Glowacz, Z. Diagnosis of the Three-Phase Induction Motor Using Thermal Imaging. *Infrared Phys. Technol.* **2017**, *81*, 7–16. [CrossRef]
166. Tse, P.W.; Wang, D. Performance Degradation Assessment of Slurry Pumps. In *Engineering Asset Management-Systems, Professional Practices and Certification. Lecture Notes in Mechanical Engineering*; Tse, P.W., Mathew, J., Wong, K., Lam, C.N., Eds.; Springer International Publishing: Cham, Switzerland, 2015; pp. 149–158.
167. Bold, S.; Urschel, S. Assessment of Approaches for Technical Diagnostic of Pump Faults with Induction Motor as Transducer. In Proceedings of the 2020 23rd International Conference on Electrical Machines and Systems (ICEMS), Hamamatsu, Japan, 24–27 November 2020; pp. 810–815. [CrossRef]
168. Koteleva, N.; Valnev, V.; Frenkel, I. Investigation of the Effectiveness of an Augmented Reality and a Dynamic Simulation System Collaboration in Oil Pump Maintenance. *Appl. Sci.* **2022**, *12*, 350. [CrossRef]
169. Pourebrahimi, A.; Mokhtar, S.; Sahami, S.; Mahmoodi, M. Status Detection and Fault Diagnosing of Rotatory Machinery by Vibration Analysis Using Data Mining. In Proceedings of the 2010 2nd International Conference on Computer Technology and Development, Cairo, Egypt, 2–4 November 2010; pp. 131–135. [CrossRef]
170. Isermann, R. *Fault-Diagnosis Applications. Model-Based Condition Monitoring: Actuators, Drivers, Machinery, Plants, Sensors, and Fault-Tolerant Systems*; Springer Science & Business Media: Berlin/Heidelberg, Germany, 2011; ISBN 9783642127663.
171. Holmberg, K.; Kivikytö-Reponen, P.; Härkisaari, P.; Valtonen, K.; Erdemir, A. Global Energy Consumption Due to Friction and Wear in the Mining Industry. *Tribol. Int.* **2017**, *115*, 116–139. [CrossRef]
172. Kamenicky, J.; Zajicek, J. Failure Rate Evaluation of the Pumps Used in Power Industry. In Proceedings of the 2015 16th International Scientific Conference on Electric Power Engineering (EPE), Kouty nad Desnou, Czech Republic, 20–22 May 2015; pp. 503–506. [CrossRef]
173. Moffatt, J.; Zaitouny, A.; Hodkiewicz, M.R.; Small, M. Detecting Asset Cascading Failures Using Complex Network Analysis. *IEEE Access* **2021**, *9*, 120624–120637. [CrossRef]
174. Wu, S.; Wei, X. A New Integrated Numerical Modeling Approach toward Piston Diaphragm Pump Simulation: Three-Dimensional Model Simplification and Characteristic Analysis. *Adv. Mech. Eng.* **2019**, *11*, 1687814019845955. [CrossRef]
175. Vellman, D.H.; Venter, J.; Pelzer, R. Automated Demand Side Management of Clear Water Pumping and Storage for Shifting Load out of Eskom's Domestic Peak in the Mining Industry. In Proceedings of the 2012 9th Industrial and Commercial Use of Energy Conference, Cape Town, South Africa, 15–16 August 2012; pp. 1–4.
176. Korzhev, A.A.; Bolshunova, O.M.; Voytyuk, I.N. Simulation of an Adjustable Synchronous Electric Motor Drive of a Pumping Unit in a Reservoir Pressure Maintenance System. *J. Phys. Conf. Ser.* **2019**, *1333*, 052014. [CrossRef]
177. Barberá, L.; Crespo, A.; Viveros, P.; Stegmaier, R. A Case Study of GAMM (Graphical Analysis for Maintenance Management) in the Mining Industry. *Reliab. Eng. Syst. Saf.* **2014**, *121*, 113–120. [CrossRef]
178. Martinez, L.B.; Crespo Marquez, A.; Gunckel, P.V.; Stegmaier, R. *Advanced Maintenance Modelling for Asset Management: Techniques and Methods for Complex Industrial Systems*; Crespo Marquez, A., Gonzalez-Prida Diaz, V., Gomez Fernandez, J.F., Eds.; Springer International Publishing AG: Berlin/Heidelberg, Germany, 2018; pp. 331–348, ISBN 9783319580456.

179. Jasinski, M.; Jasinska, E.; Jasinski, M.; Jasinski, L. Computer-Aided Appliances to Underground Machines Maintenance—Selected Issues. In Proceedings of the 10th International Conference on Electronics, Computers and Artificial Intelligence, ECAI 2018, Iasi, Romania, 28–30 June 2018.
180. NFC Offers. Asset Management Plan Cuts Maintenance Costs at Tronox Mine. *Eng. Min. J.* **2021**, *222*, 58–60.
181. Fan, C. Key Technologies for Intelligent Mining of Difficult-to-Mine Coal Seams. *J. Civ. Eng. Urban Plan.* **2024**, *6*, 69–75. [CrossRef]
182. Ren, B.; Ding, K.; Wang, L.; Wang, S.; Jiang, C.; Guo, J. Research on an Intelligent Mining Complete System of a Fully Mechanized Mining Face in Thin Coal Seam. *Sensors* **2023**, *23*, 9034. [CrossRef]
183. Singh, V.; Gangsar, P.; Porwal, R.; Atulkar, A. Artificial Intelligence Application in Fault Diagnostics of Rotating Industrial Machines: A State-of-the-Art Review. *J. Intell. Manuf.* **2021**, *34*, 931–960. [CrossRef]
184. Miao, B.; Ge, S.; Guo, Y.; Zhou, J.; Jiang, E. Construction of Digital Twin System for Intelligent Mining in Coal Mines. *Metaverse* **2022**, *3*, 16. [CrossRef]
185. Xu, Y.; Xu, X. Research on Pump Product Intelligent Fault Diagnosis System Based on Dig Data. *J. Phys. Conf. Ser.* **2021**, *2121*, 012024. [CrossRef]
186. Tian, C.; Wei, X.; Yang, C.; Su, C. Optimization Technology of Frequency Conversion Constant Pressure Control System of Mine Emulsion Pump Station in Electrical Engineering and Automation Specialty. *Adv. Civ. Eng.* **2021**, *2021*, 5569994. [CrossRef]
187. Fiscor, B.S. Managing a Digital Ecosystem for Mining New Tools Offer Mines Ways to Collect and Analyze Useful Data. *Eng. Min. J.* **2022**, 30–34.
188. Peach, R.H.; Visser, J.K. Measuring Human Factors in Maintenance: A Literature Review. *S. Afr. J. Ind. Eng.* **2020**, *31*, 104–114. [CrossRef]
189. Noroozi, A.; Khakzad, N.; Khan, F.; Mackinnon, S.; Abbassi, R. The Role of Human Error in Risk Analysis: Application to Pre- and Post-Maintenance Procedures of Process Facilities. *Reliab. Eng. Syst. Saf.* **2013**, *119*, 251–258. [CrossRef]
190. Lawrence, A.C. Importance of Human Factors in Mining Productivity. *J. S. Afr. Inst. Min. Metall.* **1974**, *74*, 399–411.
191. Mayeza, T.; Peach, R. *Maintenance Human Factors in the South African Mining Industry*; Springer: Berlin/Heidelberg, Germany, 2024; pp. 217–225.
192. Bonsu, J.; Van Dyk, W.; Franzidis, J.P.; Petersen, F.; Isafiade, A. A Systemic Study of Mining Accident Causality: An Analysis of 91 Mining Accidents from a Platinum Mine in South Africa. *J. S. Afr. Inst. Min. Metall.* **2017**, *117*, 59–66. [CrossRef]
193. Noroozi, A.; Khan, F.; Mackinnon, S.; Amyotte, P.; Deacon, T. Determination of Human Error Probabilities in Maintenance Procedures of a Pump. *Process Saf. Environ. Prot.* **2014**, *92*, 131–141. [CrossRef]

sensors

MDPI

Article

Sensor-Based Diagnostics for Conveyor Belt Condition Monitoring and Predictive Refurbishment

Ryszard Błażej, Leszek Jurdziak and Aleksandra Rzeszowska *

Faculty of Geoengineering, Mining and Geology, Wroclaw University of Science and Technology, Na Grobli 15 St., 50-421 Wrocław, Poland; ryszard.blazej@pwr.edu.pl (R.B.); leszek.jurdziak@pwr.edu.pl (L.J.)
* Correspondence: aleksandra.rzeszowska@pwr.edu.pl

Abstract: Rising raw material costs and complex global supply chains have reduced the durability and availability of conveyor belts. In response, condition-based maintenance (CBM) with in situ diagnostics has become essential. This case study from a Polish lignite mine shows how subjective visual inspections were replaced with objective, repeatable measurements of belt core condition and thickness. Shifting refurbishment decisions from the plant to the conveyor improved success rates from 70% to over 90% and optimized belt lifecycle management. Sensor-based monitoring enables predictive maintenance, reduces premature or delayed replacements, increases belt reuse, lowers costs, and supports the circular economy by extending belt core life and reducing raw material demand. The study demonstrates how real-time, sensor-based diagnostics using inductive and ultrasonic technologies supports predictive maintenance of conveyor belts, improving refurbishment efficiency and lifecycle management.

Keywords: conveyor belt; refurbishment; predicting maintenance; condition monitoring; circular economy

1. Introduction—Global Conveyor Belt Market

Belt conveyors are mechanical systems consisting of continuous belt loops used for transporting bulk or unit materials over short to long distances, including within buildings, storage yards, assembly lines, and supply chains—sometimes across hundreds of kilometers [1,2].

The transported material rests on the belt's top cover and is driven by pulling forces within the core. The system typically consists of at least two pulleys, one of which drives the loop. Due to the low belt-to-material mass ratio, conveyor belts are among the most cost-effective transportation systems [3]. Their versatility makes them widely used in mining [4], cement manufacturing [5], power generation [6,7], recycling [8], metal processing, chemicals [9], oils, gases, food and beverage processing [10], general manufacturing, supply chain, automated distribution, warehousing [11], airports [12], and other sectors. Increasing deployment of belt systems contributes to reducing production costs, supporting overall market growth.

This growth is further fueled by technological advances, rising demand for automation, and the need for condition monitoring, diagnostics, and environmental sustainability. Electric-powered conveyors are also considered more eco-friendly than fossil-fuel alternatives. Despite market expansion, challenges remain—especially for small and medium enterprises—due to high belt procurement, installation, and maintenance costs, exacerbated by rising electricity prices, particularly in the European Union. Solutions such as belt

https://doi.org/10.3390/s25113459

refurbishment [13–15], energy-saving belts [16,17] and structures [18–20] and intelligent management strategies [6,21] are gaining importance.

The market includes textile-reinforced belts, steel cord belts, and solid woven types [22,23]. Textile belts (EE and EP) are projected to remain dominant due to their cost-effectiveness. Steel cord belts are favored in mining for their superior impact resistance and strength [24–26], though rising steel prices affect market value.

While growing demand is expected in developing countries, the shift away from coal-fired power in industrialized nations presents a challenge. Conveyor belts, especially heavy-duty ones, require large quantities of raw materials, which account for 70–75% of total production costs and strongly influence product quality [27].

Market size estimates vary widely, ranging from over $3 billion in 2020 to $11.5 billion by 2030, with projected CAGR between 3.74% and 5.9%. These discrepancies stem from differing methodologies and segmentation criteria (e.g., belt types, construction materials, applications, and geography), as illustrated in Figure 1.

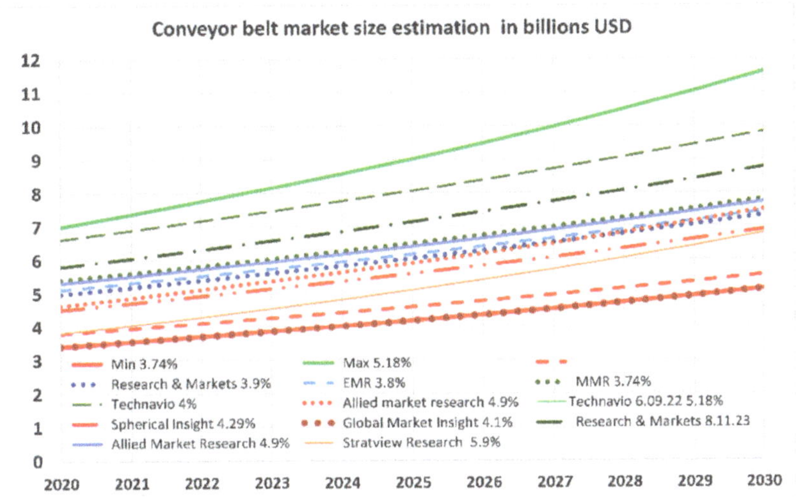

Figure 1. Global conveyor belt market size and forecasted growth measured in billion dollars. The Compound Annual Growth Rate (CAGR) of the belt market growth between 2023 and 2032 are given along with the names of market research firms and the date of report publication.

As belts are the costliest component of conveyor systems, their condition must be closely monitored. Predictive diagnostics and preventive replacement strategies not only reduce downtime but also promote material reuse and sustainability.

The pandemic, wars and the imposition of customs barriers disrupt the global supply of raw materials, causing material difficulties, extending the routes and delivery times of components for production and finished products. It also affects the increase in the prices of belts and lowers their durability when good quality components are replaced with cheaper substitutes.

In a situation of supply uncertainty, the strategy of extending the life of products through their refurbishment becomes a rational choice. It leads to cost and rare raw materials savings. The core of St type belts can be used three times, because in Poland, conveyor belts in brown coal mines are refurbished twice. In the case of the Bełchatów lignite mines, when one of two pits is being liquidated due to the lignite depletion, the mine has a large number of used belts. Their use by transferring them to other conveyors or refurbishment deepens the rationality of refurbishment, as it almost completely eliminates the need to purchase new belts. Considering that the cost of recondition is half the cost of purchasing a new belt, it is also economically justified. This study evaluates the cost-effectiveness of preventive belt replacement strategies integrated with refurbishment practices. The proposed methodology has been successfully applied in Polish lignite mines for over five decades and aims to optimize belt lifecycle management through condition-based diagnostics. Unlike conventional approaches relying on calendar age and visual cues, the proposed system uses objective, in situ diagnostics to evaluate the internal condition of steel-cord belts in real time, allowing optimized refurbishment decisions.

The innovation of this study lies in applying in situ magnetic diagnostics to quantify internal core damage in steel-cord belts, replacing subjective visual inspections with objective, data-driven assessments that improve refurbishment success rates.

2. Strategy of Belt Replacements

The market value of conveyor belts includes both the installation of new belts on investment conveyors and the replacement of worn sections on existing ones. A belt loop on long conveyors consists of multiple spliced sections (vulcanized, adhesive, or mechanical). New loops are built from the longest possible sections to minimize the number of splices, which represent structural weak points. Belts are typically delivered on single (250–400 m) or double reels (up to 700+ m). During operation, damaged segments or joints are replaced, gradually shortening section lengths and increasing the number of splices over time [28]. This process is referred to as belt renewal [29–31], which may involve new, less worn, or refurbished belts.

Belt degradation is influenced by many factors: conveyor design, belt construction, material type, working conditions, and operational regime [32–36]. Because these factors interact dynamically over time, belt lifetime is difficult to predict and should be treated as a random variable. Attempts to model belt durability based on operational statistics remain limited due to complexity, subjective replacement decisions, and inadequate metrics (e.g., calendar time instead of effective operational hours or transported mass) [37–40]. As a result, belt durability varies significantly, and its evaluation requires multicriteria analysis [41–44].

Furthermore, unit costs of belts with identical specifications may vary twofold. A lower price typically reflects lower material quality, but a high price does not guarantee longer lifespan. Without predictive tools, durability cannot be reliably assessed in advance.

While data on the new belt market is available, the refurbished belt market remains poorly quantified. Refurbishment was once common and widely offered by major service providers. Today, only selected suppliers (e.g., Fenner Dunlop Conveyor Belting (Pittsburgh, PA, USA), ContiTech AG (Hanover, Germany)) and service firms (e.g., Bestgum Sp. z o.o. (Rogowiec, Poland)) continue this service. Most users opt for new belts and consider refurbishment only after early-stage damage. Only a few, such as Polish lignite mines, apply refurbishment as a regular replacement strategy.

Failures caused by external stochastic events (e.g., punctures, cuts) may occur regardless of belt age or condition. The likelihood increases as belts wear, but accurate prediction would require extensive failure case data, which many operations do not collect or publish.

Ultimately, the decision to replace, repair, or refurbish depends on economic trade-offs. If repair downtime is shorter than replacement time, in situ repairs are preferred. Otherwise, belts are dismantled and refurbished either on-site or in specialized plants.

Figure 2 shows the expected runtimes for different replacement policies. The earliest belts to be replaced are those replaced preventively for refurbishment. The point is to ensure that overuse of the belt does not cause damage to the belt core precludes the belt from being refurbished with success. Corrosion of the cables in the core in many places with numerous punctures will reduce the adhesion of the cables to the rubber, and the need to make many repairs may increase the costs and time of refurbishment. A reconditioned belt may not pass the tests allowing it to be used, or the refurbishment costs will be too high compared to purchasing new belts.

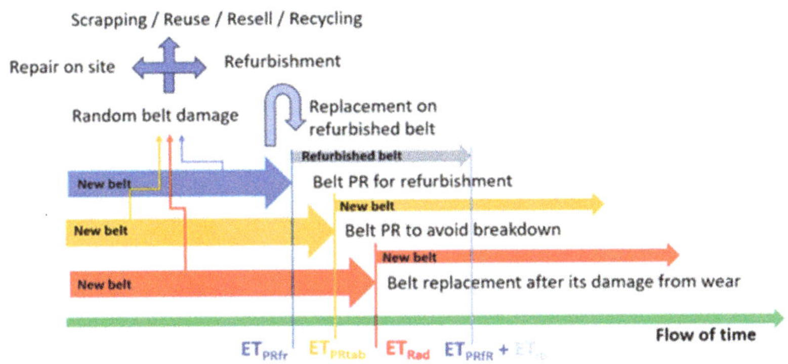

Figure 2. Expected times (ET) of belt replacements in different replacement strategies.

The belt segment can be damaged randomly at any age. In such cases, the maintenance supervisor has to decide what to do with failures. They can be repaired on site (at once, at the closest planned standstill, and later when more small failures occur). When failures require a long time to be repaired on-site and create a threat to continuous operation, it is better to replace the belt fragment or the whole segment on a new one and send the belt for refurbishment. If the belt is not worth repairing, it can be scrapped, or reused for another purpose (e.g., as seals, shock-absorbing linings, fenders, construction mats, marine dock bumpers, truck bed liners, grain pit covers, etc.). They can be resold to less demanding users or recycled. The last solution requires a special line for worn-out belt cover milling, cutting the core into smaller parts, scrubbing, separating, etc. Such lines will be required to utilize a lot of old belts when coal mines are closed shortly.

Decisions regarding sending the belt for refurbishment must be economically justified. If the cost of refurbishment the belt is C_{rb}, and the cost of purchasing a new belt is C_{nb}, then the unit cost of using a new belt will be the quotient $\frac{C_{nb}}{ET_{PRtab}}$ (where ET_{PRtab} is expected time in preventive replacement to avoid breakdown strategy), and for the refurbished belt, it will be $\frac{C_{rb}}{ET_{PRfr}}$ (where ET_{PRfr} is expected time in preventive replacement for refurbishment). Refurbishment is cost-effective if both conditions (1) and (2) are met.

$$\frac{C_{rb}}{ET_{PRfr}} < \frac{C_{nb}}{ET_{PRtab}} \tag{1}$$

$$C_{rb} < C_{nb} \cdot \frac{ET_{PRfr}}{ET_{PRtab}} \tag{2}$$

If the cost of refurbishment is 50% of the price of a new belt, then even when their lifespan is shorter than that of new belts but exceeds 50%, the use of refurbishment is cost-effective. Because not all belts sent for refurbishment can be successfully reconditioned, the relationship should be modified by introducing a refurbishment success rate indicator, showing the percentage of belts sent for refurbishment that are successfully reconditioned (*RSR*—Refurbishment Success Rate). It can be assumed that this rate depends on the time of belt usage and changes from 1 to 0.

To determine the unit cost of using 1 m of refurbished belts when $RSR(t)$ is less than 1, more belts need to be sent for refurbishment because some may not be successfully refurbished. The length of belts sent for reconditioning is obtained by dividing 1 by $RSR(t)$. Over time, as $RSR(t)$ decreases, the length of belts sent for refurbishment will increase according to the function $\frac{1}{RSR(t)}$ (Equations (3) and (4)).

$$\frac{C_{rb}}{RSR(t)\cdot ET_{PRfr}} < \frac{C_{nb}}{ET_{PRtab}} \tag{3}$$

$$C_{rb} < C_{nb}\cdot RSR(t)\cdot \frac{ET_{PRfr}}{ET_{PRtab}} \tag{4}$$

If the $RSR(t)$ indicator at the time of replacement is 0.7, and the ratio of the durability of refurbishment and new belts is, for example, 80%, the inequality will be true since 50% of $C_{rb} < C_{nb}\cdot 0.7\cdot 80\%$ and $C_{rb} < C_{nb}\cdot 56\%$. Knowing the course of the $RSR(t)$ function, you can determine the critical time, T_{rbt}, from Equations (5) and (6).

$$C_{rb} = C_{nb}\cdot RSR(T_{rbt})\cdot \frac{T_{rbt}}{ET_{PRtab}} \tag{5}$$

$$T_{rbt}\cdot RSR(T_{rbt}) = ET_{PRtab}\cdot \frac{C_{rb}}{C_{nb}} \tag{6}$$

If all belts could be successfully refurbished and incorporated into production, $RSR(t)$ would be equal to 1 (at least up to T_{rbt}), and then (7).

$$T_{rbt} = ET_{PRtab}\cdot \frac{C_{rb}}{C_{nb}} \tag{7}$$

If, thanks to the diagnostics of the belt's condition, it were possible to determine the belt wear index $I_{bc}(t)$ and its changes over time, as well as the critical state $I_{bc}(t) = I_{cr}$, after reaching which the belt should be dismantled and sent for refurbishment, then the replacement moment would be chosen not based on time, which is not a good measure of belt wear, but based on its condition, determined by reliable and objective methods. For this time, $RSR(t)$ can be calculated and checked to see if it ensures cost-effective refurbishment.

Figure 3 illustrates the proposed method for determining the optimal replacement time based on projected failure density over time, which forms the basis of the predictive replacement algorithm.

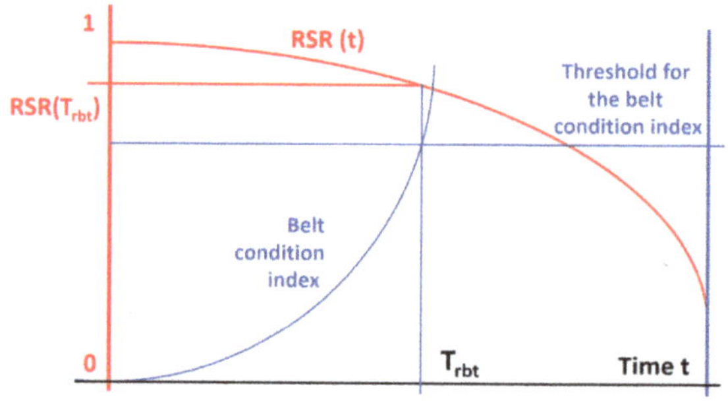

Figure 3. Method of determining the replacement time T_{rbt} based on the hypothetical deterioration of the belt condition $I_{bc}(t)$ until reaching the critical value. The intersection point determines the optimal replacement time, and this allows the calculation of $RSR(T_{rbt})$ to verify if cost-effectiveness is maintained.

The Belt Condition Index ($I_{bc}(t)$) is a quantitative measure of belt wear. At the Bełchatów lignite mine, a related metric—failure density—is commonly used. This is defined as the number of detected failures on a given belt segment divided by its length. This approach allows for meaningful comparisons of wear across belts of varying lengths and ages.

Failure density can serve multiple purposes. Locally, it helps identify specific sections of the belt that may require targeted repair or replacement. On a broader scale, it can be applied to entire belt sections to determine the optimal timing for removal and refurbishment.

Threshold values for I_{bc} and failure density are selected empirically by the mine's belt maintenance team and refurbishment plant. These teams analyze the Refurbishment Success Rate (RSR) for belts with different I_{bc} values to guide threshold selection.

The critical time for preventive replacement—denoted as T_{rbf}—is established at the point where the $I_{bc}(t)$ trend line intersects the defined threshold (see Figure 3). Validation of this threshold is also empirical: if belts with lower failure densities fail refurbishment criteria (e.g., strength tests), the threshold is adjusted downward. Conversely, if all belts at the threshold density pass refurbishment successfully, the threshold may be slightly increased.

In essence, RSR should be determined not for time but for the condition of the belt.

3. The Beginning of Conveyor Belt Refurbishment in Poland [45]

Since the mid-1960s, conveyor belts have been reconditioned in lignite mines in Poland. The first reconditioning department was set up in the Turow lignite surface mine (1a, 1b, Figure 4) The next was started up in 1977 at the Konin mine (2a, 2b, Figure 4) and in 1981 reconditioning began at the biggest and most modern lignite mine Bełchatów (3b, Figure 4).

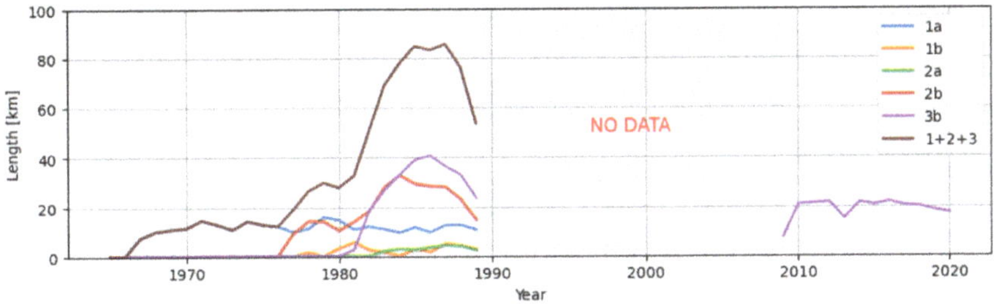

Figure 4. Reconditioning of conveyor belts at the Polish lignite surface mines: Turow—1, Konin—2, Belchatow—3. a—textile belt, b—steel cord belts. Data from [45].

Although the data in this section originate mainly from the 1980s, they are included to illustrate the long-term scale and historical significance of conveyor belt refurbishment practices in Poland's lignite industry. These insights provide a valuable context for understanding how current refurbishment strategies have evolved and how they compare to present-day practices supported by modern diagnostics.

In Figure 4, we can see that the maximal total length of reconditioned belts was close to 85 km in mid 80-ties, which gives about 14.8% of the total length (573 km) of installed belts. The lifespan of naturally worn-out belts is 49 months (4.1 years). Refurbished belts attain 85.7% of it (42.9 months = 3.6 years). With the refurbishment costs lower than 50% of a new belt price. We have profitable refurbishment, due to 85.7% is greater than 50%. Even for all belts removed the durability of refurbished belts was greater than 50% (min. 54.3%; max. 62.9%). The quality of refurbished belts has increased since the mid-80-ties. One of the reasons is that this time, the belt management computer system "Tasma" has been applied in the Belchatow mine which helped in rational belt management and maintenance decisions. The success rate of belt recondition can be as low as 58.34%

due to 50% = 87.5% · 58.34%. This provies a good margin for wrong decisions about belt replacements. Data published in the paper [45] allows for the estimation of the proportion of new and reconditioned belts to all belts supplied the mine between 1981 and 1989 to be 58.65% and 41.35%, respectively.

A similar proportion of new belts, belts after the 1st recondition and the 2nd recondition, estimated for the conditions of one of our lignite surface mines, are a bit different. For a hypothetical mine with 100 km belts installed, proportion stabilizes, as is shown in Figure 5.

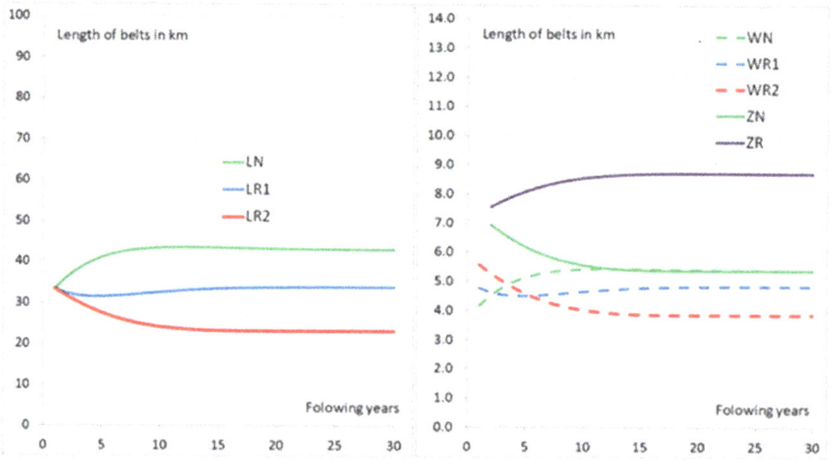

Figure 5. Stabilization process of belt length installed in the mine (**left**) and demanded (**right**) in the simulated strategy (deterministic variant) of belt replacements for recondition at the hypothetical mines having 100 km belts installed. Installed belts: new LN, after 1st reconditioning LR1, after 2nd reconditioning LR2), replaced belts (WN, WR1 and WR2), the newly purchased ones (ZN). Data from [46].

The initial state for iterative calculations of belt replacements including high refurbishment success rate $RSR = 0.9$ and $RSR_2 = 0.8$ (in reality, it is about 0.7). All types of belts new (LN), after the 1st recondition (LR1) and after the 2nd recondition (LR2) have the same length of 33.3 km. After several years of belt replacements, as is seen with new belts that are 43 km in length and refurbished belts 57 km in length. Among the replaced belts, refurbished made up 58.65% and new ones 41.35%. Real data in the past were the opposite. The new belts supplied to the mine for 10 years made up 58.65% and refurbished ones made up 41.35% (Figure 6). The differences came from the fact that demand for belts was not so stable due to changes in the length of conveyors' routes (mainly their increase from 50 km up to 112 km due to pit development—the length of belts is twice as much). The durability of reconditioned belts was much lower than that of new belts (slightly over 60% for the whole period and 85.7% in 1989) and unknown RSR which was much lower than the current value, especially after the application of diagnostics to select the best moment of belt dismantling for recondition.

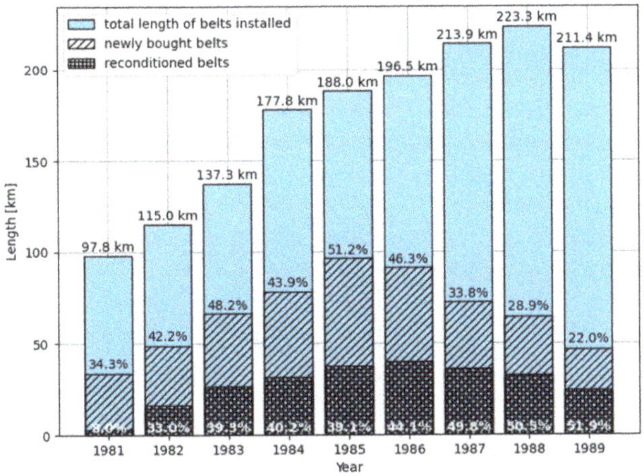

Figure 6. The 'Belchatow' open pit mine demand for belts concerning the length of belts installed. Data from [45].

4. Sensor-Based Diagnostic Systems for Predictive Belt Refurbishment

*4.1. Overview of Diagnostic Strare*gy

Various techniques have been developed to assess the condition of conveyor belts, particularly those with steel cords. Among them, X-ray imaging systems are often cited for their high diagnostic accuracy in detecting internal structural damage. However, these systems are rarely used in Polish industrial practice due to several limitations: they are expensive to install and operate, involve radiation safety concerns, and require specialized staff to interpret radiographic images manually. Additionally, the need to halt operations and isolate belt sections for scanning significantly reduces their applicability for routine monitoring.

In contrast, the DiagBelt+ system provides a non-invasive, in situ diagnostic solution that scans the belt loop while in motion, without disassembly or interruption of conveyor operation. The system uses magnetic field distortion analysis to identify discontinuities in steel cords and areas of weakened adhesion. The results are automatically visualized and quantified, enabling objective assessment by technical staff and eliminating the subjectivity inherent in visual inspection or X-ray interpretation.

BeltSonic complements DiagBelt+ by assessing the condition of the rubber cover using ultrasonic measurements. This combination provides a comprehensive evaluation of both the core and surface condition, supporting more accurate decisions about refurbishment, continued use, or disposal. Together, these systems offer a practical, scalable alternative to existing technologies, adapted to real industrial constraints.

4.2. Sensor System Architecture

The diagnostic framework employed in this study integrates two complementary sensing technologies: the DiagBelt+ system, based on inductive magnetic field sensors for core integrity assessment, and the BeltSonic system, based on ultrasonic distance sensors for measuring cover thickness and belt sagging. Both systems are designed for in situ diagnostics of moving conveyor belts, eliminating the need for belt disassembly or operational downtime.

DiagBelt+ employs a set of inductive sensors arranged within a measuring head, which is mounted on a stationary frame surrounding the conveyor belt. The system also includes permanent magnet strips that magnetize the steel cords embedded in the belt's core. As the belt moves, magnetic field disturbances caused by damage (e.g., cord breakage, corrosion, splice defects) are detected by the coils and converted into voltage signals.

The measurement resolution is typically one coil per 1–2 cords, allowing high spatial granularity. A tachometer mounted on the belt frame ensures synchronization of spatial measurements. The resulting magnetic signal is processed to produce a two-dimensional defect map of the entire belt loop. Parameters such as failure density (defects per meter), splice locations, and damage types are extracted using proprietary algorithms.

The BeltSonic module consists of two ultrasonic heads placed on opposing sides of the belt. Each head contains high-frequency transducers (typically 40–60 kHz), which measure the distance to the belt surface by analyzing the time-of-flight of reflected ultrasonic pulses. Knowing the fixed distance between the heads, the system computes the local thickness of the belt.

This differential measurement technique enables continuous scanning of belt cover thickness, sagging, and longitudinal/transverse deformations. Measurements can be repeated periodically, allowing trend analysis of belt wear over time. The system outputs a wear map of the belt, showing average and minimum cover thickness and percentage material loss relative to nominal specifications.

Both sensor systems are connected to local data acquisition (DAQ) modules, which digitize and preprocess the signals. Sensor data are transmitted to a central diagnostic computer. Custom software (DiagBelt+ Data Analysis v. 1.0) aggregates and visualizes the data, performing threshold analysis, condition indexing, and degradation trend forecasting.

A schematic of the integrated sensor system and data flow is presented in Figure 7. This diagram illustrates the physical arrangement of the sensors, data flow pathways, and analytical processing blocks that support decision-making.

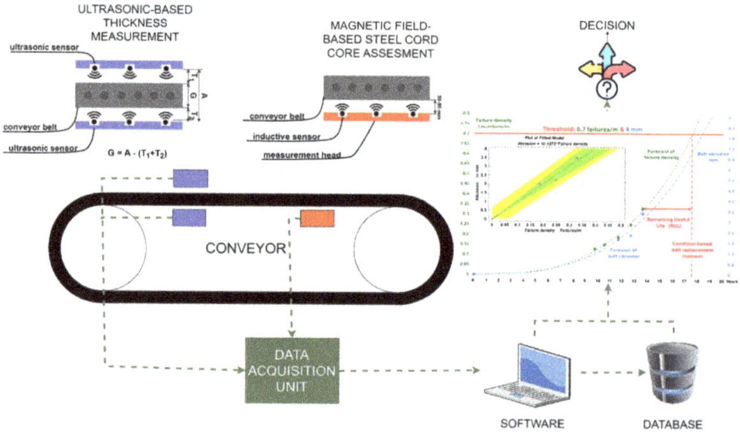

Figure 7. Overall architecture of the sensor-based conveyor belt diagnostic system.

The diagnostic outputs generated from sensor data—such as Belt Segment Condition Index (BSCindex), failure density profiles, and cover thickness loss—are used in a predictive maintenance algorithm. Each belt segment is evaluated against empirically defined thresholds to determine whether it is:

- fit for continued operation,
- eligible for refurbishment, or
- in need of scrapping or replacement.

This sensor-driven approach replaces subjective, visual assessments with quantitative diagnostics, significantly improving decision accuracy and enabling condition-based refurbishment planning. Results from full-scale deployment in the Bełchatów lignite mine confirm the practical value of this architecture: refurbishment success rates increased from 70% to over 90%, while belt utilization time was significantly extended in many cases.

4.3. Data Interpretation and Condition Indices

To accurately assess the technical condition of conveyor belts and make informed decisions regarding refurbishment at the right time, it is necessary to have information about both the condition of the belt's core and the thickness of its covers [13]. The DiagBelt+ system, developed at the Wrocław University of Science and Technology, utilizes a magnetic method to assess the technical condition of the core with steel cords, while the BeltSonic system uses a differential measurement method to determine the thickness of the measured object. Both of these systems are installed around the moving conveyor belt, as shown in Figure 8. The conveyor belt core diagnostic system (DiagBelt+ [47]) requires the installation of permanent magnet strips, which are intended to magnetize the core links of the belt. It also includes a measuring head equipped with magnetic coils that record changes in the magnetic field at the location of discontinuities in the links. On the other hand, the thickness measurement system (BeltSonic [48]) has two measuring heads installed parallel on both sides of the conveyor belt. Ultrasonic sensors installed in them measure the distance to the supporting and running covers of the belt, and knowing the distance between the sensors, allows for a precise determination of the thickness of the belt at a given location.

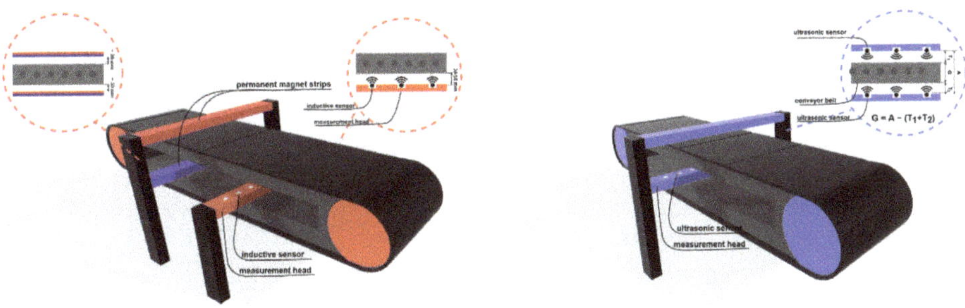

(a) (b)

Figure 8. Schematic diagram of the measurement system installation on the conveyor belt. (**a**) DiagBelt+ system (**b**) BeltSonic system.

The BeltSonic system is designed for measuring the thickness and assessing changes in the transverse and longitudinal profiles of various types of conveyor belts. The operating principle of this system is based on a differential measurement. Measurement heads, placed above and below the examined object, have ultrasonic sensors that measure the distance from the sensor to the obstacle, i.e., the cover of the conveyor belt. By knowing the distance between the measurement heads and readings from both sensors (located above and below the belt), the thickness of the examined object can be determined unambiguously. The operation of the system and its design parameters have been detailed in works [49].

4.4. Diagnostic Maps and Visualization

Measurements using the BeltSonic system enable the determination of the average and minimum thickness of the examined belt section, as well as the calculation of the percentage loss of its surface. Conducting measurements at specified time intervals allows for monitoring the rate of belt wear [33,40]. Analyzing the results of measurements for individual sections (or shorter fragments of the conveyor belt) and comparing the measured thickness with the nominal thickness of the conveyor belt provides an image of the wear of individual sections of the examined object. An exemplary contour map of a performed measurement and a wear measure map (average percentage loss of the original cover thickness calculated for the entire surface of the section between connections) are presented in Figure 9 for an extended loop of a belt consisting of twenty sections.

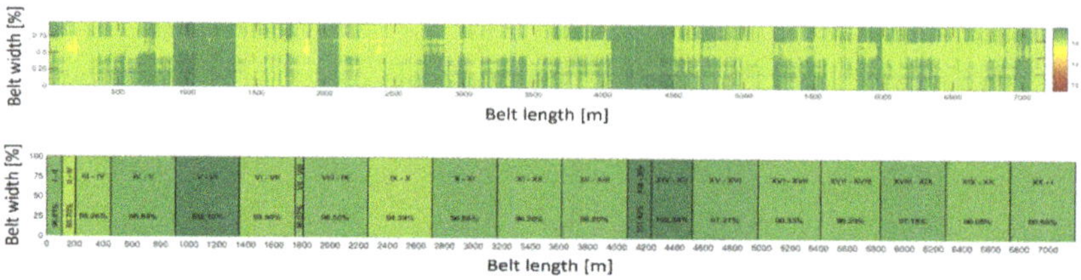

Figure 9. Conveyor belt loop map with sections—assessment of the percentage loss relative to the original thickness of the covers of the conveyor belt section—BeltSonic system.

The DiagBelt+ system is designed to detect and record changes in the magnetic field that occur at the belt splice or in areas of core damage. The most common failures include corrosion of the links, cuts in the links, or their absence on a specific section of the belt (e.g., a few centimeters long). It consists of a measurement head, which is equipped with coils to read changes in the magnetic field, two strips of permanent magnets to magnetize the core, a tachometer, and a data acquisition module. A detailed description of the system's operation can be found in works [13,50].

Thanks to measurements taken by the DiagBelt+ system, it is possible to obtain an image of failures, which, in turn, allows assessing the degree of damage to the belt's core and verifying the correctness of the joint's execution and its technical condition. By analyzing the number of failures, their density on a given belt section, and the characteristics of failures (their shape and belonging to specific categories), it is possible to determine which section of the belt requires repair. Figure 10 shows signals recorded by the DiagBelt+ system, which are then used for damage analysis.

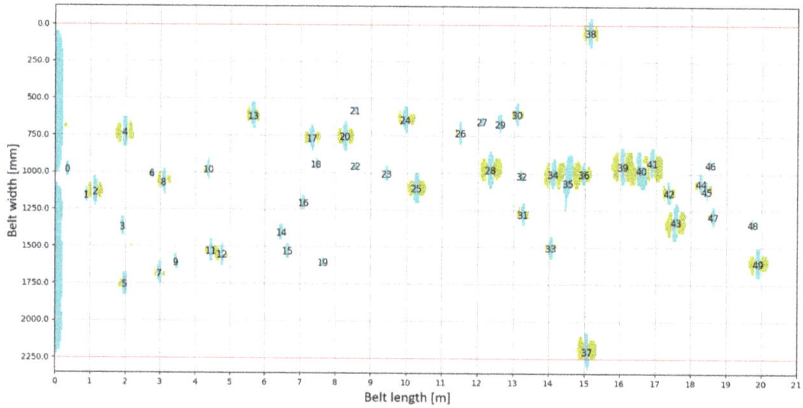

Figure 10. Visualization of measurement data—an example of a two-dimensional image from the DiagBelt+ system. Cyan indicates a signal value of +1, and yellow indicates a value of −1 (associated signal). The numbers mark consecutive identified belt failures.

The diagnostic system also includes software that, based on measurement data, generates an assessment report of the technical condition of the examined object. This program takes into account various parameters and drawings, such as the number of failures, the quantity of splices, and the length of the belt. After analyzing the data, the system generates a map of the technical condition of the examined conveyor belt, covering all sections of the belt loop. On this map, by setting thresholds beforehand, it is possible to visually assess the condition of a particular section, as the failures represented on the map correspond to the density of failures. An example of such a map is presented in Figure 11.

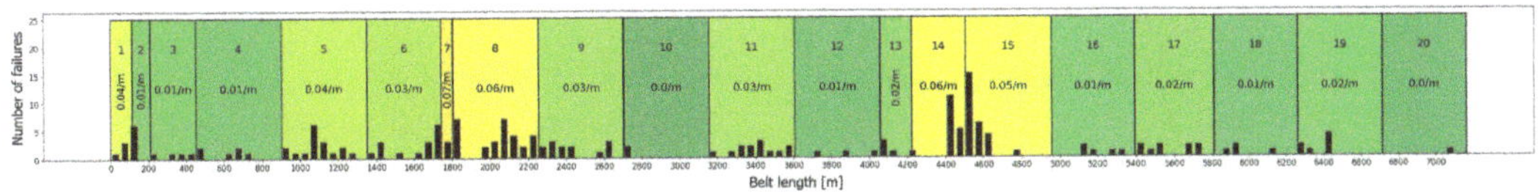

Figure 11. Map of aggregated wear measure (here, average density of failures per 1 m for the entire section of the belt) for the extended loop of the conveyor belt—assessment of the technical condition of the belt core—image from the DiagBelt+ system.

Analysis of the distribution of registered failures on histograms, visible in the report, allows for determining which part of the belt is most susceptible to failures.

The DiagBelt+ system represents a shift from visual, experience-based judgments to quantitative magnetic measurements of core damage. This transition enables repeatable and data-driven decisions that significantly reduce diagnostic uncertainty.

The data used for evaluating belt refurbishment success rates (RSR) and service life were collected and verified by the conveyor belt management team at the Bełchatów lignite mine. The team tracks each belt section's usage and monitors refurbishment outcomes using internal protocols provided by the regeneration plant. These protocols include detailed reasons for unsuccessful refurbishments. The mine utilizes a belt management software system ("Taśma"), implemented in the mid-1980s, which stores historical data on the installation, dismantling, and operating time of each belt segment. Prior to adopting DiagBelt+, refurbishment decisions were based on belt age and visual inspections. The introduction of in situ scanning has replaced subjective assessments with objective, measurable diagnostics that significantly improve decision quality and traceability.

The DiagBelt+ system was developed and refined as part of an industrial–academic project focused on modernizing belt maintenance strategies. The Bełchatów lignite mine (KWB Bełchatów), Poland's largest open-pit operation, played a key role in testing and implementing the solution. The mine adopted the system for full-scale deployment in 2023, integrating it into its belt condition monitoring and replacement decision process.

The DiagBelt+ system was launched in mid-2023 and has since been used to analyze approximately 106.5 km of conveyor belts (widths B1800 and B2250, ST 3150 strength class) across more than 20 excavator–conveyor–stacker systems at the Bełchatów lignite mine. Based on the diagnostic results, 62% of the belts were deemed fit for continued operation, 22% were directed for refurbishment, and 16% were scrapped. Diagnostic data enabled optimized decisions, leading to extended service life for 25.5 km of belts and early replacement of 6.1 km due to advanced wear (as of July 2024). From August 2024 to February 2025, an additional 36.9 km of belts across 15 conveyors were scanned. As a result, predicted service life was extended for 21.8 km of belts and shortened for 3.96 km. These decisions reduced the amount of belt waste and significantly cut costs related to new belt purchases, directly improving operational efficiency and ecological performance. The DiagBelt+ system was developed as part of a collaborative industrial–academic project aimed at modernizing conveyor belt maintenance strategies through sensor-based diagnostics. The Bełchatów lignite mine, the largest open-pit operation in Poland, was an active partner in this initiative. In 2023, the mine fully implemented the DiagBelt+ system into its operational belt management process. This real-world deployment played a key role in validating the method's effectiveness and integrating it into decision-making protocols related to refurbishment, reuse, or disposal of steel-cord belts.

These results confirm the practical advantage of the proposed diagnostic method: compared to earlier decisions based on visual inspection, DiagBelt+ increased the refurbishment success rate from 70% to over 90%, and enabled informed decisions that extended the operation time of 25.5 km of belts and avoided premature disposal.

The DiagBelt+ system underwent extensive validation prior to industrial implementation, including laboratory tests and field trials. The magnetic diagnostic approach it uses is based on technology that has been in use internationally since the late 1970s, and was refined to achieve high resolution imaging—typically one sensor per 1–2 steel cords.

As part of the implementation of the NCBR POIR.01.01.01-00-1194/19-00 project, a series of belts operating in a brown coal mine in Poland were examined. This process served as a ground truth validation for the DiagBelt+ system, confirming that the magnetic signals correspond precisely to actual core defects observed during physical inspection. The belts, directed for refurbishment during this project, were first scanned using the DiagBelt+ diagnostic system, and then redirected to the refurbishment process. On the refurbishment station, milling of the supporting and running cover of the belt was carried out (2 mm above the breaker). Visible failures on the belt were located in the form of a signal. Figures 12 and 13 show a fragment of the belt where the occurrence of damage is visible on the remaining, unmilled cover, as well as in the read measurement signal. Failures in Figure 12 are marked with numbering according to the DiagBelt+ system (Figure 13). It should be noted that not all failures visible on the remaining cover are core defects, and these were not marked or registered in the magnetic signal. This confirmed that the system reliably detects internal steel cord damage and distinguishes it from surface-level imperfections not relevant to refurbishment quality.

Figure 12. Belt photo after milling of covers.

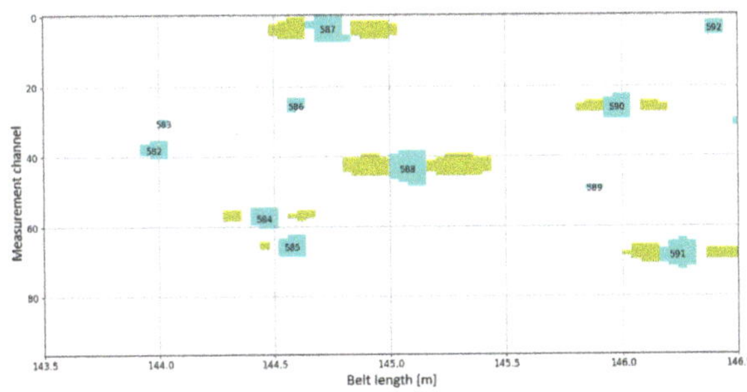

Figure 13. Corresponding magnetic signal. Cyan indicates a signal value of +1, and yellow indicates a value of –1 (associated signal).

With a relatively shallow milling of the carrying cover (with preserving the breaker), only very large failures were visible on the milled surface of the belt. On the most damaged sections of both belts, they were identified. The DiagBelt+ system identified a greater number of failures, but some of them may have remained undiscovered during milling when they were not related to piercing the cover to the core with the links. This could include cracked wires and crushed links.

4.5. Decision Algorithm and Practical Implementation

Figure 14 depicts the flowchart of the belt replacement strategy utilizing a diagnostic system.

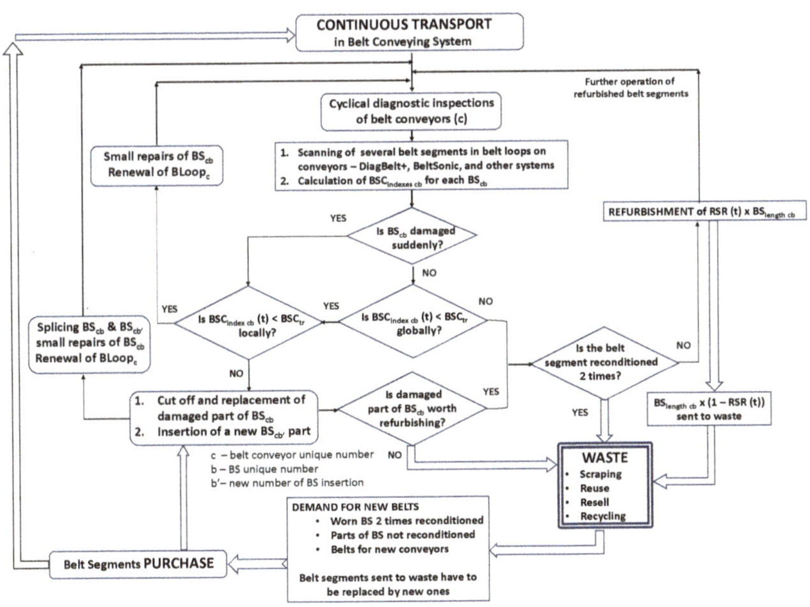

Figure 14. A flow chart of condition-based preventive belt replacement strategy for belt reconditions with the calculation of BSC_{index} for each belt segment (BS) in a belt loop (BL) on each belt conveyor (BC) in the belt conveying system (BCS).

The decision-making algorithm illustrated in Figure 14 has been applied in practice to support real-time maintenance decisions based on sensor data. Below are several representative cases from the Bełchatów lignite mine demonstrating how the system influenced refurbishment planning:

- Case 1—Belt life extended beyond visual estimate.
  A 2250 mm-wide ST3150 belt with an initial BSCindex of 0.95 defects/m was visually classified for replacement. However, DiagBelt+ scans showed uniform core condition and low deterioration dynamics. The algorithm projected that the threshold value of 1.5 defects/m would not be exceeded for another 8000 operating hours. As a result, the refurbishment was postponed, and the belt remained in use, reducing unnecessary downtime and saving over 1000 m of new belt material.
- Case 2—Belt removed earlier than expected.
  Another segment, 1800 mm-wide with BSCindex at 1.4 defects/m, showed rapid deterioration (slope > 0.1 defects/m per 1000 h). Though visually acceptable, the algorithm forecasted the index would exceed 1.7 within weeks. The belt was preemptively removed and sent for refurbishment, avoiding sudden failure and unplanned stoppage of the KTZ system.
- Case 3—Refurbishment confirmed on schedule.
  A 2250 mm belt planned for refurbishment after 12,000 h of operation had a BSCindex of 1.45 with moderate degradation rate. DiagBelt+ data confirmed that the predicted moment of crossing the 1.5 threshold aligned with the scheduled plan. The refurbishment proceeded as originally planned, validating the previous decision and demonstrating how sensor data can also confirm traditional schedules when appropriate.

The main difference between condition-based replacement using visual inspection and inspections with diagnostic tools is that the subjective and incomplete (based on the external view) assessment of the belt condition is replaced by the reliable assessment of the belt core condition by the magnetic system. Calendar time (the degree of filling the belt service life), which is sensitive to many factors, including a change in the conveyor load, is replaced by an objective measure of the belt condition BSC_{index} (e.g., the density of damage per 1 m of belt segment). After crossing the threshold by measured BSC_{index} the belt segment will be sent to the refurbishment plant. The threshold (BSC_{tr}), identical to all conveyors (regardless of their operating condition, length, belt speed, load, etc.), will be optimally selected, taking into account the changes in the recondition success rate ($RSR(t)$), costs of refurbishment and belt segment durability after reconditioning. $RSR(t)$ is not known as a priori but can be calculated as a posteriori after sending several belt segments to refurbishment with known BSC_{index} and attained RSR during refurbishment.

For 50 years, a preventive belt replacement strategy to refurbishment was used in the Bełchatów mine based on the belt condition. The belt condition was determined during the visual inspection taking into account their age in comparison to selected service life for a given conveyors (expected time of their operation) working in specific conditions. The use of a subjective visual assessment and operating time, which is a very inaccurate measure of belt condition, led to frequent recondition failures. The belts were pulled off the conveyors too late, because, although the belt did not look worn from the outside, the core condition was poor. Numerous cables and corrosion intersections, invisible from the outside, prevented successful reconditioning. The RSR indicator was about 70%. The solution was to implement magnetic diagnostics of the core condition using the Diagbelt+ system. Determining the core condition became possible on the conveyor, which not only reduced the costs, but above all, enabled an objective assessment of the invisible core and better determination of the moment of dismantling. Currently, the mine is in the phase of intensive scanning of belts directed for refurbishment so that an optimal selection of $BSIT$ will be possible to minimize the costs of using belts in the mine. Savings generated by avoiding dismantling of belts too early turned out to be a big positive surprise. Many belts working on less loaded conveyors have a little used belt core, although they have long filled the temporary Resurs (expected operational time). They would normally be directed to refurbishment, and it would probably be successful. However, greater savings allow further use of the belt for the coming months and years, because the belt loop can continue working without incurring costs of refurbishment. The success of refurbishment depends more on the core condition and less on the condition of the covers and the edges. The use of magnetic diagnostics NDT core condition makes this task easier. This will increase the refurbishment success rate (RSR) up to 90%, extend belt service life, reduce the need for new belt purchases, and improve cost-efficiency. The cost of reconditioning at Bestgum plants is estimated at 50% of the purchase price of the new belt. Leading belt manufacturers are estimated to make up 60–65% of purchase costs.

5. Discussion and Conclusions

The work aims to demonstrate the profitability of preventive replacements for belt regeneration. Lignite mines in Poland have successfully employed this strategy for over 50 years. The current market size exceeds $4 billion and is growing at a rate of 3 to almost 6% annually. Various estimates indicate that the production and delivery of belts consume significant resources, with approximately 75% of production costs attributed to raw materials. The remaining costs include energy, which is subject to price increases, and labor.

Many cheaper components are not approved for use in Europe but can be used outside the EU. The pandemic has highlighted extended supply chain timelines and geographical reach (due to carbon leakage), increasing the risk of deadline and quality failures and leading to price hikes, sometimes doubling. Refurbishment has once again become an attractive method to meet growing demand and enhance the security of supply and transport system continuity.

In Poland, belts are refurbished on-site, a practice that originated due to the lack of foreign exchange for imports and continues today due to extended delivery times and related uncertainties. Heavy-weight belts, being the most expensive, undergo refurbishment to maximize their utility. It is considered environmentally wasteful to discard them after a single use. Therefore, in Poland, heavy-weight belts are reconditioned, often twice (with the core being used three times), and in the past, even three times. Textile belts were also reconditioned in the past [51], a practice now rare due to the lower price of belts and the less valuable and durable core. Water ingress into the core causes delamination of the interlayers, excluding the belt from refurbishment.

Correct refurbishment is possible when the core is not excessively damaged. The use of core diagnostics and assessment of its condition on conveyors significantly reduces errors in decision-making during refurbishment. With the implementation of diagnostics using the DiagBelt+ system and the ability to test belt abrasion with the BeltSonic system, the probability of errors decreases, leading to an improved refurbishment success rate (RSR) and increased economic efficiency in the reconditioning process.

In the 1980s, the proportions of new and remanufactured belts on conveyors were 60% new and 40% remanufactured. Current simulations suggest a reversal of these proportions, with 60% remanufactured and 40% new. The Bełchatów mine, following the closure of one of its workings, practically abandoned the purchase of new belts. Instead, it effectively assesses the condition of dismantled belts and makes informed decisions on whether to reuse, refurbish, or discard them. To ensure the environmental effectiveness of the full belt use cycle, efforts are made to resell used belts, use them for other purposes (reuse), or recycle them to recover all components [52,53]. The refurbishment plant already recovers milled rubber from the reconditioning process to produce mats, carpets, and paving stones for parking lots. As used belts become more available, recycling rates will likely need to increase. Rubber from conveyor belts can be recycled in a similar way to rubber from tires [54].

The use of both measurement systems, DiagBelt+ for assessing the technical condition of the core and BeltSonic for evaluating the cover thickness, provides a comprehensive understanding of the technical state of the examined conveyor belt. This approach allows for the continuous monitoring of changes in its condition over time through periodic measurements. This enables the implementation of predictive belt replacement, based on the diagnosed condition, following any preventive replacement strategy. This strategy allows for choosing (for refurbishment only) belts that actually show a significant number of failures but still qualify for refurbishment. Utilizing the refurbishment capabilities of dismantled belt fragments that have reached a level of wear qualifying them for refurbishment increases the working time of full sections, minimizing the risk of emergency replacement and production losses associated with downtime and waiting for the delivery of new belts.

Traditional methods of assessing the condition of belts, such as visual inspection combined with the calendar age of the belt's operation, proved to be ineffective from the mine's perspective. As a result of the project implementation and the use of the system, the mine reports significant benefits, avoiding the need to send belts for refurbishment that, despite exceeding the calendar age of operation, still maintain a good technical condition.

The core contribution of this study is demonstrating that quantified, sensor-based diagnostics enable significantly more accurate refurbishment decisions, supporting both cost optimization and sustainability in belt lifecycle management.

Author Contributions: Conceptualization, L.J. and R.B.; methodology, L.J.; software, L.J. and A.R.; validation, R.B.; formal analysis, L.J. and A.R.; investigation, R.B.; resources, R.B.; data curation, R.B. and A.R.; writing—original draft preparation, L.J. and A.R.; writing—review and editing, L.J. and R.B. and A.R.; visualization, L.J. and A.R.; supervision, R.B. All authors have read and agreed to the published version of the manuscript.

Funding: This research received no external funding.

Institutional Review Board Statement: Not applicable.

Informed Consent Statement: Not applicable.

Data Availability Statement: The data presented in this study are available on request from the corresponding author.

Conflicts of Interest: The authors declare no conflicts of interest.

References

1. Kawalec, W. Przenośniki Taśmowe Dalekiego Zasięgu Do Transportu Węgla Brunatnego. *Transp. Przemysłowy I Masz. Robocze* **2009**, *1*, 6–13.
2. Bogacz, P.; Cieślik, Ł.; Osowski, D.; Kochaj, P. Analysis of the Scope for Reducing the Level of Energy Consumption of Crew Transport in an Underground Mining Plant Using a Conveyor Belt System Mining Plant. *Energies* **2022**, *15*, 7691. [CrossRef]
3. Mathaba, T.; Xia, X. Optimal and Energy Efficient Operation of Conveyor Belt Systems with Downhill Conveyors. *Energy Effic.* **2017**, *10*, 405–417. [CrossRef]
4. Krol, R.; Kawalec, W.; Gladysiewicz, L. An Effective Belt Conveyor for Underground Ore Transportation Systems. *IOP Conf. Ser. Earth Environ. Sci.* **2017**, *95*, 042047. [CrossRef]
5. Vijay, A.; Mathew, M.; Rohith, P.K. Development of an Economical Digital Control Method for a Continuously Running Conveyor Belt. In Proceedings of the 2017 International Conference on Inventive Systems and Control (ICISC), Coimbatore, India, 19–20 January 2017; pp. 1–5.
6. Luo, J.; Huang, W.; Zhang, S. Energy Cost Optimal Operation of Belt Conveyors Using Model Predictive Control Methodology. *J. Clean. Prod.* **2015**, *105*, 196–205. [CrossRef]

7. Zhang, S.; Tang, Y. Optimal Scheduling of Belt Conveyor Systems for Energy Efficiency. With Application in a Coal-Fired Power Plant. In Proceedings of the 2011 Chinese Control and Decision Conference (CCDC), Mianyang, China, 23–25 May 2011; pp. 1434–1439.
8. Oliveira Neto, R.; Gastineau, P.; Cazacliu, B.G.; Le Guen, L.; Paranhos, R.S.; Petter, C.O. An Economic Analysis of the Processing Technologies in CDW Recycling Platforms. *Waste Manag.* **2017**, *60*, 277–289. [CrossRef]
9. Salawu, G.; Glen, B. Improving the Efficiency of a Conveyor System in an Automated Manufacturing Environment Using a Model-Based Approach. *Int. J. Mech. Eng. Robot. Res.* **2023**, *12*, 107–112. [CrossRef]
10. Rashid, S.; Bashir, O.; Majid, I. Different Mechanical Conveyors in Food Processing. In *Transporting Operations of Food Materials Within Food Factories*; Elsevier: Amsterdam, The Netherlands, 2023; pp. 253–263.
11. Martin, H. *Warehousing and Transportation Logistics: Systems, Planning, Application and Cost Effectiveness*; Kogan Page Publishers: London, UK, 2018.
12. Draganová, K.; Semrád, K.; Spodniak, M.; Cúttová, M. Innovative Analysis of the Physical-Mechanical Properties of Airport Conveyor Belts. *Transp. Res. Procedia* **2020**, *51*, 20–27. [CrossRef]
13. Błażej, R.; Jurdziak, L.; Kirjanów-Błażej, A.; Bajda, M.; Olchówka, D.; Rzeszowska, A. Profitability of Conveyor Belt Refurbishment and Diagnostics in the Light of the Circular Economy and the Full and Effective Use of Resources. *Energies* **2022**, *15*, 7632. [CrossRef]
14. Bajda, M.; Hardygóra, M. Analysis of the Influence of the Type of Belt on the Energy Consumption of Transport Processes in a Belt Conveyor. *Energies* **2021**, *14*, 6180. [CrossRef]
15. Jurdziak, L.; Błażej, R. Economic Analysis of Steel Cord Conveyor Belts Replacement Strategy in Order to Undertake Profitable Refurbishment of Worn out Belts. In Proceedings of the International Multidisciplinary Scientific GeoConference: SGEM 17, Albena, Bulgaria, 29 June–5 July 2017; pp. 283–290.
16. Bajda, M.; Hardygóra, M. Examination and Assessment of the Impact of Working Conditions on Operating Parameters of Selected Conveyor Belts. *Min. Sci.* **2022**, *29*, 165–178. [CrossRef]
17. Youssef, G. Development of Conveyor Belts Design for Reducing Energy Consumption in Mining Application. Master's Thesis, Ain Shams University, Cairo, Egypt, 2016.
18. Opasiak, T.; Marglewicz, J.; Gąska, D.; Haniszewski, T. Rollers for Belt Conveyors in Terms of Rotation Resistance and Energy Efficiency. *Transp. Probl.* **2022**, *17*, 57–68. [CrossRef]
19. Masaki, M.S.; Zhang, L.; Xia, X. A Design Approach for Multiple Drive Belt Conveyors Minimizing Life Cycle Costs. *J. Clean. Prod.* **2018**, *201*, 526–541. [CrossRef]
20. Kawalec, W.; Król, R. Sustainable Development Oriented Belt Conveyors Quality Standards. In *Proceedings of the 14th International Scientific Conference: Computer Aided Engineering, Wroclaw, Poland, 20–23 June 2018*; CAE 2018. Lecture Notes in Mechanical Engineering; Rusiński, E., Pietrusiak, D., Eds.; Springer: Cham, Switzerland, 2019; pp. 327–336.
21. Zhang, M.; Jiang, K.; Cao, Y.; Li, M.; Hao, N.; Zhang, Y. A Deep Learning-Based Method for Deviation Status Detection in Intelligent Conveyor Belt System. *J. Clean. Prod.* **2022**, *363*, 132575. [CrossRef]
22. Subba Rao, D.V. *The Belt Conveyor*; Rao, D.V.S., Ed.; CRC Press: Boca Raton, FL, USA, 2020; ISBN 9781003089315.
23. Dick, J.S. *Testing Rubber Products for Performance: An Overview of Commercial Rubber Product Performance Requirements for the Non-Product Specialists*; ASTM International: West Conshohocken, PA, USA, 2023.
24. Ambriško, Ľ.; Marasová, D. Experimental Research of Rubber Composites Subjected to Impact Loading. *Appl. Sci.* **2020**, *10*, 8384. [CrossRef]
25. Andrejiova, M.; Grincova, A.; Marasova, D. Analysis of Tensile Properties of Worn Fabric Conveyor Belts with Renovated Cover and with the Different Carcass Type. *Eksploat. I Niezawodnosc* **2020**, *22*, 472–481. [CrossRef]
26. Bajda, M.; Błażej, R.; Jurdziak, L. Partial Replacements of Conbeyor Belt Loop Analysis with Regard to Its Reliability. In Proceedings of the 17th International Multidisciplinary Scientific GeoConference, SGEM 2017, Albena, Bulgaria, 29 June–5 July 2017.
27. Agg-Net. The Competitive Conveyor Belt Market. Available online: https://www.agg-net.com/resources/articles/materials-handling/the-competitive-conveyor-belt-market (accessed on 28 May 2025).
28. Bajda, M.; Hardygóra, M. Laboratory Tests of Operational Durability and Energy-Efficiency of Conveyor Belts. In *IOP Conference Series: Earth and Environmental Science*; Institute of Physics: Philadelphia, PA, USA, 2019; Volume 261.
29. Ambriško, Ľ.; Marasová, D.; Grendel, P.; Lukáč, S. Application of Logistics Principles When Designing the Process of Transportation of Raw Materials. *Acta Montan. Slovaca* **2015**, *20*, 141–147.
30. Knežo, D.; Andrejiová, M.; Kimáková, Z.; Radchenko, S. Determining of the Optimal Device Lifetime Using Mathematical Renewal Models. *TEM J.* **2016**, *5*, 121–125. [CrossRef]
31. Grinčová, A.; Andrejiová, M.; Grendel, P. Application the Renewal Theory to Determining the Models of the Optimal Lifetime for Conveyor Belts. *Appl. Mech. Mater.* **2014**, *683*, 97–101. [CrossRef]
32. Parmar, P.; Burduk, A.; Jurdziak, L. *Ishikawa Diagram Indicating Potential Causes for Damage Occurring to the Rubber Conveyor Belt Operating at Coal Mining Site*; Springer: Cham, Switzerland, 2024; pp. 704–713.

33. Webb, C.; Sikorska, J.; Khan, R.N.; Hodkiewicz, M. Developing and Evaluating Predictive Conveyor Belt Wear Models. *Data-Centric Eng.* **2020**, *1*, e13. [CrossRef]
34. Fedorko, G.; Molnár, V.; Živčák, J.; Dovica, M.; Husáková, N. Failure Analysis of Textile Rubber Conveyor Belt Damaged by Dynamic Wear. *Eng. Fail. Anal.* **2013**, *28*, 103–114. [CrossRef]
35. Fedorko, G.; Molnár, V.; Ferková, Ž.; Peterka, P.; Krešák, J.; Tomašková, M. Possibilities of Failure Analysis for Steel Cord Conveyor Belts Using Knowledge Obtained from Non-Destructive Testing of Steel Ropes. *Eng. Fail. Anal.* **2016**, *67*, 33–45. [CrossRef]
36. Bortnowski, P.; Kawalec, W.; Król, R.; Ozdoba, M. Types and Causes of Damage to the Conveyor Belt—Review, Classification and Mutual Relations. *Eng. Fail. Anal.* **2022**, *140*, 106520. [CrossRef]
37. Ilanković, N.; Živanić, D.; Zuber, N. The Influence of Fatigue Loading on the Durability of the Conveyor Belt. *Appl. Sci.* **2023**, *13*, 3277. [CrossRef]
38. Sikorska, J.Z. Variable Selection for Conveyor-Belt Mean Wear Rate Prediction. *Insights Min. Sci. Technol.* **2021**, *2*, 555594. [CrossRef]
39. Temerzhanov, A.; Stolpovskikh, I.; Sładkowski, A. Analysis of Reliability Parameters of Conveyor Belt Joints. *Transport Problems* **2012**, *7*, 107–112.
40. Webb, C.; Hodkiewicz, M.; Khan, N.; Muller, S.; Wilson, R. Conveyor Belt Wear Life Modelling. In Proceedings of the CEED Seminar Proceedings 2013, Zagreb, Croatia, 20–21 May 2013.
41. Al-Attar, R.T.; AL-Khafaji, M.S.; AL-Ani, F.H. Fuzzy-Based Multi-Criteria Decision Support System for Maintenance Management of Wastewater Treatment Plants. *Civ. Environ. Eng.* **2021**, *17*, 654–672. [CrossRef]
42. Patyk, M.; Bodziony, P.; Krysa, Z. A Multiple Criteria Decision Making Method to Weight the Sustainability Criteria of Equipment Selection for Surface Mining. *Energies* **2021**, *14*, 3066. [CrossRef]
43. Roumpos, C.; Partsinevelos, P.; Agioutantis, Z.; Makantasis, K.; Vlachou, A. The Optimal Location of the Distribution Point of the Belt Conveyor System in Continuous Surface Mining Operations. *Simul. Model. Pract. Theory* **2014**, *47*, 19–27. [CrossRef]
44. Bajda, M.; Błażej, R.; Jurdziak, L. Analysis of Changes in the Length of Belt Sections and the Number of Splices in the Belt Loops on Conveyors in an Underground Mine. *Eng. Fail. Anal.* **2019**, *101*, 436–446. [CrossRef]
45. Jurdziak, L.; Kawalec, W. Komputerowo Wspomagana Gospodarka Taśmami w KWB Turów. *Górnictwo Odkryw.* **1990**, 3–4.
46. Blazej, R.; Jurdziak, L. Condition-Based Conveyor Belt Replacement Strategy in Lignite Mines with Random Belt Deterioration. *IOP Conf. Ser. Earth Environ. Sci.* **2017**, *95*, 042051. [CrossRef]
47. Available online: https://Diagbeltplus.Pwr.Edu.Pl/ (accessed on 30 March 2025).
48. Available online: https://Beltsonic.Pwr.Edu.Pl/ (accessed on 30 March 2025).
49. Kirjanów-Błażej, A.; Jurdziak, L.; Błażej, R.; Rzeszowska, A. Calibration Procedure for Ultrasonic Sensors for Precise Thickness Measurement. *Measurement* **2023**, *214*, 112744. [CrossRef]
50. Blazej, R.; Jurdziak, L.; Kirjanow-Blazej, A.; Kozlowski, T. Identification of Damage Development in the Core of Steel Cord Belts with the Diagnostic System. *Sci. Rep.* **2021**, *11*, 12349. [CrossRef]
51. Ruuth, E.; Sanchis-Sebastiá, M.; Larsson, P.T.; Teleman, A.; Jiménez-Quero, A.; Delestig, S.; Sahlberg, V.; Salén, P.; Sanchez Ortiz, M.; Vadher, S.; et al. Reclaiming the Value of Cotton Waste Textiles: A New Improved Method to Recycle Cotton Waste Textiles via Acid Hydrolysis. *Recycling* **2022**, *7*, 57. [CrossRef]
52. Idriss, L.K.; Gamal, Y.A.S. Properties of Rubberized Concrete Prepared from Different Cement Types. *Recycling* **2022**, *7*, 39. [CrossRef]
53. Abdulrahman, A.S.; Jabrail, F.H. Treatment of Scrap Tire for Rubber and Carbon Black Recovery. *Recycling* **2022**, *7*, 27. [CrossRef]
54. Li, J.; Xiao, X.; Chen, Z.; Xiao, F.; Amirkhanian, S.N. Internal De-Crosslinking of Scrap Tire Crumb Rubber to Improve Compatibility of Rubberized Asphalt. *Sustain. Mater. Technol.* **2022**, *32*, e00417. [CrossRef]

MDPI AG
Grosspeteranlage 5
4052 Basel
Switzerland
Tel.: +41 61 683 77 34

Sensors Editorial Office
E-mail: sensors@mdpi.com
www.mdpi.com/journal/sensors

www.ingramcontent.com/pod-product-compliance
Lightning Source LLC
Chambersburg PA
CBHW041946260926
51850CB00018B/973
9783725854615